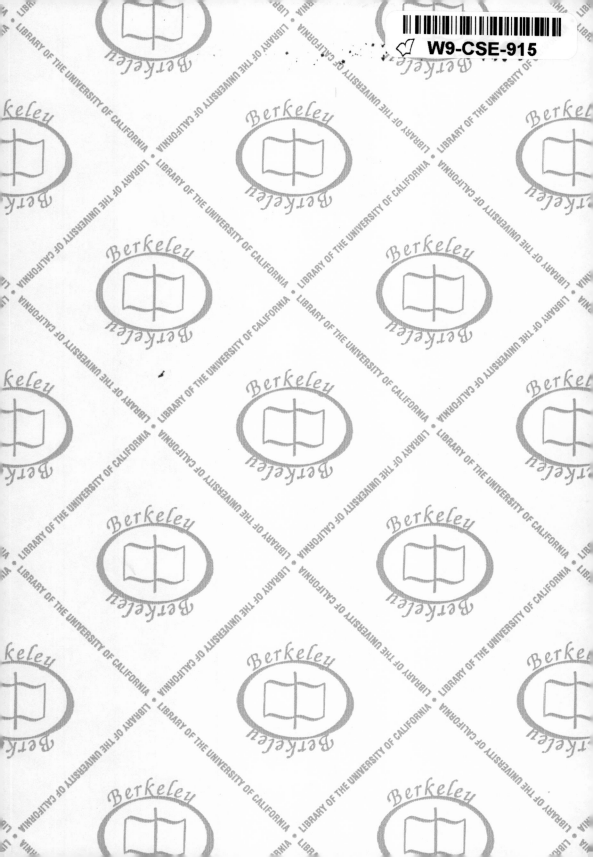

Dieter Meschede

Optics, Light and Lasers

The Practical Approach to Modern Aspects of
Photonics and Laser Physics

Dieter Meschede

Optics, Light and Lasers

The Practical Approach to Modern Aspects of
Photonics and Laser Physics

WILEY-
VCH

WILEY-VCH Verlag GmbH & Co. KGaA

Author

Prof. Dr. rer. nat. Dieter Meschede
Institute for Applied Physics
University of Bonn

Translators
Dr. Dorothea Gauer and Dr. Markus Mauerer

"Optik, Licht und Laser" was originally published
by Teubner in 1999.

Library of Congress Card No.: applied for
British Library Cataloging-in-Publication Data:
A catalogue record for this book is available from
the British Library

Bibliographic information published by
Die Deutsche Bibliothek
Die Deutsche Bibliothek lists this publication in
the Deutsche Nationalbibliografie; detailed bibli-
ographic data is available in the
Internet at <http://dnb.ddb.de>.

Cover Picture
The cover picture shows coherent white light which
is generated by the propagation of a 100-femto-
second pulse at 800 nm through a tapered fibre.
The light emerging from this fibre is recollimated.
A holographic 8-fold beamsplitter causes both
spectral and spatial dispersion of the optical wave.
The central spot corresponds to the zeroth order
and is therefore not diffracted. Further discus-
sions of white laser light as a recent research topic
can be found in section 8.5.6.
© Group of Prof. Dr. H. Giessen, Institute of
Applied Physics, University of Bonn

Printed in the Federal Republic of Germany
Printed on acid-free paper

Printing Strauss Offsetdruck GmbH,
Mörlenbach
Bookbinding Großbuchbinderei J. Schäffer
GmbH & Co. KG, Grünstadt

ISBN 3-527-40364-7

Preface

Though being taught as a subfield of classical electrodynamics, optics is now once again considered to be an important branch of physical sciences for the 21st century. Furthermore, optics implicitly exists due to its propaedeutic contributions to the theory of classical fields and quantum mechanics. In lecture halls today we can easily demonstrate coherence phenomena with laser light sources. It is hence appropriate also in lecturing to devote more room to the concepts of optics created since the 1960s.

This textbook attempts to link the central topics of optics that were established 200 years ago to the most recent research topics such as laser cooling or holography. To compromise between depth and breadth, it is assumed that the reader is familiar with the formal concepts of electrodynamics and also basic quantum mechanics. In scientific education, this textbook may serve as a reference for the foundations of modern optics: classical optics, laser physics, laser spectroscopy, nonlinear optics as well as applied optics may profit. Teaching will be complemented through materials presented by new media such as the internet. Nevertheless, the author strongly believes that conventional textbooks will continue to be a prime source of learning. Novel materials and complements will be made available, however, through the www.uni-bonn.de/iap/ollE website.

Contents

1 Light rays

1.1 Light rays in human experience

The formation of an image is one of our most fascinating emotional experiences. Even in ancient times it was realized that our 'vision' is the result of rectilinearly propagating light rays, because everybody was aware of the sharp shadows of illuminated objects. Indeed, rectilinear propagation may be influenced by certain optical instruments, e.g. by mirrors or lenses. Following the successes of Tycho Brahe (1546–1601), knowledge about *geometrical optics* made for the consequential design and construction of magnifiers, microscopes and telescopes. All these instruments serve as aids to vision. Through their assistance, 'insights' have been gained that added to our world picture of natural science, because they enabled observations of the world of both micro- and macro-cosmos.

Thus it is not surprising that the terms and concepts of optics had tremendous impact on many areas of natural science. Even such a giant instrument as the new Large Hadron Collider (LHC) particle accelerator in Geneva is basically nothing other than an admittedly very elaborate microscope, with which we are able to observe the world of elementary particles on a subnuclear length scale. Perhaps as important for the humanities is the wave theoretical description of optics, which spun off from the development of quantum mechanics.

In our human experience, rectilinear propagation of light rays – in a homogeneous medium – stands in the foreground. But it is a rather newer understanding that our ability to see pictures is caused by an optical image in the eye.

Fig. 1.1: *Light rays.*

Nevertheless, we can understand the formation of an image with the fundamentals of ray optics. That is why this textbook starts with a chapter on ray optics.

Optics, Light and Laser. Dieter Meschede
Copyright © 2004 Wiley-VCH Verlag GmbH & Co. KGaA
ISBN: 3-527-40364-7

1.2   Ray optics

When light rays spread spherically into all regions of a homogeneous medium, in general we think of an idealized, point-like and isotropic luminous source at their origin. Usually light sources do not fulfil any of these criteria. Not until we reach a large distance from the observer may we cut out a nearly parallel beam of rays with an aperture. Therefore, with an ordinary light source, we have to make a compromise between intensity and parallelism, to achieve a beam with small divergence. Nowadays optical demonstration experiments are nearly always performed with laser light sources, which offer a nearly perfectly parallel, intense optical beam to the experimenter.

When the rays of a beam are confined within only a small angle with a common optical axis, then the mathematical treatment of the propagation of the beam of rays may be greatly simplified by linearization within the so-called 'paraxial approximation'. This situation is met so often in optics that properties such as those of a thin lens, which go beyond that situation, are called 'aberrations'.

The direction of propagation of light rays is changed by refraction and reflection. These are caused by metallic and dielectric interfaces. Ray optics describes their effect through simple phenomenological rules.

1.3   Reflection

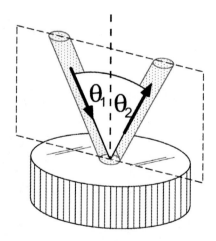

Fig. 1.2: *Reflection at a planar mirror.*

We observe reflection, or mirroring of light rays not only on smooth metallic surfaces, but also on glass plates and other dielectric interfaces. Modern mirrors may have many designs. In everyday life they mostly consist of a glass plate coated with a thin layer of evaporated aluminium. But if the application involves laser light, more often dielectric multilayer mirrors are used; we will discuss these in more detail in the chapter on interferometry (Chapter 5). For ray optics, the type of design does not play any role.

1.3.1   Planar mirrors

We know intuitively that at a planar mirror like in Fig. 1.2 the *angle of incidence* θ_1 is identical with the *angle of reflection* θ_2 of the reflected beam,

$$\theta_1 = \theta_2, \tag{1.1}$$

and that incident and reflected beams lie within a plane together with the surface normal. Wave optics finally gives us a more rigid reason for the laws of reflection. Thereby also details like, for example, the intensity ratios for dielectric reflection (Fig. 1.3) are explained, which cannot be derived by means of ray optics.

1.4 Refraction

At a planar dielectric surface, like e.g. a glass plate, reflection and transmission occur concurrently. Thereby the transmitted part of the incident beam is 'refracted'. Its change of direction can be described by a single physical quantity, the 'index of refraction' (also: refractive index). It is higher in an optically 'dense' medium than in a 'thinner' one.

In ray optics a general description in terms of these quantities is sufficient to understand the action of important optical components. But the refractive index plays a key role in the context of the macroscopic physical properties of dielectric matter and their influence on the propagation of macroscopic optical waves as well. This interaction is discussed in more detail in the chapter on light and matter (Chapter 6).

1.4.1 Law of refraction

At the interface between an optical medium '1' with refractive index n_1 and a medium '2' with index n_2 (Fig. 1.3) Snell's law of refraction (Willebrord Snell, 1580–1626) is valid,

$$n_1 \sin \theta_1 = n_2 \sin \theta_2, \tag{1.2}$$

where θ_1 and θ_2 are called the angle of incidence and angle of emergence at the interface. It is a bit artificial to define two absolute, material-specific refractive indices, because according to Eq. (1.2) only their ratio $n_{12} = n_1/n_2$ is determined at first. But considering the transition from medium '1' into a third material '3' with n_{13}, we realize that, since $n_{23} = n_{21}n_{13}$, we also know the properties of refraction at the transition from '2' to '3'. We can prove this relation, for example by inserting a thin sheet of material '3' between '1' and '2'. Finally, fixing the refractive index of vacuum to $n_{\text{vac}} = 1$ – which is argued within the context of wave optics – the specific and absolute values for all dielectric media are determined.

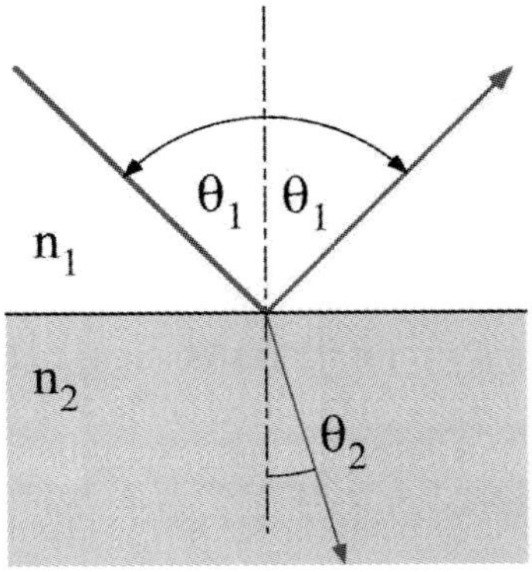

Fig. 1.3: *Refraction and reflection at a dielectric surface.*

In Tab. 1.1 on p. 9 we collect some physical properties of selected glasses. The refractive index of most glasses is close to $n_{\text{glass}} = 1.5$. Under usual atmospheric conditions the refractive index in air varies between 1.000 02 and 1.000 05. Therefore, using $n_{\text{air}} = 1$, the refraction properties of the most important optical interface, i.e. the glass–air interface, may be described adequately in terms of ray optics. Nevertheless, small deviations and variations of the refractive index may play an important role in everyday optical phenomena in the atmosphere (for example, fata morgana, p. 6).

1.4.2   Total internal reflection

According to Snell's law, at the interface between a dense medium '1' and a thinner medium '2' ($n_1 > n_2$), the condition (1.2) can only be fulfilled for angles smaller than the critical angle θ_c,

$$\theta < \theta_c = \sin^{-1}(n_2/n_1). \tag{1.3}$$

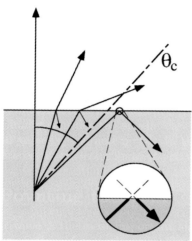

For $\theta > \theta_c$ the incident intensity is totally reflected at the interface. We will see in the chapter on wave optics that light penetrates into the thinner medium for a distance of about one wavelength with the so-called 'evanescent' wave, and that the point of reflection does not lie exactly at the interface (Fig. 1.4). The existence of the evanescent wave enables the application of the so-called 'frustrated' total internal reflection, e.g. for the design of polarizers (Chapter 3.5.4).

Fig. 1.4: *Total internal reflection at a dielectric surface. The point of reflection of the rays does not lie exactly within the interface, but slightly beyond (the Goos–Haenchen effect [36, 88]).*

1.5   Fermat's principle: the optical pathlength

As long as light rays propagate in a homogeneous medium, they seem to follow the shortest geometric path from the source to a point, making their way in the shortest possible time. If refraction occurs along this route, then the light ray obviously no longer moves on the geometrically shortest path.

The French mathematician Pierre de Fermat (1601–1665) postulated in 1658 that in this case the light ray should obey a *minimum principle*, moving from the source to another point along the path that is *shortest in time*.

For an explanation of this principle, one cannot imagine a better one than that given by the American physicist Richard P. Feynman (1918–1988), who visualized

Fermat's principle with a human example: One may imagine Romeo discovering his great love Juliet at some distance from the shore of a shallow, leisurely flowing river, struggling for her life in the water. Without thinking, he runs straight towards his goal – although he might have saved valuable time if he had taken the longer route, running the greater part of the distance on dry land, where he would have achieved a much higher speed than in the water.

Considering this more formally, we determine the time required from the point of observation to the point of the drowning maiden as a function of the geometric pathlength. Thereby we find that the shortest time is achieved exactly when a path is chosen that is refracted at the water–land boundary. It fulfils the refraction law (1.2) exactly, if we substitute the indices of refraction n_1 and n_2 by the inverse velocities in water and on land, i.e.

$$\frac{n_1}{n_2} = \frac{v_2}{v_1}. \qquad \text{low } v_\perp \longrightarrow \text{ lov } r.$$

According to Fermat's minimum principle, we have to demand the following. The propagation velocity of light in a dielectric c_n is reduced in comparison with the velocity in vacuum c by the refractive index n:

$$c_n = c/n.$$

Now the *optical pathlength* along a trajectory $\mathcal{C}$, where the refractive index n depends on the position $\mathbf{r}$, can be defined in general as

$$\mathcal{L}_{\text{opt}} = c \int_\mathcal{C} \frac{ds}{c/n(\mathbf{r})} = \int_\mathcal{C} n(\mathbf{r})\, ds. \qquad (1.4)$$

With the tangential unit vector $\mathbf{e_t}$, the path element $ds = \mathbf{e_t} \cdot d\mathbf{r}$ along the path can be calculated.

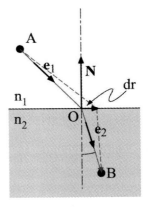

Fig. 1.5: *Fermat's principle and refraction at a dielectric surface.*

Example: Fermat's principle and refraction
As an example of the use of the integral principle, we will again consider refraction at a dielectric surface and therefore vary the length of the optical path between the points A and B in Fig. 1.5 ($\mathbf{r}_{\text{AO}}$ = vector from A to O etc., $\mathbf{e}_{1,2}$ = unit vectors),

$$\begin{aligned}
\mathcal{L}_{\text{opt}} &= n_1\mathbf{e_1} \cdot \mathbf{r}_{\text{AO}} + n_2\mathbf{e_2} \cdot \mathbf{r}_{\text{OB}}, \\
d\mathcal{L}_{\text{opt}} &= (n_1\mathbf{e_1} - n_2\mathbf{e_2}) \cdot d\mathbf{r}.
\end{aligned}$$

Now varying only the distance along the surface ($\mathbf{N}$ = surface normal) and taking into account $d\mathbf{r} \perp \mathbf{N}$, we can specify the condition

$$(n_1\mathbf{e_1} - n_2\mathbf{e_2}) \times \mathbf{N} = 0,$$

which is a vectorial formulation of Snell's law (1.2) reproducing it at once.

1.5.1   Inhomogeneous refractive index

In general, the index of refraction of a body is not spatially homogeneous, but has underlying, continuous, even though small, fluctuations like the material itself, which affect the propagation of light rays: $n = n(\mathbf{r})$. We observe such fluctuations in, for example, the flickering of hot air above a flame. From the phenomenon of mirages, we know that efficient reflection may arise like in the case of grazing incidence at a glass plate, even though the refractive index decreases only a little bit towards the hot bottom.

Again using the idea of the integral principle, this case of propagation of a light ray may also be treated by applying Fermat's principle. The contribution of a path element ds to the optical pathlength is $d\mathcal{L}_{\mathrm{opt}} = n\,ds = n\mathbf{e}_{\mathrm{t}} \cdot d\mathbf{r}$, where $\mathbf{e}_{\mathrm{t}} = d\mathbf{r}/ds$ is the tangential unit vector of the trajectory. On the other hand $d\mathcal{L}_{\mathrm{opt}} = \boldsymbol{\nabla}\mathcal{L}_{\mathrm{opt}} \cdot d\mathbf{r}$ is valid in accordance with Eq. (1.4), which yields the relation

$$n\mathbf{e}_{\mathrm{t}} = n\frac{d\mathbf{r}}{ds} = \boldsymbol{\nabla}\mathcal{L}_{\mathrm{opt}} \quad \text{and} \quad n^2 = (\boldsymbol{\nabla}\mathcal{L}_{\mathrm{opt}})^2,$$

which is known as the *eikonal equation* in optics. We get the important *ray equation* of optics, by differentiating the eikonal equation after the path[1],

$$\frac{d}{ds}\left(n\frac{d\mathbf{r}}{ds}\right) = \boldsymbol{\nabla}n. \tag{1.5}$$

A linear equation may be reproduced for homogeneous materials ($\boldsymbol{\nabla}n = 0$) from (1.5) without difficulty.

Example: Fata morgana

As a short example we will treat reflection at a hot film of air near the ground, which induces a decrease in air density and thereby a reduction of the refractive index. (Another example is the propagation of light rays in a gradient waveguide – section 1.7.3.) We may assume in good approximation that for calm air the index of refraction increases with distance y from the bottom, e.g. $n(y) = n_0(1 - \varepsilon\,e^{-\alpha y})$. Since the effect is small, $\varepsilon \ll 1$ is valid in general, while the scale length α is of the order $\alpha = 1\,\mathrm{m}^{-1}$. We look at Eq. (1.5) for $\mathbf{r} = (y(x), x)$ for all individual components and find for the x coordinate with constant C

$$n\frac{dx}{ds} = C.$$

We may use this result as a partial parametric solution for the y coordinate,

$$\frac{d}{ds}\left(n\frac{dy}{ds}\right) = \frac{d}{dx}\left(n\frac{dy}{dx}\frac{dx}{ds}\right)\frac{dx}{ds} = \frac{d}{dx}\left(C\frac{dy}{dx}\right)\frac{C}{n} = \frac{\partial n(y)}{\partial y}.$$

---

[1]Thereby we apply $d/ds = \mathbf{e}_{\mathrm{t}} \cdot \boldsymbol{\nabla}$ and

$$\frac{d}{ds}\boldsymbol{\nabla}\mathcal{L} = (\mathbf{e}_{\mathrm{t}} \cdot \boldsymbol{\nabla})\boldsymbol{\nabla}\mathcal{L} = \frac{1}{n}(\boldsymbol{\nabla}\mathcal{L} \cdot \boldsymbol{\nabla})\boldsymbol{\nabla}\mathcal{L} = \frac{1}{2n}\boldsymbol{\nabla}(\boldsymbol{\nabla}\mathcal{L})^2 = \frac{1}{2n}\boldsymbol{\nabla}n^2.$$

The constant may be chosen to be $C = 1$, because it is only scaling the x coordinate. Since $2n\,\partial n/\partial y = \partial n^2/\partial y$ and $n^2 \simeq n_0^2(1 - 2\varepsilon\,e^{-\alpha y})$, we get for $\varepsilon \ll 1$

$$\frac{d^2y(x)}{dx^2} = \frac{1}{2}\frac{\partial}{\partial y}n^2(y) = n_0^2\varepsilon\alpha\,e^{-\alpha y}.$$

This equation can be solved by fundamental methods and it is convenient to write the solution in the form

$$y = y_0 + \frac{1}{\alpha}\ln\left[\cosh^2(\kappa(x - x_0))\right] \overset{\kappa(x-x_0)\gg 1}{\longrightarrow} y_0 + \frac{2\kappa}{\alpha}(x - x_0).$$

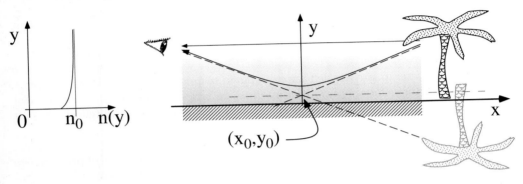

Fig. 1.6: *Profile of the refractive index and optical path for a fata morgana.*

For large distances from the point of reflection at $x = x_0$ we find straight propagation as expected. The maximum angle $\phi = \arctan(2\kappa/\alpha)$, where reflection is still possible, is defined by $\kappa \leq n_0\alpha(\varepsilon/2)^{1/2}$. As in Fig. 1.6 the observer registers two images – one of them is upside down and corresponds to a mirror image. The curvature of the light rays declines quickly with increasing distance from the bottom and therefore may be neglected for the 'upper' line of sight. At (x_0, y_0) a 'virtual' point of reflection may be defined.

1.6 Prisms

The technically important rectangular reflection is achieved with an angle of incidence of $\theta_i = 45°$. For ordinary glasses ($n \simeq 1.5$), this is above the angle of total internal reflection $\theta_c = \sin^{-1}1.5 = 42°$. Glass prisms are therefore often used as simple optical elements, which are applied for beam deflection. More complicated prisms are realized in many designs for multiple reflections, where they have advantages over the corresponding mirror combinations due to their minor losses and more compact and robust designs.

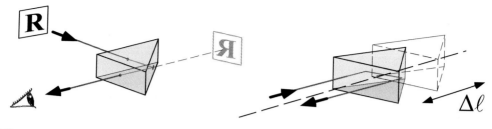

Fig. 1.7: *Reflection or 90° prism. This prism is used for rectangular beam deflection. It may also be used for the design of a retroreflector, whereby an optical delay $\Delta t = \Delta\ell/c$ is realized by simple adjustment.*

Often used designs are the Porro prism and the retroreflector (Fig. 1.8) – other names for the latter are 'corner cube reflector', 'cat's eye' or 'triple mirror'. The Porro prism and its variants are applied for example in telescopes to create upright images. The retroreflector not only plays an important role in optical distance measurement techniques and interferometry, but also enables functioning of security reflectors - cast in plastics - in vehicles.

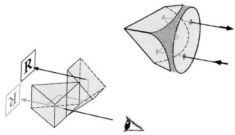

Fig. 1.8: *The Porro prism (left) is combined out of two rectangular prisms, which rotate the image plane of an object such that in combination with lenses one gets an upright image. The retroreflector (right) throws back every light ray independently of its angle of incidence, but causing a parallel shift.*

in more detail.

We may also regard cylindrical glass rods as a variant of prims where total internal reflection plays an important role. In such a rod (see Fig. 1.11) a light ray is reflected back from the surface to the interior again and again, without changing its path angle relative to the rod axis. Such fibre rods are used, for example, to guide light from a source towards a photodetector. In miniaturized form they are applied as *waveguides* in optical telecommunications. Their properties will be discussed in the section on beam propagation in waveguides (Section 1.7) and later on in the chapter on wave optics (Chapter 3.3)

1.6.1   Dispersion

Prisms played a historical role in the spectral decomposition of white light into its constituents. The refractive index and thus also the angle of deflection δ in Fig. 1.9 actually depend on the wavelength, $n = n(\lambda)$, and therefore light rays of different colours are deflected with different angles. Under *normal dispersion* blue wavelengths are refracted more strongly than red, $n(\lambda_{\text{blue}}) > n(\lambda_{\text{red}})$.

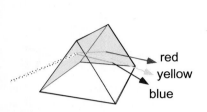
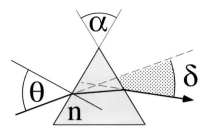

Fig. 1.9: *Refraction and dispersion at a symmetrical prism. The index of refraction n can be calculated from the minimum angle of deflection* $\delta = \delta_{\min}$ *in a simple manner.*

Refractive index and dispersion are very important technical quantities for the application of optical materials. The refractive index is tabulated in manufacturers' data sheets for various wavelengths, and (numerous different) empirical formulae are used for the wavelength dependence. The constants from Tab. 1.1 are valid for the formula which is also called *Sellmeier equation*:

$$n^2 = 1 + \frac{B_1 \lambda^2}{\lambda^2 - C_1} + \frac{B_2 \lambda^2}{\lambda^2 - C_2} + \frac{B_3 \lambda^2}{\lambda^2 - C_3} \quad (\lambda \text{ in } \mu m). \tag{1.6}$$

Tab. 1.1: *Optical properties of selected glasses.*

Name	Boron crown	Heavy flint glass		Barium crown	Flint glass
Abbreviation	BK7	SF11	LaSF N9	BaK 1	F 2
Abbé number A	64.17	25.76	32.17	57.55	36.37
Refractive index n for selected wavelengths					
$\lambda = 486.1\,\text{nm}$	1.5224	1.8065	1.8690	1.5794	1.6321
$\lambda = 587.6\,\text{nm}$	1.5168	1.7847	1.8503	1.5725	1.6200
$\lambda = 656.3\,\text{nm}$	1.5143	1.7760	1.8426	1.5695	1.6150
Dispersion constants of refractive index (see Eq. 1.6)					
B_1	1.0396	1.7385	1.9789	1.1237	1.3453
B_2	0.2379	0.3112	0.3204	0.3093	0.2091
B_3	1.0105	1.1749	1.9290	0.8815	0.9374
C_1	0.0060	0.0136	0.0119	0.0064	0.0100
C_2	0.0200	0.0616	0.0528	0.0222	0.0470
C_3	103.56	121.92	166.26	107.30	111.89
Density ρ (g cm^{-3})					
	2.51	4.74	4.44	3.19	3.61
Expansion coefficient $\Delta\ell/\ell$ $(-30$ to $+70°\text{C})$ $\times 10^6$					
	7.1	6.1	7.4	7.6	8.2

Strain birefringence: typically $10\,\text{nm cm}^{-1}$.

Homogeneity of the refractive index from melt to melt: $\delta n/n = \pm 1 \times 10^{-4}$.

By geometrical considerations we find that the angle of deflection δ in Fig. 1.9 depends not only on the angle of incidence θ but also on the aperture angle α of the symmetrical prism and of course on the index of refraction, n,

$$
\begin{aligned}
\delta &= \alpha - \theta + \arcsin\left[\cos\alpha\sin\theta - \sin(\alpha\sqrt{n^2 - \sin^2\theta})\right], \\
\delta_{\min} &= \alpha - 2\theta_{\text{symm}}.
\end{aligned}
$$

The minimum deflection angle $\delta_{\min}$ is achieved for symmetrical transit through the prism and enables a precise determination of the refractive index. The final result is expressed straightforwardly by the quantities α and $\delta_{\min}$,

$$
n = \frac{\sin\left[(\alpha + \delta_{\min})/2\right]}{\sin(\alpha/2)}.
$$

For quantitative estimation of the dispersive power K of glasses, the Abbe number A may be used. This relates the refractive index at a yellow wavelength (at $\lambda = 587.6\,\text{nm}$, the D line of helium) to the change of the refractive index, estimated from the difference of the refractive indices at a blue ($\lambda = 486.1\,\text{nm}$, Fraunhofer line F of hydrogen) and a red wavelength ($\lambda = 656.3\,\text{nm}$, Fraunhofer line C of hydrogen),

$$
A = K^{-1} = \frac{n_{\text{D}} - 1}{n_{\text{F}} - n_{\text{C}}}.
$$

According to the above, a large Abbe number means weak dispersion, and a small Abbe number means strong dispersion. The Abbe number is also important when correcting chromatic aberrations (see Chapter 4.5.3).

The index of refraction describes the interaction of light with matter, and we will come to realize that it is a complex quantity, which describes not only the properties of dispersion but also those of absorption as well. Furthermore, it is the task of a microscopic description of matter to determine the dynamic polarizability and thus to establish the connection to a macroscopic description.

1.7   Light rays in waveguides

The transmission of messages via light signals is a very convenient method that has a very long history of application. For example, in the 19th century, mechanical pointers were mounted onto high towers and were observed with telescopes to realize transmission lines of many hundreds of miles. An example of a historic relay station from the 400 mile Berlin-Cologne-Coblenz transmission line is shown in Fig.1.10. Basically, in-air transmission is also performed nowadays, but with laser light. But it is always affected by its scattering properties even at small distances, because turbulence, dust and rain can easily inhibit the propagation of a free laser beam.

Ideas for guiding optical waves have been in existence for a very long time. In analogy to microwave techniques, for example, at first hollow tubes made of copper were applied, but their attenuation is too large for transmission over long distances. Later on periodical lens systems have been used for the same purpose, but due to high losses and small mechanical flexibility they also failed.

The striking breakthrough happened to 'optical telecommunication' through the development of low-loss *waveguides*, which are nothing other than elements for guiding light rays. They can be distributed like electrical cables, provided that adequate transmitters and receivers are available. With overseas cables, significantly shorter signal transit times and thus higher comfort for phone calls can be achieved than via geostationary satellites, where there is always a short but unpleasant and unnatural break between question and answer.

Therefore, propagation of light rays in dielectric waveguides is an important chapter in modern optics. Some basics may yet be understood by the methods of ray optics.

Fig. 1.10: *Historic station No. 51 of the Berlin–Cologne–Coblenz optic-mechanical 'sight' transmission line on the tower of the St. Pantaleon church, Cologne. Picture from Weiger (1840).*

1.7.1 Ray optics in waveguides

Total internal reflection in an optically thick medium provides the fundamental physical phenomenon for guiding light rays within a dielectric medium. Owing to this effect, for example, in cylindrical homogeneous glass fibres, rays whose angle with the cylinder axis stays smaller than the angle of total internal reflection θ_c are guided from one end to the other. Guiding of light rays in a homogeneous glass cylinder is affected by any distortion of the surface, and a protective cladding could even suppress total internal reflection.

Therefore, various concepts have been developed, where the optical waves are guided in the centre of a waveguide through variation of the index of refraction. These waveguides may be surrounded by cladding and entrenched like electrical cables.

We will present the two most important types. Step-index fibres consist of two homogeneous cylinders with different refractive indices (Fig. 1.11). To achieve beam guiding, the higher index of refraction must be in the cladding. Gradient-index fibres with continuously changing (in good approximation, parabolic) refractive index are more sophisticated to manufacture, but they have technical advantages like, for example, a smaller group velocity dispersion.

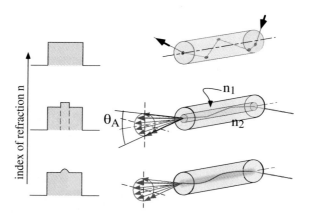

Fig. 1.11: *Profiles of the refractive index and ray path in optical waveguides. Upper: waveguide with homogeneous refractive index. Centre: waveguide with stepped profile of refractive index (step-index fibre). Lower: waveguide with continuous profile of refractive index (gradient-index fibre).*

Excursion: Manufacturing waveguides

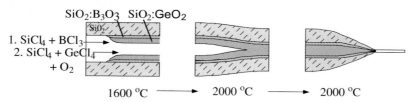

Fig. 1.12: *Manufacturing of waveguides. The preform is manufactured with appropriate materials with distinct indices of refraction, which are deposited on the inner walls of a quartz tube by a chemical reaction.*

The starting material is an ordinary tube made of quartz glass. It rotates on a lathe and is blown through on the inside by a gas mixture (chlorides like highly purified $SiCl_4$, $GeCl_4$, etc.). A oxyhydrogen burner heats a small zone of only a few inches up to about 1600°C, in which the desired materials are deposited as oxides on the inner walls (*chemical vapour deposition, CVD*). Thus by multiple repetition a refractive index profile is established, before the tube is melted at about 2000°C to a massive glass rod of about 10 mm diameter, a so-called preform. In the last step a fibre pulling machine extracts the glass fibre out of a crucible with viscous material. Typical cross-sections are 50 and 125 μm, which are coated with a cladding for protection.

1.7.2   Step-index fibres

The principle of total internal reflection is applied in *step-index fibres* (Fig. 1.13), which consist of a *core* with refractive index n_1 and a *cladding* with $n_2 < n_1$. The relative difference in the index of refraction

$$\Delta = \frac{n_1 - n_2}{n_1} \qquad (1.7)$$

is not more than 1–2%, and the light rays are only guided if the angle α towards the fibre axis is shallow enough to fulfil the condition for total internal reflection.

For example, for quartz glass ($n_2 = 1.45$ at $\lambda = 1.55\,\mu\mathrm{m}$), whose core index of refraction has been enhanced by GeO_2 doping up to $n_1 = n_2 + 0.015$, according to $\theta_c = \sin^{-1}(n_2/n_1)$ one finds the critical angle $\theta_c = 81.8°$. The complementary beam angle relative to the fibre axis, $\alpha_G = 90° - \theta_c$, can be approximated by

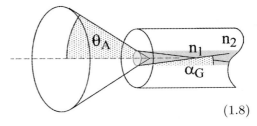

Fig. 1.13: *Critical angle in a step-index fibre.*

$$\alpha_G \simeq \sin\alpha_G \simeq \sqrt{2\Delta}, \qquad (1.8)$$

since $n_2/n_1 = 1 - \Delta$, and thus is set in relation to Δ, which yields $\alpha \leq 8.2°$ for this case.

When light rays cross the axis of a fibre, propagation takes place in a cut plane, which is called the *meridional plane*. *Skewed rays* pass through the axis and are guided on a polygon around the circle. It can be shown that the rays must confine an angle $\alpha < \alpha_G$ with the z axis to be guided by total internal reflection.

Numerical aperture of an optical fibre

To guide a light ray in an optical fibre, the angle of incidence at the incoupler must be chosen small enough. The maximum aperture angle θ_a of the acceptance cone can be calculated according to the refraction law, $\sin\theta_a = n_1 \sin\alpha_G = n_1 \cos\theta_c$. The sine of the aperture angle is called the numerical aperture (NA). According to Eq. (1.8) and $\cos\theta_c \simeq \sqrt{2\Delta}$ it can be related with the physical parameters of the optical fibre,

$$\mathrm{NA} = n_1\sqrt{2\Delta}. \qquad (1.9)$$

This yields, for example, NA = 0.21 for the quartz glass fibre mentioned above, which is a useful and typical value for standard waveguides.

Propagation velocity

Light within the core of the waveguide propagates along the trajectory with a velocity $v(r(z)) = c/n(r(z))$. Along the z axis the beam propagates with a reduced velocity, $\langle v_z \rangle = v\cos\alpha$, which can be calculated for small angles $\alpha(z)$ to the z axis according to

$$\langle v_z \rangle \simeq \frac{c}{n_1}\left(1 - \frac{1}{2}\alpha^2\right). \qquad (1.10)$$

In Chapter 3.3 on the wave theory of light, we will see that the propagation velocity is identical with the group velocity.

1.7.3   Gradient-index fibres

Beam guiding can also be performed by means of a *gradient-index fibre* (GRIN), where the quadratic variation of the index of refraction is important. To determine the curvature of a light ray induced by the refractive index, we apply the ray equation (1.5). This is greatly simplified in the paraxial approximation $(ds \simeq dz)$ and for a cylindrically symmetric fibre,

$$\frac{d^2 r}{dz^2} = \frac{1}{n} \frac{dn}{dr}.$$

A parabolic profile of the refractive index with a difference of the refractive index of $\Delta = (n_1 - n_2)/n_1$,

$$n(r{\le}a) = n_1 \left[1 - \Delta \left(\frac{r}{a}\right)^2\right] \qquad \text{and} \qquad n(r{>}a) = n_2, \tag{1.11}$$

decreases from the maximum value n_1 at $r = 0$ to n_2 at $r = a$. One ends up with the equation of motion of a harmonic oscillator,

$$\frac{d^2 r}{dz^2} + \frac{2\Delta}{a^2} r = 0,$$

and realizes immediately that the light ray performs oscillatory motion about the z axis. The period is

$$\Lambda = \frac{2\pi a}{\sqrt{2\Delta}}, \tag{1.12}$$

and a light ray is described with a wavenumber $K = 2\pi/\Lambda$ according to

$$r(z) = r_0 \sin(2\pi z/\Lambda).$$

The maximum elongation allowed is $r_0 = a$, because otherwise the beam loses its guiding. Thereby also the maximum angle $\alpha_G = \sqrt{2\Delta}$ for crossing the axis occurs. It is identical with the critical angle for total internal reflection in a step-index fibre and yields also the same relation to the numerical aperture (Eq. (1.9)). As in the case of a step-index fibre, the propagation velocity of the light ray is of interest. Using the approximation eq. (1.10) we calculate the average velocity during an oscillation period with $\tan \alpha \simeq \alpha = dr(z)/dz$,

$$\langle v_z \rangle = \left\langle \frac{c \, \cos \alpha(r(z))}{n(r(z))} \right\rangle = \frac{c}{n_1} \left\langle \frac{1 - \Delta(r_0/a)^2 \cos^2(Kz)}{1 - \Delta(r_0/a)^2 \sin^2(Kz)} \right\rangle \quad,$$

and find after a short conversion the remarkable result

$$\langle v_z \rangle = \frac{c}{n_1} \left[1 - \left(\frac{\Delta}{2}\right)^2 \left(\frac{\alpha}{\alpha_G}\right)^2\right] \quad,$$

which actually means that, because $\Delta \ll 1$, the propagation velocity within a gradient-index fibre depends much less on the angle α than that within a step-index fibre. As we will see, this circumstance plays an important role for signal propagation in waveguides (see Chapter 3.3).

1.8   Lenses and curved mirrors

The formation of an image plays a major role in optics, and *lenses* and *curved mirrors* are essential parts in optical devices. First we will discuss the effect of these components on the propagation of rays; owing to its great importance we have dedicated an extra chapter (Chapter 4) to the formation of images.

1.8.1   Lenses

We define an *ideal* lens as an optical element that merges all rays of a point-like source into one point again. An image where all possible object points are transferred into image points is called a *stigmatic* image (from the Greek: *stigma*, point). The source may even be far away and illuminate the lens with a parallel bundle of rays. In this case the point of merger is called the *focal point* or *focus*. In Fig. 1.14, we consider a beam of parallel rays that passes through the lens and is merged in the focal point. According to Fermat's principle, the optical pathlength must be equal for all possible paths, which means that they are independent from the distance of a partial beam from the axis. Then the propagation of light must be delayed most on the symmetry axis of the lens and less and less at the outer areas!

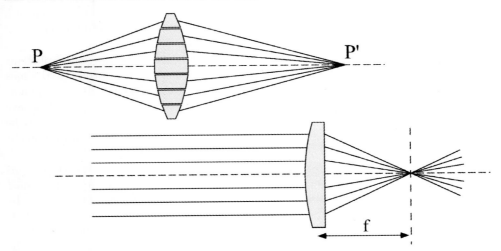

Fig. 1.14: *Upper: Stigmatic lens imaging. All rays starting at object point P are merged again at image point P'. The light rays are delayed more near the axis of the lens body than in the outermost areas, so that all rays make the same optical pathlength to the image point. A lens may be figured as a combination of several prisms. Lower: A parallel beam of rays originating from a source at infinite distance is focused at the focal point at focal distance f.*

For a simplified analysis, we neglect the thickness of the lens body, consider the geometrical increase of the pathlength from the lens to the focal point at a distance f

and expand the term as a function of distance r from the axis,

$$\ell(r) = \sqrt{f^2 + r^2} \simeq f \left(1 + \frac{r^2}{2f^2}\right). \tag{1.13}$$

To compensate for the quadratic increase of the optical pathlength $\ell(r)$, the delay by the path within the lens glass – i.e. the thickness – must also vary quadratically. This is actually the condition for spherical surfaces, which have been shown to be extremely successful for convergent lenses! The result is the same with much more mathematical effort, if one explores the properties of refraction at a lens surface assuming that a lens is constructed of many thin prisms (Fig. 1.14). In the chapter on lens aberrations, we will deal with the question of which criteria should be important for the choice of a planar convex or biconvex lens.

1.8.2   Concave mirrors

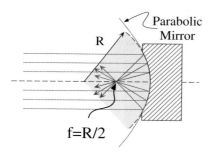

Among curved mirrors, concave or parabolic mirrors play the most important role. They are very well known from huge astronomical telescopes (see Chapter 4), because we entered the fascinating world of the cosmos with their aid. But they are used much more often in laser resonators (Chapter 5.6).

Taking into account the tangential plane at the intercept of the surface normal at the lens surface, we can transfer the conditions of planar reflection to curved mirror surfaces. Concave mirrors mostly have axial symmetry, and the effect on a parallel beam of rays within one cut plane is visualized in Fig. 1.15.

Fig. 1.15: *Path of rays for a concave mirror. For near-axis incident light, spherical mirrors are used.*

The reflected partial rays meet at the focal point or focus on the mirror axis, as they do in the case of a lens. It is known from geometry that the reflection points must then lie on a parabola. Near the axis, parabolic mirrors may in good approximation be substituted by spherical mirrors, which are much easier to manufacture. On the left-hand side of Fig. 1.16 the geometrical elements are shown, out of which the dependence of the focal length (defined here by the intersection point with the optical axis) on the axis distance y_0 of a parallel incident beam may be calculated,

$$f = R - \frac{R}{2\cos\alpha} \simeq \frac{R}{2}\left[1 - \frac{1}{2}\left(\frac{y_0}{R}\right)^2 + \cdots\right].$$

In general we neglect the quadratic correction, which causes an aperture error and is investigated in more detail in Chapter 4.5.2.

In laser resonators a situation often occurs in which spherical mirrors are simultaneously used as deflection mirrors, e.g. in the 'bowtie resonator' in Fig. 7.35. Then the

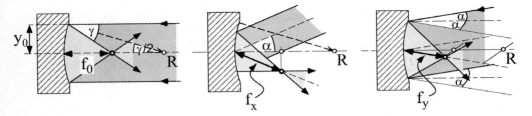

Fig. 1.16: *Focusing an incident beam that is parallel to (left) and oblique to (centre: top view; right: side view) the optical axis.*

focal width of the rays within the ray plane (f_x) and within the plane perpendicular to that one (f_y) will differ from $f_0 = R/2$,

$$f_x = \frac{R}{2\cos\alpha} = \frac{f_0}{\cos\alpha} \qquad \text{and} \qquad f_y = \frac{R\cos\alpha}{2} = f_0\cos\alpha.$$

The geometrical situations in the top view (Fig. 1.16, centre) are easy to see. In the side view one looks at the projection onto a plane perpendicular to the direction of emergence. The projections of the radius and focal length are reduced to $R\cos\alpha$ and $f\cos\alpha$, respectively. The difference between the two planes occurring here is called astigmatic aberration and sometimes can be compensated by simple means (see for example p. 126).

1.9 Matrix optics

As a result of its rectilinear propagation, a free light ray may be treated like a straight line. In optics, systems with axial symmetry are especially important, and an individual light ray may be described sufficiently well by the distance from and angle to the axis (Fig. 1.17). If the system is not rotationally symmetric, for example after passing through a cylindrical lens, then we can deal with two independent contributions in the x and y directions with the same method.

The modification of the beam direction by optical components – mirrors, lenses, dielectric surfaces – is described by a trigonometric and therefore not always simple relation. For near-axis rays, these functions can often be linearized, and thus the mathematical treatment is simplified enormously. This becomes obvious, for example, for a linearized form of the law of refraction (1.2):

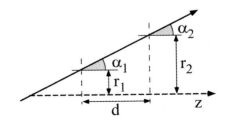

Fig. 1.17: *Key variables of an optical ray for simple translation.*

$$n_1\theta_1 = n_2\theta_2. \tag{1.14}$$

Here we have made use of this approximation already with the application of Fermat's principle for ideal lenses. Near-axis rays allow the application of spherical surfaces for lenses, which are much easier to manufacture than mathematical ideal surfaces. Furthermore, ideal systems are only 'ideal' for selected ray systems, otherwise they suffer from image aberrations like other systems.

When treating the modification of a light ray by optical elements in this approximation by linear transformation, matrices are a convenient mathematical tool for calculating the fundamental properties of optical systems. The development of this method made for the denomination *matrix optics*. The introduction of transformation matrices for ray optics may be visualized very easily, but they achieved striking importance, because they do not change their form when treating near-axis rays according to wave optics (see section 2.3.2). Furthermore this formalism is also applicable for other types of optics like 'electron optics', or the even more general 'particle optics'.

1.9.1   Paraxial approximation

Let us consider the propagation of a light ray at a small angle α to the z axis. The beam is fully determined by the distance r from the z axis and the slope $r' = \tan \alpha$. Within the so-called *paraxial approximation*, we now linearize the tangent of the angle and substitute it by its argument, $r' \simeq \alpha$, and then merge r with r' to end up with a vector $\mathbf{r} = (r, \alpha)$. At the start a light ray may have a distance to the axis and a slope of $\mathbf{r}_1 = (r_1, \alpha_1)$. Having passed a distance d along the z axis, then

$$
\begin{aligned}
r_2 &= r_1 + \alpha_1 d, \\
\alpha_2 &= \alpha_1,
\end{aligned}
$$

holds. One may use 2×2 matrices to write the translation clearly,

$$
\mathbf{r}_2 = \mathbf{T} \, \mathbf{r}_1 = \begin{pmatrix} 1 & d \\ 0 & 1 \end{pmatrix} \mathbf{r}_1. \tag{1.15}
$$

A bit more complicated is the modification by a refracting optical surface. For that purpose we look at the situation of Fig. 1.18, where two optical media with refractive indices n_1 and n_2 are separated by a spherical interface with radius R. If the radius vector subtends an angle ϕ with the z axis, then the light ray is obviously incident on the surface at an angle $\theta_1 = \alpha_1 + \phi$ and is related to the angle of emergence by the law of refraction.

In paraxial approximation according to Eq. (1.2), $n_1 \theta_1 \simeq n_2 \theta_2$ and $\phi \simeq r_1/R$ is valid, and one finds

$$
n_1 \left(\alpha_1 + \frac{r_1}{R} \right) = n_2 \left(\alpha_2 + \frac{r_2}{R} \right).
$$

The linearized relations may be described easily by the refraction matrix $\mathbf{B}$,

$$
\begin{pmatrix} r_2 \\ \alpha_2 \end{pmatrix} = \mathbf{B} \begin{pmatrix} r_1 \\ \alpha_1 \end{pmatrix} = \begin{pmatrix} 1 & 0 \\ (n_1 - n_2)/n_2 R & n_1/n_2 \end{pmatrix} \begin{pmatrix} r_1 \\ \alpha_1 \end{pmatrix}. \tag{1.16}
$$

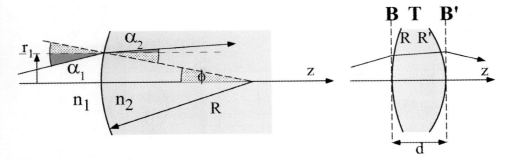

Fig. 1.18: *Modification of a light ray at curved refracting surfaces.*

1.9.2 ABCD matrices

The most important optical elements may be specified by their transformations, also called *ABCD matrices*,

$$
\begin{pmatrix} r_2 \\ \alpha_2 \end{pmatrix} = \begin{pmatrix} A & B \\ C & D \end{pmatrix} \begin{pmatrix} r_1 \\ \alpha_1 \end{pmatrix}, \tag{1.17}
$$

which we collect in Tab. 1.2 for look-up purposes and will be presented in the following in more detail.

According to Fig. 1.18 the effect of a lens on a light ray is characterized by a refraction $\mathbf{B}$ at the entrance, a translation $\mathbf{T}$ in the glass and one further refraction $\mathbf{B'}$ at the exit. Now the matrix method shows its strength, because the effect of a lens can easily be expressed as a product $\mathbf{L} = \mathbf{B'TB}$ of three operations,

$$
\begin{pmatrix} r_2 \\ \alpha_2 \end{pmatrix} = \mathbf{L} \begin{pmatrix} r_1 \\ \alpha_1 \end{pmatrix} = \mathbf{B'TB} \begin{pmatrix} r_1 \\ \alpha_1 \end{pmatrix}. \tag{1.18}
$$

Before we discuss the lens and some more examples in detail, we have to fix some conventions, which in general are used in matrix optics:

1. The ray direction goes from left to right in the positive direction of the z axis.

2. The radius of a convex surface is positive, $R > 0$, and that of a concave surface is negative, $R < 0$.

3. The slope is positive when the beam moves away from the axis, and negative when it moves towards the axis.

4. An object distance or image distance is positive (negative) when lying in front of (behind) the optical element.

5. Object distances are defined to be positive (negative) above (below) the z axis.

6. Reflective optics is treated by flipping the ray path after every element.

Tab. 1.2: *Important ABCD matrices.*

Operation		ABCD-Matrix
Translation		$\begin{pmatrix} 1 & d \\ 0 & 1 \end{pmatrix}$
Refraction (planar surface)		$\begin{pmatrix} 1 & 0 \\ 0 & n_1/n_2 \end{pmatrix}$
Refraction (curved surface)		$\begin{pmatrix} 1 & 0 \\ \dfrac{n_1-n_2}{n_2 R} & \dfrac{n_1}{n_2} \end{pmatrix}$
Lenses Curved Mirrors (focal length f)		$\begin{pmatrix} 1 & 0 \\ -1/f & 1 \end{pmatrix}$
Optical Fiber GRIN (length ℓ)		$\begin{pmatrix} \cos K\ell & K^{-1}\sin K\ell \\ -K\sin K\ell & \cos K\ell \end{pmatrix}$

1.9.3   Lenses in air

Now we will explicitly calculate the lens matrix **L** according to eq. (1.18) and we take into account the index of refraction $n_{\mathrm{air}} = 1$ in Eqs. (1.15) and (1.16). The expression

$$\mathbf{L} = \begin{pmatrix} 1 - \dfrac{n-1}{n}\dfrac{d}{R} & \dfrac{d}{n} \\ (n-1)\left[\dfrac{1}{R'} - \dfrac{1}{R} - \dfrac{d(n-1)^2}{RR'n}\right] & 1 + \dfrac{n-1}{n}\dfrac{d}{R'} \end{pmatrix}$$

makes a complicated and not very convenient expression at first sight. Though it may allow the treatment of very thick lenses, by far the most important are the predominantly used 'thin' lenses, whose thickness d is small compared to the radii of curvature R, R' of the surfaces. With d/R, $d/R' \ll 1$ or by direct multiplication $\mathbf{B'B}$, we find the much simpler form

$$\mathbf{L} \simeq \begin{pmatrix} 1 & 0 \\ (n-1)\left(\dfrac{1}{R'} - \dfrac{1}{R}\right) & 1 \end{pmatrix}$$

and introduce the symbol $\mathcal{D}$ for the *refractive power* in the lensmaker's equation ,

$$\mathcal{D} = -(n-1)\left(\frac{1}{R'} - \frac{1}{R}\right) = \frac{1}{f}. \tag{1.19}$$

Thus the ABCD matrix for thin lenses becomes very simple

$$\mathbf{L} = \begin{pmatrix} 1 & 0 \\ -\mathcal{D} & 1 \end{pmatrix} = \begin{pmatrix} 1 & 0 \\ -\frac{1}{f} & 1 \end{pmatrix}, \tag{1.20}$$

where the sign is chosen such that convergent lenses have a positive refractive power. The refractive power is identical with the inverse focal length, $\mathcal{D} = 1/f$. The refractive power $\mathcal{D}$ is measured in units of dioptres ($1\,\mathrm{dpt} = 1\,\mathrm{m}^{-1}$).

To support the interpretation of Eq. (1.20), we consider a bundle of rays that originates from a point source G on the z axis (Fig. 1.19). Such a bundle of rays can be described at a distance g from the source according to

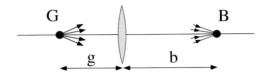

Fig. 1.19: *Point image formation with a lens.*

$$\begin{pmatrix} r \\ \alpha \end{pmatrix} = \alpha \begin{pmatrix} g \\ 1 \end{pmatrix}. \tag{1.21}$$

We calculate the effect of the lens in the form

$$\mathbf{L} \begin{pmatrix} r \\ \alpha \end{pmatrix} = \alpha \begin{pmatrix} g \\ 1 - g/f \end{pmatrix} = \alpha' \begin{pmatrix} -b \\ 1 \end{pmatrix}. \tag{1.22}$$

The lens transforms the incident bundle of rays into a new bundle, which again has the form (1.21). It converges for $\alpha' < 0$ to the axis, crosses it at a distance $b > 0$ (convention 4) behind the lens, and creates there an image of the point source. If $b < 0$, then the virtual image of the point source lies in front of the lens and the lens has the properties of a dispersive lens.

By comparison of coefficients, we obtain the relation between object distance g and image distance b from Eq. (1.22) for lens imaging:

$$\frac{1}{f} = \frac{1}{g} + \frac{1}{b}. \tag{1.23}$$

This equation is the known basis for optical imaging. We refer to this topic again in Chapter 4 in more detail.

Example: ABCD matrix of an imaging system
For imaging by an arbitrary ABCD system, we must claim that a bundle of rays (r_1, α_1) is again merged at a point at a certain distance $d = d_1 + d_2$:

$$\begin{pmatrix} r_2 \\ \alpha_2 \end{pmatrix} = \begin{pmatrix} 1 & d_1 \\ 0 & 1 \end{pmatrix} \begin{pmatrix} A & B \\ C & D \end{pmatrix} \begin{pmatrix} 1 & d_2 \\ 0 & 1 \end{pmatrix} \begin{pmatrix} r_1 \\ \alpha_1 \end{pmatrix}.$$

For stigmatic imaging r_2 must be independent of α_1 and by calculation one finds the condition $d_1 D + d_2 A + d_1 d_2 C + B = 0$, which for $B = 0$ can be fulfilled by suitable choice of d_1 and d_2, even if $C < 0$. Thus the ABCD matrix takes exactly the form that we know already from lenses and lens systems.

1.9.4   Lens systems

The matrix method enables us to explore the effect of a system consisting of two lenses with focal lengths f_1 and f_2 at a distance d. We multiply the ABCD matrices according to Eqs. (1.20) and (1.15) and get the matrix $\mathbf{M}$ of the system

$$\mathbf{M} = \mathbf{L}_2\mathbf{T}\mathbf{L}_1 = \begin{pmatrix} 1 & 0 \\ -1/f_2 & 1 \end{pmatrix} \begin{pmatrix} 1 & d \\ 0 & 1 \end{pmatrix} \begin{pmatrix} 1 & 0 \\ -1/f_1 & 1 \end{pmatrix}$$

$$= \begin{pmatrix} 1 - \dfrac{d}{f_1} & d \\ -\left(\dfrac{1}{f_2} + \dfrac{1}{f_1} - \dfrac{d}{f_1 f_2}\right) & 1 - \dfrac{d}{f_2} \end{pmatrix}. \tag{1.24}$$

The system of two lenses substitutes a single lens with focal length given by

$$\frac{1}{f} = \frac{1}{f_2} + \frac{1}{f_1} - \frac{d}{f_1 f_2}. \tag{1.25}$$

We consider the following two interesting extreme cases.

(i) $d \ll f_{1,2}$: Two lenses that are mounted directly next to each other, with no space between them, add their refractive powers, $\mathbf{M} \simeq \mathbf{L}_2\mathbf{L}_1$ with $\mathcal{D} = \mathcal{D}_1 + \mathcal{D}_2$. This circumstance is used for example when adjusting eyeglasses, when refractive powers are combined until the required correction is found. Obviously a biconvex lens can be constructed out of two planar convex lenses, expecting that the focal length of the system is divided by two.

(ii) $d = f_1 + f_2$: If the focal points coincide, a telescope is realized. A parallel bundle of rays with radius r_1 is widened or collimated into a new bundle of parallel rays with a new diameter $(f_2/f_1)r_1$. The refractive power of the system vanishes according to eq. (1.25), $\mathcal{D} = 0$. Such systems are called *afocal*.

A thin lens is one of the oldest optical instruments, and thin lenses may have many different designs due to their various applications. But since lens aberrations are of major interest, we will dedicate a specific chapter to the various designs (Chapter 4.5.1).

1.9.5   Periodic lens systems

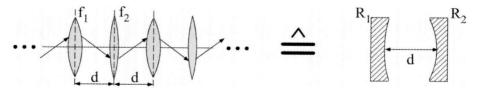

Fig. 1.20: *Periodic lens system and equivalence to a two-mirror resonator.*

Periodic lens systems had already been analysed in early times to realize optical light transmission lines. For such an application it is important that a light ray does not leave the system even after long distances. We consider a periodic variant of the lens system with focal lengths f_1 and f_2 at a distance d. For that purpose we add one more identical translation to the transformation matrix from Eq. (1.24), which yields a system equivalent to a system of two concave mirrors (Fig. 1.20):

$$\begin{pmatrix} A & B \\ C & D \end{pmatrix} = \begin{pmatrix} 1 & 0 \\ -1/f_2 & 1 \end{pmatrix} \begin{pmatrix} 1 & d \\ 0 & 1 \end{pmatrix} \begin{pmatrix} 1 & 0 \\ -1/f_1 & 1 \end{pmatrix} \begin{pmatrix} 1 & d \\ 0 & 1 \end{pmatrix}$$

$$= \begin{pmatrix} 1 & d \\ -1/f_2 & 1 - d/f_2 \end{pmatrix} \begin{pmatrix} 1 & d \\ -1/f_1 & 1 - d/f_1 \end{pmatrix}.$$

Now for n-fold application the individual element will cause total transformation

$$\begin{pmatrix} A_n & B_n \\ C_n & D_n \end{pmatrix} = \begin{pmatrix} A & B \\ C & D \end{pmatrix}^n.$$

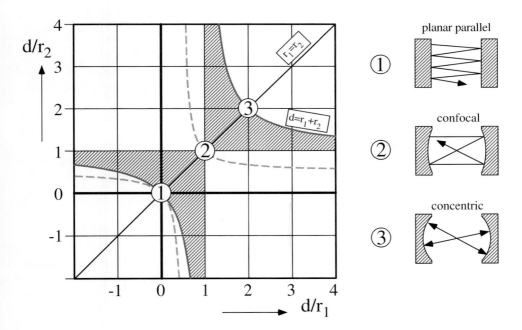

Fig. 1.21: *Stability diagram for lenses and optical resonators according to the condition (1.27). Stable resonator configurations are within the hatched area. The dashed lines indicate the positions of confocal resonators, $d = (r_1 + r_2)/2$. Symmetric planar parallel, confocal and concentric resonators are at the positions circled 1, 2 and 3.*

Introducing

$$\cos \Theta = \frac{1}{2}(A + D) = 2 \left(1 - \frac{d}{2f_1} \right) \left(1 - \frac{d}{2f_2} \right) - 1, \tag{1.26}$$

this matrix can be evaluated algebraically. Thus one calculates

$$\begin{pmatrix} A & B \\ C & D \end{pmatrix}^n = \frac{1}{\sin\Theta} \begin{pmatrix} A\sin n\Theta - \sin(n-1)\Theta & B\sin n\Theta \\ C\sin n\Theta & D\sin n\Theta - \sin(n-1)\Theta \end{pmatrix}.$$

The angle Θ must remain real, to avoid the matrix coefficients increasing to infinity. Otherwise the light ray would actually leave the lens system. Thus from the properties of the cosine function,

$$-1 \leq \cos\Theta \leq 1,$$

and in combination with Eq. (1.26) we get

$$0 \leq \left(1 - \frac{d}{2f_1}\right)\left(1 - \frac{d}{2f_2}\right) \leq 1. \tag{1.27}$$

This result defines a stability criterion for the application of a waveguide consisting of lens systems, and the corresponding important stability diagram is shown in Fig. 1.21. We will deal with this in more detail later on, because multiple reflection between concave mirrors of an optical resonator can be described in this way as well (Chapter 5.6).

1.9.6   ABCD matrices for waveguides

According to Section 1.7 and with the aid of the wavenumber constant $K = 2\pi/\Lambda$ (Eq. (1.12)) a simple ABCD matrix for the transformation of a ray by a graded-index fibre of length ℓ can be specified:

$$\mathbf{G} = \begin{pmatrix} \cos K\ell & K^{-1}\sin K\ell \\ -K\sin K\ell & \cos K\ell \end{pmatrix}. \tag{1.28}$$

With short pieces of fibre ($K\ell < \pi/4$) also thin lenses can be realized, and it can be shown that the focal point lies at $f = K^{-1}\cot K\ell$. These components are called *GRIN lenses*.

1.10   Ray optics and particle optics

Traditional optics, which deals with light rays and is the topic of this textbook, was conceptually in every respect a role model for 'particle optics', which started around the year 1900 with the exploration of electron beams and radioactive rays. Since ray optics describes the propagation of light rays, it is convenient to look for analogies in the trajectories of particles. We will see in the chapter on coherence and interferometry (Chapter 5) that the wave aspects of particle beams are widely coined in terms of the ideas of optics as well.

To re-establish the analogy explicitly, we refer to considerations about Fermat's principle (p. 4), because there a relation between the velocity of light and the index of refraction is described. This relation is particularly simple if a particle moves in a conservative potential (potential energy $E_{\mathrm{pot}}(\mathbf{r})$), like for example an electron in an electric field. As a result of energy conservation,

$$E_{\mathrm{kin}}(\mathbf{r}) + E_{\mathrm{pot}}(\mathbf{r}) = E_{\mathrm{tot}},$$

we can immediately infer from $E_{\text{kin}} = mv^2/2$ that

$$v(\mathbf{r}) = \sqrt{\frac{2}{m}[E_{\text{tot}} - E_{\text{pot}}(\mathbf{r})]},$$

if the particles do not move too fast and we can adopt classical Newtonian mechanics (in a particle accelerator, the special theory of relativity has to be applied.).

We can define an effective relative index of refraction by

$$\frac{v(\mathbf{r}_1)}{v(\mathbf{r}_2)} = \frac{n_{\text{eff}}(\mathbf{r}_2)}{n_{\text{eff}}(\mathbf{r}_1)} = \frac{\sqrt{[E_{\text{tot}} - E_{\text{pot}}(\mathbf{r}_2)]}}{\sqrt{[E_{\text{tot}} - E_{\text{pot}}(\mathbf{r}_1)]}}.$$

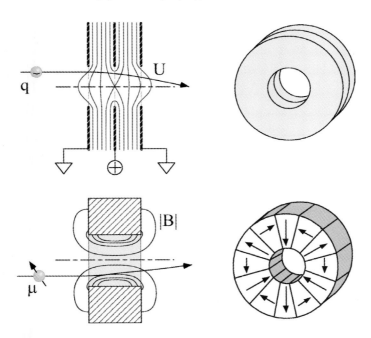

Fig. 1.22: *Lenses for particle optics. Upper: So-called 'single lens' for electron optics with equipotential surfaces qU [81]. The potential is created by symmetric positioning of three conducting electrodes, the two outer ones lying on the same potential. Lower: Magnetic lens for atom optics with equipotential surfaces $|\mu \cdot B|$ [53]. An axial magnetic hexapole is formed out of circle segments, which are manufactured from a homogeneously magnetized permanent magnet (e.g. NdFeB or SmCo). The strength of the magnetic field rises as a square function of the radial distance.*

As in the case of light, it must satisfy an additional condition, to be defined absolutely. For example, we may claim that $n_{\text{eff}} = 1$ for $E_{\text{pot}} = 0$. But then it is obvious that n_{eff} depends extremely on the velocity outside of the potential – particle optics has properties that are very much chromatic! The fundamental reason for this difference is the different relation between kinetic energy E and momentum p for light and for particles having mass, which is also called the *dispersion relation*:

Light $E = pc,$
Particles $E = p^2/2m.$

In charged particle beams a narrow velocity distribution can be created by acceleration, which makes the difference not very pronounced. But the broadness of the velocity distribution in thermal beams of neutral atoms induces significant problems. Indeed, this velocity distribution can be manipulated by so-called supersonic jets or by laser cooling (see Chapter 11.6) in such a way that even 'atom optics' can be established [53]. We present some important devices of electron and atom optics in Fig. 1.22.

2 Wave optics

At the beginning of the 19th century a few phenomena were known that could not be reconciled with simple rectilinear, ray-like propagation of light, and made wave theory necessary. The beginning is marked by Huygens' principle (after the Dutch mathematician and physicist C. Huygens, 1629–1695), an explanation of wave propagation often used up to now and very intuitive. About 100 years later T. Young (1773–1829) from England and A. P. Fresnel (1788–1827) from France developed a very successful wave theory, which could explain all the phenomena of interference known at that time. After G. Kirchhoff (1824–1887) had given a mathematical formulation of Huygens' principle, the final breakthrough occurred with the famous Maxwell's equations, which will serve also here as a systematic basis for the wave theory of light.

The development of a common theoretical description of electric and magnetic fields by the Scottish physicist J. C. Maxwell (1831–1879) had a crucial influence not only on physics, but also on the science and technology of the 20th century. Maxwell's equations, which had at first been found through empirical knowledge and aesthetic considerations, caused for example Heinrich Hertz in 1887 to excite radio waves for the first time, thereby laying the foundation for modern telecommunications techniques.

2.1 Electromagnetic radiation fields

Electromagnetic fields are defined by two vector fields,[1]

$$\mathbf{E}(\mathbf{r}, t), \quad \text{electric field,}$$
$$\text{and} \quad \mathbf{H}(\mathbf{r}, t), \quad \text{(magnetic) } H \text{ field.}$$

They are caused by electric charges and currents.

2.1.1 Static fields

Charges are the *sources* of electric fields. The formal relation between field strength and charge density ρ and total charge Q, respectively, in a volume with surface S is described by Gauss's law in differential or integral form,

$$\mathbf{\nabla} \cdot \mathbf{E} = \rho/\epsilon_0 \quad \text{or} \quad \oint_S \mathbf{E} \cdot \mathbf{df} = Q/\epsilon_0. \tag{2.1}$$

---

[1] We follow the recent general literature, where usually the notation $\mathbf{B}(\mathbf{r}, t)$ stands for the magnetic field or B field, while $\mathbf{H}(\mathbf{r}, t)$ is simply called the (magnetic) H field.

Optics, Light and Laser. Dieter Meschede
Copyright © 2004 Wiley-VCH Verlag GmbH & Co. KGaA
ISBN: 3-527-40364-7

Furthermore, an electrostatic field is irrotational (curl-free), which means that $\nabla \times \mathbf{E} = 0$, and it may be described as the gradient of a scalar electrostatic potential $\Phi(\mathbf{r})$,

$$\mathbf{E}(\mathbf{r}) = -\nabla\Phi(\mathbf{r}).$$

The sources of the magnetic field are not charges, because it is known that

$$\nabla \cdot \mathbf{H} = 0, \tag{2.2}$$

but instead *curls*, which are caused by currents (current density $\mathbf{j}$, total current I crossing a surface with contour $\mathcal{C}$). According to Stokes' law

$$\nabla \times \mathbf{H} = \mathbf{j} \qquad \text{or} \qquad \oint_{\mathcal{C}} \mathbf{H} \cdot \mathbf{dl} = I \tag{2.3}$$

is valid. The field strength of the H-field may be described as the curl of a vector potential $\mathbf{A}(\mathbf{r})$,

$$\mathbf{H}(\mathbf{r}) = \frac{1}{\mu_0}\nabla \times \mathbf{A}(\mathbf{r}, t).$$

2.1.2   Dielectric media

The considerations of the preceding section are only valid for free charges and currents. But usually these are bound to materials, which we can roughly divide into two classes, *conductors* and *insulators*. In conducting materials charges can move freely; in insulators they are bound to a centre, but an external field causes a macroscopic dielectric *polarization* through displacement of charge.[2] For example, polar molecules may be oriented in a water bath, or a charge asymmetry may be induced in initially symmetric molecules (Fig. 2.1). In a homogeneous sample, negative and positive charges compensate, and there is left only an effective charge density at the border of the polarized volume. If the polarization varies continuously, then the compensation is cancelled and one gets an effective charge density

$$\rho_{\mathrm{pol}} = -\nabla \cdot \mathbf{P}(\mathbf{r}, t).$$

Of course, polarization charges must be accounted for as well, and therefore in dielectric matter it holds that

$$\nabla \cdot \mathbf{E} = \frac{1}{\epsilon_0}\left(\rho_{\mathrm{free}} + \rho_{\mathrm{pol}}\right).$$

In many important optical materials the polarization charge is proportional to the external field strength, and the coefficient is called the *dielectric susceptibility* χ,

$$\mathbf{P} = \epsilon_0\chi\mathbf{E}.$$

We introduce the *dielectric displacement* with the relative dielectric constant $\kappa = 1 + \chi$,

$$\mathbf{D} = \epsilon_0\mathbf{E} + \mathbf{P} = \epsilon_0\kappa\mathbf{E}, \tag{2.4}$$

and thus we can write much more simply

$$\nabla \cdot \mathbf{D} = \rho. \tag{2.5}$$

---

[2]More precisely it is a polarization density.

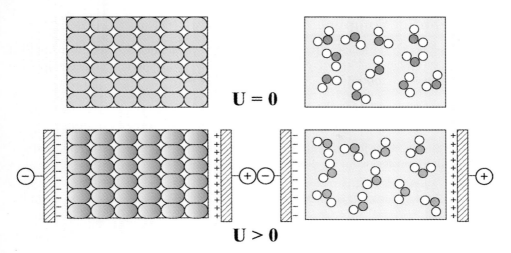

Fig. 2.1: *In a solid (left), charges are separated. In a glass with polar molecules (right), existing dipoles are oriented.*

An analogue to the dielectric polarization, namely a magnetic polarization $\mathbf{M}(\mathbf{r}, t) = \chi_{\text{mag}}\mathbf{H}(\mathbf{r}, t)$, may occur, which is in general called magnetization. Magneto-optical effects (e.g. the Faraday effect) may be of less importance than dielectric phenomena, but on the other hand they play a significant role in optical applications. In most of the cases we treat here, the assumption is justified that for the magnetic permeability of vacuum, $\mu_{\text{mag}} = 1 + \chi_{\text{mag}} = 1$ is valid.

2.1.3   Dynamic fields

The magnetic or B field and the H field are related by

$$\mathbf{B}(\mathbf{r}, t) = \mu_0\mathbf{H}(\mathbf{r}, t),$$

differing from each other only by μ_0, the permeability of vacuum. It is known that a change of the magnetic field within a circuit loop induces a voltage. Thus we formulate the law of induction as the third Maxwell equation,

$$\nabla \times \mathbf{E} = -\frac{\partial}{\partial t}\mathbf{B} \qquad \text{or} \qquad \oint_C \mathbf{E} \cdot \mathbf{dl} = -\frac{\partial}{\partial t}\oint_S \mathbf{B}\,\mathbf{df}, \tag{2.6}$$

In analogy, a changing dielectric field strength causes a displacement current, $\mathbf{j}_{\text{dis}} = \epsilon_0(\partial/\partial t)\mathbf{E}$, and a time-dependent polarization causes a polarization current, $\mathbf{j}_{\text{pol}} = (\partial/\partial t)\mathbf{P}$. This yields the complete fourth Maxwell equation for time-varying fields, if we consider these contributions in Eq. (2.3) (with $(\partial/\partial t)\mathbf{D} = \mathbf{j}_{\text{dis}} + \mathbf{j}_{\text{pol}}$):

$$\nabla \times \mathbf{H} = \mathbf{j} + \frac{\partial}{\partial t}\mathbf{D}. \tag{2.7}$$

2.1.4   Fourier components

Electric and magnetic fields with harmonic time development are central to optical wave theory. When talking of *Fourier components* of an electromagnetic field we mean the Fourier amplitudes $\mathcal{E}$, $\mathcal{H}$:[3]

$$
\begin{aligned}
E(\mathbf{r}, t) &= \mathfrak{Re}\{\mathcal{E}(\omega, \mathbf{k})\, e^{-i(\omega t - \mathbf{kr})}\}, \\
H(\mathbf{r}, t) &= \mathfrak{Re}\{\mathcal{H}(\omega, \mathbf{k})\, e^{-i(\omega t - \mathbf{kr})}\}.
\end{aligned}
$$

In general the relation for an amplitude in position and time space, $\mathcal{A}(\mathbf{r}, t)$, and the corresponding Fourier or $(\omega, \mathbf{k})$ dimension can be stated as

$$
\begin{aligned}
\mathcal{A}(\mathbf{r}, \omega) &= \frac{1}{(2\pi)^{1/2}} \int \mathbf{A}(\mathbf{r}, t)\, e^{-i\omega t}\, dt, \\
\mathcal{A}(\mathbf{k}, t) &= \frac{1}{(2\pi)^{3/2}} \int \mathbf{A}(\mathbf{r}, t)\, e^{i\mathbf{kr}}\, d^3 r.
\end{aligned}
$$

Of course, time and space variables may be Fourier-transformed simultaneously. It is particularly convenient to describe monochromatic fields, which have a fixed harmonic frequency $\omega = 2\pi\nu$, by Fourier components. Applying Maxwell's equations to this, the differential equations result in vector equations. We collect an overview of all variants in Tab. 2.1 and add the Coulomb–Lorentz force, which acts on a charge q at the point $\mathbf{r}$ and with velocity $\mathbf{v} = d\mathbf{r}/dt$.

2.1.5   Maxwell's equations for optics

For most applications in optics we can assume that no free charges and currents exist. It is the task of a microscopic theory to calculate the dynamical dielectric function $\epsilon(\omega) = \epsilon_0 \kappa(\omega) = \epsilon_0[1 + \chi(\omega)]$ from Eq. (2.4). For simple cases we will discuss this question in the chapter on the interaction of light with matter (Chapter 6). First of all we substitute the dielectric function $\epsilon_0 \kappa$ by phenomenological means by the index of refraction n,

$$
\epsilon_0 \kappa = \epsilon_0 n^2,
$$

which can depend on frequency ω and on position $\mathbf{r}$, and find a set of Maxwell's equations, meaningful for optics, which features high symmetry:

$$
\begin{aligned}
\boldsymbol{\nabla} \cdot n^2 \mathbf{E} &= 0, & \boldsymbol{\nabla} \times \mathbf{E} &= -\mu_0 \frac{\partial}{\partial t} \mathbf{H}, \\
\boldsymbol{\nabla} \cdot \mathbf{H} &= 0, & \boldsymbol{\nabla} \times \mathbf{H} &= \epsilon_0 \frac{\partial}{\partial t} n^2 \mathbf{E}.
\end{aligned}
\tag{2.8}
$$

Since we are particularly interested in the motion of charged, polarized matter, we must add the Lorentz force. These five equations are also called the *Maxwell–Lorentz equations*. They are specified in Tab. 2.1 in differential and integral form.

---

[3]We will write dynamic electromagnetic fields mainly in complex notation. Thereby the physical fields should always be considered as the real parts, even when this is not expressed explicitly like here.

Tab. 2.1: Summary: Maxwell–Lorentz equations.

In vacuum	In matter	In $(\omega, \mathbf{k})$ space
Charges are sources of electric field:		
$\boldsymbol{\nabla} \cdot \mathbf{E} = \rho/\epsilon_0$	$\boldsymbol{\nabla} \cdot \mathbf{D} = \rho$	$i\mathbf{k} \cdot \mathcal{D} = \rho/\epsilon_0$
No magnetic charges exist:		
$\boldsymbol{\nabla} \cdot \mathbf{B} = 0$		$i\mathbf{k} \cdot \mathcal{B} = 0$
Law of induction:		
$\boldsymbol{\nabla} \times \mathbf{E} = -\frac{\partial}{\partial t}\mathbf{B}$		$i\mathbf{k} \times \mathcal{E} = i\omega\mathcal{B}$
Currents are curls of the magnetic field:		
$c^2 \boldsymbol{\nabla} \times \mathbf{B} = \frac{1}{\epsilon_0}\mathbf{j} + \frac{\partial}{\partial t}\mathbf{E}$	$\boldsymbol{\nabla} \times \mathbf{H} = \mathbf{j} + \frac{\partial}{\partial t}\mathbf{D}$	$i\mathbf{k} \times \mathcal{H} = \mathbf{j} - i\omega\mathcal{D}$
Lorentz force:		
$m\frac{d^2}{dt^2}\mathbf{r} = q(\mathbf{E} + \mathbf{v} \times \mathbf{B})$		–

2.1.6 Continuity equation and superposition principle

We can draw two important conclusions from Maxwell's equations:

1. Charges are conserved, as can be found easily by applying the divergence to Eq. (2.3) and applying Eq. (2.1) for the *continuity equation*:

$$\boldsymbol{\nabla} \cdot \mathbf{j} = -\frac{\partial}{\partial t}\rho.$$

2. The *superposition principle* is a consequence of the linearity of the Maxwell equations. Two independent electromagnetic fields $\mathbf{E}_1$ and $\mathbf{E}_2$ are superpositioned linearly to yield a superposition field $\mathbf{E}_{\text{sup}}$,

$$\mathbf{E}_{\text{sup}} = \mathbf{E}_1 + \mathbf{E}_2. \tag{2.9}$$

The superposition field is important as a basis for the discussion of interference.

2.1.7 The wave equation

Electromagnetic fields propagate in vacuum ($n_{\text{vac}} = 1$) at the velocity of light, and they are a direct consequence of Maxwell's equations. In vacuum, there exist neither currents, $\mathbf{j} = 0$, nor charges, $\rho = 0$. This simplifies Maxwell's equations (2.1) and (2.6) significantly,

$$\boldsymbol{\nabla} \cdot \mathbf{E} = 0 \quad \text{and} \quad \boldsymbol{\nabla} \times \mathbf{H} = \epsilon_0 \frac{\partial}{\partial t}\mathbf{E}.$$

With the vector identity $\mathbf{\nabla}\times(\mathbf{\nabla}\times\mathbf{E}) = \mathbf{\nabla}(\mathbf{\nabla}\cdot\mathbf{E}) - \nabla^2\mathbf{E}$ and using $c = 1/\sqrt{\mu_0\epsilon_0}$, we find the wave equation in vacuum,

$$\left(\nabla^2 - \frac{1}{c^2}\frac{\partial^2}{\partial t^2}\right)\mathbf{E}(\mathbf{r}, t) = 0. \tag{2.10}$$

The corresponding one-dimensional wave equation can be written in the form

$$\left(\frac{\partial}{\partial z} - \frac{1}{c}\frac{\partial}{\partial t}\right)\left(\frac{\partial}{\partial z} + \frac{1}{c}\frac{\partial}{\partial t}\right)\mathbf{E}(z, t) = 0,$$

and by some straightforward algebra one finds solutions of the form

$$\mathbf{E}(z, t) = \mathbf{E}(z \pm ct).$$

The solutions propagate with the *phase velocity* c, the value of which in vacuum is called the velocity of light c. The velocity of light is one of the most important universal constants. Its numerical value was measured ever more precisely up to 1983, since when it has been set by definition once and for all to the vacuum value of

velocity of light, $c = 299.792\,458\,\mathrm{m\,s}^{-1}$.

Excursion: Velocity of light and theory of relativity
According to our direct experience, light propagates 'instantaneously'. The Danish astronomer Olaf Roemer (1644–1710) discovered in 1676 that the phases of Jupiter's moons get shorter when the planet approaches the Earth, and longer when it moves away from Earth. From that, he concluded that the propagation of light rays does not occur on an unmeasurably short time scale, but with a finite velocity, which he determined to be $225\,000\,\mathrm{km\,s}^{-1}$.

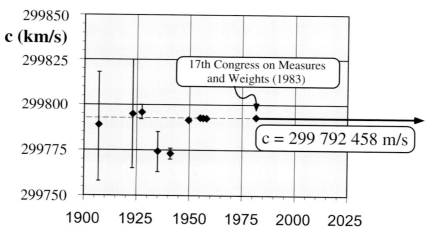

Fig. 2.2: *Values of the velocity of light before and after the 17th Congress on Weights and Measures (1983). The diamonds indicate the measured values of various laboratories including uncertainties [29].*

Since 1983 the value of the velocity of light has been fixed once and for all by international convention. At first sight it may seem surprising that one may just define a physical universal

constant. But it must be considered that velocities are determined by the physical quantities *time* and *distance*, and therefore independent measurements of time and distance are
always necessary. Time measurements can be performed by comparison with an atomic time
standard (*atomic clock*) with extreme precision, but for distance measurements such a measuring unit is not available. Therefore the procedure has been inverted and now – at least in
principle – any distance measurement is derived from a much more precise time measurement:

One metre is the distance that light covers in vacuum within $1/299\,792\,458$ s.

The velocity of light played a central role in the discovery of the special theory of relativity.
In a famous interference experiment the US physicists Michelson and Morley registered in
1886 that from the position of an observer light *always* propagates with the same velocity,
which is independent of the motion of the light source itself. One of the consequences of
this theory is that no particle or object, nor even any *action* of physical origin, can move or
propagate faster than the velocity of light c.

The theory of relativity epitomizes an outstanding intersection point of classical and modern physics. Owing to the theory, it is necessary that the equations of mechanical motion
are modified for very high velocities. From the very start Maxwell's equations, describing
the propagation of light, have been consistent with the theory of relativity. This property is
called 'relativistic invariance'.

The wave equation is simplified more if monochromatic waves with harmonic development are permitted. We use complex numbers, because in that way many waveforms
can be discussed formally in a clear manner.

In general only the real part of the complex amplitude is considered as a physically
real quantity. Inserting into eq.(2.10)

$$\mathbf{E}(\mathbf{r}, t) = \Re\{\mathbf{E}(\mathbf{r})\, e^{-i\omega t}\}$$

yields, with $\omega^2 = c^2 \mathbf{k}^2$, the *Helmholtz equation*, which depends only on the position $\mathbf{r}$:

$$(\nabla^2 + \mathbf{k}^2)\, \mathbf{E}(\mathbf{r}) = 0. \tag{2.11}$$

In homogeneous material (i.e. for constant index of refraction n), the wave equation
(2.10) experiences only one modification due to (2.8). The propagation is defined by
another phase velocity, $c \to c/n$, otherwise the wave propagates exactly as in vacuum.
One gets

$$\left[\nabla^2 - \left(\frac{n}{c}\right)^2 \frac{\partial^2}{\partial t^2}\right] \mathbf{E}(\mathbf{r}, t) = 0. \tag{2.12}$$

In theoretical electrodynamics also the dynamic electric and magnetic fields are of-
ten and slightly more elegantly derived from a common vector potential $\mathbf{A}(\mathbf{r}, t) = \mathbf{A}_0\, e^{-i(\omega t - \mathbf{k}\mathbf{r})}$, which on its part fulfils the Helmholtz equation (2.11):

$$
\begin{aligned}
\mathbf{E} &= \frac{\partial}{\partial t}\mathbf{A} &= i\omega\mathbf{A}, \\
\mathbf{H} &= \frac{1}{\mu_0}\nabla \times \mathbf{A} &= \frac{i}{\mu_0}\mathbf{k}\times\mathbf{A}.
\end{aligned}
\tag{2.13}
$$

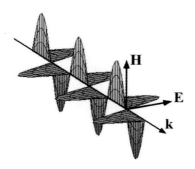

Fig. 2.3: *In isotropic space, the directions of the electric (**E**) and magnetic (**H**) fields of an electromagnetic wave (here: linearly polarized) are perpendicular to each other and to the propagation direction with wavevector **k** as well.*

For a complete definition of the potential **A**, an additional condition to ensure so-called gauge invariance is necessary. For this purpose the so-called *Coulomb gauge* ($\nabla \cdot \mathbf{A} = 0$) is a suitable choice, but in other situations alternatives like the Lorentz gauge for relativistic problems might offer advantages. From $\nabla \cdot \mathbf{E} = 0$ and Eq. (2.13) it follows that radiation fields are transverse in free space (i.e. they are orthogonal with respect to the wavevector **k**) (Fig. 2.3),[4]

$$\mathbf{E} \cdot \mathbf{k} = \mathbf{H} \cdot \mathbf{k} = 0.$$

Furthermore (2.13) may yield the useful relation

$$\mathbf{H} = \frac{1}{\mu_0 c}\mathbf{e_k} \times \mathbf{E}.$$

This shows that the **E** and **H** field are also perpendicular to each other.

2.1.8   Energy and momentum

The instantaneous energy density U of an electromagnetic field is

$$U = \tfrac{1}{2}(\epsilon_0|\mathbf{E}|^2 + \mu_0|\mathbf{H}|^2) = \epsilon_0|\mathbf{E}|^2. \tag{2.14}$$

The total energy $\mathcal{U}$ of an electromagnetic field is obtained by integration over the corresponding volume V,

$$\mathcal{U} = \epsilon_0 \int_V |\mathbf{E}(\mathbf{r})|^2 \, d^3r.$$

Formally a 'photon' oscillating with frequency ω has energy $\mathcal{U} = \hbar\omega$. From that one can get the average field strength $\langle|\mathbf{E}|\rangle = \sqrt{\hbar\omega/\epsilon_0 V}$, which corresponds to one photon. This quantity is important if for example one wants to describe the coupling of an atom to the field oscillation of an optical resonator.

Electromagnetic waves transport momentum and energy. The momentum current density is described by the *Poynting vector* **S**,

$$\mathbf{S} = \mathbf{E} \times \mathbf{H} = c\epsilon_0\mathbf{e_k}|\mathbf{E}|^2, \tag{2.15}$$

---

[4]Static fields of charge distributions are called *longitudinal*, because then according to Eq. (2.1) it holds that $\nabla \cdot \mathbf{E} = \rho(\mathbf{r}) \neq 0$. Indeed longitudinal and transverse properties depend on the calibration.

which is proportional to the energy current density, since $\mathcal{E} = pc$. In an experiment the intensity $I = c\langle U \rangle$ of an electromagnetic wave averaged over one period $T = 2\pi/\omega$ is measured most easily. It is related to the electric field amplitude $\mathcal{E}_0$ which for $E(t) = \mathcal{E}_0 \cos \omega t$ yields

$$I = \tfrac{1}{2} c \epsilon_0 \mathcal{E}_0^2.$$

2.2   Wave types

Now we want to present some limiting cases of simple and important wave types.

2.2.1   Planar waves

Planar waves are the characteristic solution of the Helmholtz equation (2.11) in Cartesian coordinates (x, y, z):

$$\left(\frac{\partial^2}{\partial x^2} + \frac{\partial^2}{\partial y^2} + \frac{\partial^2}{\partial z^2} + \mathbf{k}^2 \right) \mathbf{E}(\mathbf{r}) = 0. \tag{2.16}$$

Planar waves are vector waves with constant polarization vector $\boldsymbol{\epsilon}$ and amplitude $\mathcal{E}_0$,

$$\mathbf{E}(\mathbf{r}, t) = \mathfrak{Re}\{\mathcal{E}_0 \boldsymbol{\epsilon}\, e^{-i(\omega t - \mathbf{k}\mathbf{r})}\}.$$

In general they have two independent, orthogonal polarization directions $\boldsymbol{\epsilon}$, which we will discuss later in Chapter 2.4. Through the wavevector we define by $\mathbf{k} \cdot \mathbf{r} = \text{const}$ planes with identical phase $\Phi = \omega t - \mathbf{k}\mathbf{r}$ (Fig. 2.4).

2.2.2   Spherical waves

In our experience, light propagates into all directions of space, while the intensity declines. Because of this, it would be convenient to describe ray propagation by spherical waves as indicated in Fig. 2.4(a). In spherical coordinates (r, θ, ϕ) the Helmholtz equation (2.11) can be written as

$$\left(\frac{1}{r} \frac{\partial}{\partial r} r \frac{\partial}{\partial r} + \frac{1}{r \sin \theta} \frac{\partial}{\partial \theta} r \sin \theta \frac{\partial}{\partial \theta} + \frac{1}{r^2} \frac{\partial^2}{\partial \phi^2} + \mathbf{k}^2 \right) \mathbf{E}(\mathbf{r}) = 0. \tag{2.17}$$

But since electromagnetic waves have vector character, we have to look for solutions for 'vector' spherical waves. These are known and common, but are mathematically too complex in our case. But the problems are simplified, because in optics often only a small solid angle in a distinct direction is of practical importance. There the polarization of the light field varies only to a small extent and in good approximation we can apply the simplified, scalar solution of this wave equation. An isotropic, spherical wave has the form

$$E(\mathbf{r}, t) = \mathfrak{Re} \left\{ \mathcal{E}_0 \frac{e^{-i(\omega t - \mathbf{k}\mathbf{r})}}{|\mathbf{k}\mathbf{r}|} \right\}. \tag{2.18}$$

The amplitude of the spherical wave decreases inversely with the distance $E \propto r^{-1}$, and its intensity with the square of the inverse distance $I \propto r^{-2}$. With the scalar spherical wave approximation, the wave theory of diffraction can be described in good approximation according to Kirchhoff and Fresnel (see Section 2.5).

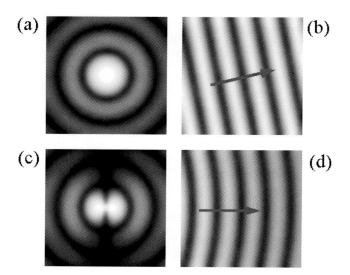

Fig. 2.4: *Snapshot of important wave types: (a) the isotropic (scalar) spherical wave has a simple structure; but it cannot describe electromagnetic waves correctly, which are always vector fields; (b) a planar wave with wavevector; (c) the dipole wave corresponds to a spherical wave with anisotropic intensity distribution; (d) yet at a distance of only a few wavelengths from the source the dipole wave is very similar to a planar wave.*

2.2.3   Dipole waves

Dipole radiators are the most important sources of electromagnetic radiation. This is true for radio waves at wavelengths in the range of metres or kilometres, which are radiated by macroscopic antennas, and for optical wavelengths as well, where induced dipoles of microscopic atoms or solids take over the role of antennas. A positive and a negative charge $\pm q$ at a separation $\mathbf{x}$ have a dipole moment $\mathbf{d}(t) = q\mathbf{x}(t)$. Dipoles can be induced by an external field displacing the centre-of-mass charge of the positive and negative charge distributions, for example of a neutral atom. Charge oscillations $\mathbf{x} = \mathbf{x}_0\, e^{-i\omega t}$ cause an oscillating dipole moment,

$$\mathbf{d}(t) = \mathbf{d}_0\, e^{-i\omega t}, \tag{2.19}$$

which radiates a dipole wave and forms the simplest version of a vector spherical wave. Let us assume that the distance of observation is large compared with the wavelength $r \gg \lambda = 2\pi c/\omega$. Under these circumstances we are located in the *far field* of the radiation field.

Although the separation $|\mathbf{x}|$ between the charges is small compared with the wavelength, we may describe the intensity distribution with the results of the *Hertzian dipole*.[5] The simplest form is shown by a linear dipole along the z axis, $\mathbf{d} = d_0\, e^{-i\omega t}\mathbf{e}_z$,

---

[5]The Hertzian dipole has vanishing spatial extent ($\mathbf{x} \to 0$), but a non-zero dipole moment $\mathbf{d}$.

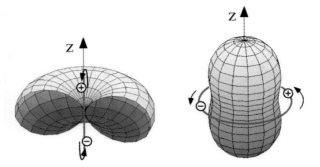

Fig. 2.5: *Angular distribution of the intensity ($\propto |\mathbf{E}|^2$) of a linearly and a circularly oscillating dipole.*

and the field amplitude is stated in spherical coordinates (r, θ, ϕ):

$$\mathbf{E}_{\text{lin}} = k^2 d_0 \sin \theta \, \frac{e^{-i(\omega t - kr)}}{r} \, \mathbf{e}_\theta.$$

The faces of constant phase are spherical faces again. Only the angle factor $\sin \theta$, which specifies exactly the component perpendicular to the direction of propagation, creates the antenna characteristics of a dipole. For a circular dipole, $\mathbf{d} = d_0 \, e^{-i\omega t}(\mathbf{e}_x + i\mathbf{e}_y)$, we find

$$\mathbf{E}_{\text{circ}} = k^2 d_0 \cos \theta \, \frac{e^{-i(\omega t - kr)}}{r} \, (\cos \theta \, \mathbf{e}_\theta + i\mathbf{e}_\varphi).$$

In Fig. 2.5 the intensity distribution of oscillating dipoles is shown. In contrast to a circular dipole, for a linear one directions occur into which no energy is radiated. The dipole character can be observed very nicely with the *Tyndall effect* by relatively simple means. One needs only a linearly polarized laser beam and a plexiglass rod (Fig. 2.6). The double refraction of the plexiglass rod causes a modulation of the polarization plane, and the observer, standing at the side, sees a periodic increase and decrease of the scattered light in the plexiglass rod.

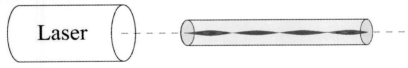

Fig. 2.6: *Tyndall effect in a plexiglass rod. By birefringence the plane of polarization gets modulated. Therefore an observer at the side sees a periodic increase and decrease of the scattered light intensity.*

2.3   Gaussian beams

Now we want to establish the connection between ray optics and wave optics, i.e. we want to describe in particular laser beam propagation through the methods of wave optics. Observation of a laser beam yields characteristic properties which we will use to construct the so called Gaussian principal mode of laser beam propagation: Laser beams are extremely well bundled, i.e. they do not seem to change over distances of many metres, and they are axially symmetric. They truly form rays without any effort. Along the propagation direction z a light ray behaves very similarly to a planar wave with constant amplitude $\tilde{\mathcal{A}}_0$, which is a known solution of the wave equations (2.10) and (2.16),

$$E(z,t) = \tilde{\mathcal{A}}_0 \, e^{-i(\omega t - kz)}.$$

On the other hand we know that, at large distances from a source, also laser light behaves more like another known solution of Eqs. (2.10) and (2.18), which is divergent like the spherical or dipole wave discussed before with an amplitude inversely decreasing with distance from the source,

$$E(z,t) = \mathcal{A}_0 \frac{e^{-i(\omega t - kz)}}{kz}. \tag{2.20}$$

A wave which combines the properties of plane waves and spherically diverging waves could have the structure

$$E(z,t) = \mathcal{A}_0 \frac{e^{-i(\omega t - kz)}}{k(z - iz_0)}, \tag{2.21}$$

where $z = z_0$ separates the region where the wave will behave more like a plane wave ($|z| < z_0$) and more like a spherically divergent wave ($|z| > z_0$). Here z_0 is a real number while the origin of the imaginary term will become transparent later. We will use this ad hoc approach to 'construct' the fundamental mode of coherent beam propagation. The 3d extension of the wave will be introduced by replacing $kz \to \mathbf{kr}$ and expanding $\mathbf{kr}$ in the vicinity of the z-axis.

2.3.1   The Gaussian principal mode or TEM$_{00}$ mode

We now consider a cutout of a spherical wave close to the z axis ('paraxial') and separate longitudinal (z coordinate) and transverse contributions. Rays with axial symmetry depend only on the transverse coordinate ρ, and when substituting $\mathbf{kr} = kr$ we may furthermore use the approximation $r = \sqrt{z^2 + \rho^2} \simeq z + \rho^2/2z$ within the so-called *Fresnel approximation* for $\rho \ll z, r$:

$$E(\mathbf{r}) = \frac{\mathcal{A}(\mathbf{r})}{|\mathbf{kr}|} e^{i\mathbf{kr}} \simeq \frac{\mathcal{A}(z,\rho)}{kz} \exp\left(i\frac{k\rho^2}{2z}\right) e^{ikz}. \tag{2.22}$$

This form of course much resembles eq.(2.20) where the spatial phase is transversely modulated, respectively curved, with the Fresnel factor $\exp(ik\rho^2/2z)$.

The linear substitution $z \to z - iz_0$ is similar to a coordinate transformation and simply realizes one more solution which also introduces a phase shift for small z due to the imaginary term iz_0. With this substitution we already arrive at the *Gaussian principal mode*,[6] if we use a constant amplitude $\mathcal{A}_0$:

$$E(z,\rho) \simeq \frac{\mathcal{A}_0}{k(z - iz_0)} \exp\left(i\frac{k\rho^2}{2(z - iz_0)}\right) e^{ikz}. \tag{2.23}$$

---

[6]The notion of 'mode', which appears here for the first time, is derived from the Latin *modus*, meaning measure or melody.

Gaussian modes propagate differently in free, isotropic space than, for example, waves in a dielectric waveguide, which depend on the inhomogeneous optical properties of the waveguide. In isotropic space the electric and magnetic fields, as well, are transverse to the direction of propagation and the waveforms are called transverse electric and magnetic (TEM) mode with indices (m, n). The basic solution is called the TEM_{00} mode. It is by far the most important form of all used wave types, and therefore will be analysed in more detail, before we consider the higher modes in Section 2.3.3.

The presentation of the field distribution in Eq. (2.23) is not yet very transparent. Therefore we introduce the new quantities $R(z)$ and $w(z)$ through

$$\frac{1}{z - iz_0} = \frac{z + iz_0}{z^2 + z_0^2} = \frac{1}{R(z)} + i\frac{2}{kw^2(z)}. \tag{2.24}$$

The decomposition of the Fresnel factor into real and imaginary parts creates two factors, a complex phase factor that describes the curvature of the phase front, and a real factor that specifies the envelope of the beam profile:

$$\exp\left(i\frac{k\rho^2}{2(z - iz_0)}\right) \quad \rightarrow \quad \exp\left(i\frac{k\rho^2}{2R(z)}\right)\exp\left(-\frac{\rho}{w(z)}\right)^2.$$

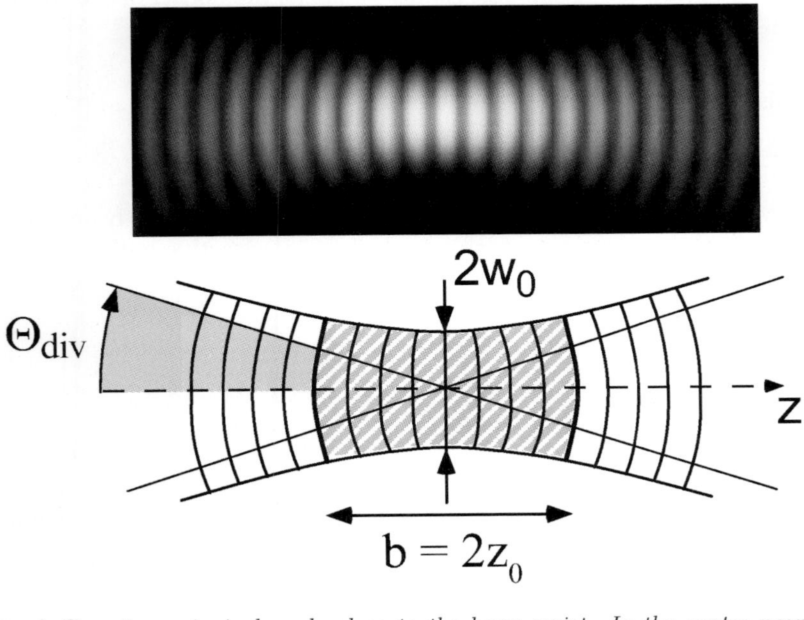

Fig. 2.7: *A Gaussian principal mode close to the beam waist. In the centre nearly planar wavefronts are achieved, while outside the waves quickly adopt a spherical form. In the lower part the Rayleigh zone is hatched.*

The form of the Gaussian principal mode in Fig. 2.7 is thoroughly characterized by the parameter couple (w_0, z_0). The following ideas and notations have been established to lend physical meaning to important parameters.

Rayleigh zone, confocal parameter b:

$$b = 2z_0$$

The Gaussian wave exhibits its largest variation for $-z_0 \leq z \leq z_0$, within the so-called Rayleigh length z_0 from Eq. (2.24). This region is called the Rayleigh zone and is also characterized with the confocal parameter $b = 2z_0$. The Rayleigh zone marks positions in the *near field* of the smallest beam cross-section or focal point ('focus'). At $z \ll z_0$ a nearly planar wave propagates and the wavefront changes only marginally. The Rayleigh zone is the shorter, the more the beam is focused. In the context of images we also use the notion *depth of focus* (see Chapter 4.3.3). In the *far field* ($z \gg z_0$) the propagation is again similar to a spherical or dipole wave, respectively.

Radius of wavefronts $R(z)$:

$$R(z) = z[1 + (z_0/z)^2] \tag{2.25}$$

Within the Rayleigh zone $R(z) \rightarrow \infty$ holds at $z \ll z_0$, whereas in the far field $R(z) \simeq z$. The largest curvature or the smallest radius occurs at the border of the Rayleigh zone with $R(z_0) = 2z_0$.

Beam waist $2w_0$:

$$w_0^2 = \lambda z_0/\pi$$

The beam waist $2w_0$, or beam waist radius w_0, specifies the smallest beam cross-section at $z = 0$. If the wave propagates within a medium of refractive index n, then λ must be substituted by λ/n. The diameter of the beam waist is then $w_0^2 = \lambda z_0/\pi n$.

Beam radius $w(z)$:

$$w^2(z) = w_0^2 \left[1 + \left(\frac{z}{z_0} \right)^2 \right]$$

Within the Rayleigh zone the beam radius $w(z)$ stays approximately constant. But in the far field it increases linearly according to $w(z) \simeq w_0 z/z_0$.

Divergence Θ_{div}:

$$\Theta_{\text{div}} = \frac{w_0}{z_0} = \sqrt{\frac{\lambda}{\pi z_0 n}}$$

In the far field ($z \gg b$) the divergence can be determined from the relation $\Theta(z) = w(z)/z$, $z \rightarrow \infty$.

Gouy phase $\eta(z)$:

$$\eta(z) = \tan^{-1}(z/z_0) \tag{2.26}$$

Passing through the focus the Gaussian wave receives a bit more curvature, i.e. shorter wavelength than a planar wave. For illustration, alternatively to (2.24), we can make the substitution

$$\frac{i}{z - iz_0} = -\frac{1}{z_0} \frac{w_0}{w(z)} e^{-i\tan^{-1}(z/z_0)}$$

(the imaginary factor establishes the common convention, to find a real amplitude or vanishing phase at $z = 0$). By this function the small deviation from the linear phase evolution of the planar wave can be described, $-\pi/2 \leq \eta(z) \leq \pi/2$. This extra phase is known by the name *Gouy phase*; half of it is collected within the Rayleigh zone. In travelling through the focus the phase is effectively inverted, which is reminiscent of two partial rays exchanging relative positions when crossing at a focal point.

With these notations the total result of the Gaussian principal mode or TEM$_{00}$ mode is the following:

$$E(\rho, z) = \mathcal{A}_0 \frac{w_0}{w(z)} e^{-[\rho/w(z)]^2} e^{ik\rho^2/2R(z)} e^{i[kz - \eta(z)]}. \tag{2.27}$$

The first factor describes the transverse amplitude distribution, the second (Fresnel) factor the spherical curvature of the wavefronts, and the last one the phase evolution along the z axis. In physics and optical techniques in most applications a Gaussian principal mode or TEM$_{00}$ mode is used.

Example: Intensity of a TEM$_{00}$ mode

The intensity distribution within a plane perpendicular to the propagation direction corresponds to the known Gaussian distribution,

$$I(\rho, z) = \frac{c\epsilon_0}{2} E E^* = \frac{c\epsilon_0}{2} |\mathcal{A}_0|^2 \left(\frac{w_0}{w(z)}\right)^2 e^{-2[\rho/w(z)]^2},$$

with the axial peak value

$$I(0, z) = \frac{c\epsilon_0}{2} |\mathcal{A}_0|^2 \left(\frac{w_0}{w(z)}\right)^2.$$

In general the 'cross-section' of a Gaussian beam is specified as the width $2w(z)$, where the intensity has dropped to $1/e^2$ or 13% of the peak value. Some 87% of the total power is concentrated within this radius.

Along the z axis the intensity follows a Lorentzian profile $1/[1+(z/z_0)^2]$. It declines from its peak value $I(0,0) = (c\epsilon_0/2)|\mathcal{A}_0|^2$ (see Fig. 2.7) and reaches half of this value at $z = z_0$. The confocal parameter b then is also a measure for the longitudinal half-width of the focal zone.

Owing to energy conservation, the total energy current density $P = 2\pi \int I(\rho, z)\rho \, d\rho$ of a Gaussian wave cannot change, as can be verified by explicit integration,

$$\frac{P}{c\epsilon_0} = 2\pi\mathcal{A}_0^2 w_0^2 \int_0^\infty \frac{\rho \, d\rho}{w^2(z)} e^{-2[\rho/w(z)]^2} = \pi w_0^2 \mathcal{A}_0^2.$$

2.3.2   The ABCD rule for Gaussian modes

The usefulness of Gaussian modes for analysis of an optical beam path is supported particularly by the simple extension of the ABCD rule (Section 1.9.2), known from ray optics. At every position z on the beam axis a Gaussian beam may be characterized either by the pair of parameters (w_0, z_0) or alternatively by the real and imaginary parts of $q(z) = z - iz_0$ according to Eq. (2.24). We know that both parameters of a light ray are transformed linearly according to Eq. (1.17) and that for every optical element a distinct type of matrix $\mathbf{T}$ with elements $ABCD$ exists. The parameters of the Gaussian beam are transformed by linear operations with coefficients that are identical to the ones from ray optics:

$$q_1 = \hat{\mathbf{T}} \otimes q_0 = \frac{Aq_0 + B}{Cq_0 + D}. \tag{2.28}$$

Now it is not very difficult to show that these operations may be applied multiple times and that the total effect $\hat{\mathbf{T}}$ corresponds to the matrix product $\hat{\mathbf{T}}_2 \hat{\mathbf{T}}_1$:

$$q_2 = \hat{\mathbf{T}}_2 \otimes (\hat{\mathbf{T}}_1 \otimes q_0) = \frac{A_2 \dfrac{A_1 q_0 + B_1}{C_1 q_0 + D_1} + B_2}{C_2 \dfrac{A_1 q_0 + B_1}{C_1 q_0 + D_1} + D_2} = \frac{(A_2 A_1 + B_2 C_1)q_0 + ...}{...}.$$

Thus we can describe the effects of all elements by the known matrices from Tab. 1.2.

Example: Focusing with a thin lens

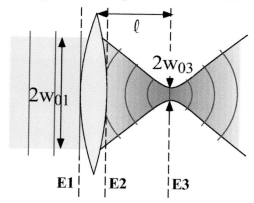

Fig. 2.8: *Focusing a Gaussian beam with a thin lens of focal length f.*

As an important and instructive example we now choose the effect of a thin lens of focal length f, with which a Gaussian beam with TEM$_{00}$ mode has to be focused, and make a comparison with the predictions from ray optics. Let us look at the parameters of the wave in planes 1 (directly in front of the lens), 2 (directly after the lens) and 3 (in the focus).

    Plane 1. A Gaussian beam with large beam waist $2w_{01}$ and infinitely large radius of curvature $R(z=0) = \infty$ is very close to our expectations of a planar wave. Then, the Rayleigh length is also very large, since $z_{01} = \pi w_{01}^2 / \lambda$; for example, it is 124 m long for a beam diameter of only 1 cm and a wavelength of 632 nm! Let us assume that the beam waist of the incident beam is at $z = 0$ and due to that $q(z)$ is purely imaginary,

$$q_1 = -iz_{01} = -i\frac{\pi w_{01}^2}{\lambda}.$$

Plane 2. The beam radius is not changed by the thin lens at once ($w_{02} = w_{01}$), but the radius of curvature is changed, and is now given by $1/R_2 = -1/f$:

$$\frac{1}{q_2(z=0)} = -\frac{1}{f} + i\frac{1}{z_{01}}.$$

Formal application of the lens transformation from Tab. 1.2 and with Eq. (2.28) would have yielded the same result.

Plane 3. For the translation from the lens to the new focus we get

$$q_3(\ell) = q_2(0) + \ell,$$

but the ℓ position of plane 3 is initially unknown and must be determined from the condition that at the focus $q_3 = i\lambda/\pi w_{03}^2$ is purely imaginary. For that purpose we determine the real and imaginary parts of q_2,

$$q_2 = -\frac{f}{1 + (f/z_{01})^2}\left(1 + i\frac{f}{z_{01}}\right).$$

Obviously the real part of q_3 is compensated exactly at

$$\ell = \frac{f}{1 + (f/z_{01})^2} = \frac{f}{1 + (\lambda f/\pi w_{01}^2)^2},$$

which means that we again find planar waves there. According to ray optics we would have expected the focus to be located exactly at $\ell = f$. But if the focal length is short compared with the Rayleigh length of the incident beam, $f \ll z_{01}$, or equivalently, which is usually the case, $\lambda f/w_{01}^2 \ll 1$, then the position of the focal point will differ only marginally from that.

More interesting is the question of how large the diameter of the beam is in the focus. We know that ray optics does not answer that, and we have to take into account diffraction at the aperture of the lens. At first we calculate the Rayleigh parameter

$$\frac{1}{z_{03}} = \frac{1}{f}\frac{1 + (f/z_{01})^2}{f/z_{01}},$$

and then determine the ratio of the beam diameter at the lens and in the focus,

$$\frac{w_{03}}{w_{01}} = \left(\frac{z_{03}}{z_{01}}\right)^{1/2} = \frac{f/z_{01}}{\sqrt{1 + (f/z_{01})^2}}. \tag{2.29}$$

In the form

$$w_{03} = \frac{\lambda f}{\pi w_{01}}\frac{1}{\sqrt{1 + (\lambda f/\pi w_{01}^2)^2}} \simeq \frac{\lambda f}{\pi w_{01}},$$

the first factor yields the Rayleigh criterion for the resolving power of a lens, known also from diffraction theory, which will be treated once more in the section on diffraction (Section 2.5, Eq. (2.48)).

2.3.3   Higher Gaussian modes

For a more formal treatment of the Gaussian modes we now also decompose the Helmholtz equation (2.11) into transverse and longitudinal contributions,

$$\nabla^2 + k^2 \;=\; \frac{\partial^2}{\partial z^2} + \nabla_{\mathrm{T}}^2 + k^2,$$

$$\nabla_{\mathrm{T}}^2 \;=\; \frac{\partial^2}{\partial x^2} + \frac{\partial^2}{\partial y^2},$$

and apply it to the electric field from Eq. (2.22). Assuming that the amplitude $\mathcal{A}$ is varying only very slowly on a wavelength scale,

$$\frac{\partial}{\partial z}\mathcal{A} = \mathcal{A}' \ll k\mathcal{A},$$

we find the approximation

$$\frac{\partial^2}{\partial z^2}\mathcal{A}\, e^{ik\rho^2/2z}\,\frac{e^{ikz}}{kz} \simeq \left(2ik\mathcal{A}' - k^2\mathcal{A}\right)e^{ik\rho^2/2z}\,\frac{e^{ikz}}{kz},$$

and finally get the *paraxial Helmholtz equation,*

$$\left(\nabla_{\mathrm{T}}^2 + 2ik\frac{\partial}{\partial z}\right)\mathcal{A}(\rho, z) = 0. \tag{2.30}$$

Obviously this is valid for $\mathcal{A} = \mathrm{const}$. This is not surprising, because by this we have just verified that, close to the z axis, the applied spherical wave fulfils the paraxial Helmholtz equation. The most fundamental solution of the paraxial Helmholtz equation we have already found by intuition and construction (Eq. (2.23), p. 38), but the Gaussian principal mode is only one particular, although important, solution. We look for the higher solutions as variants of the principal solution, known from Eq. (2.27),

$$E(x, y, z) = \frac{\mathcal{A}(x, y, z)}{z - iz_0}\exp\left(i\frac{k(x^2 + y^2)}{2(z - iz_0)}\right)e^{ikz},$$

and initially we want to use Cartesian coordinates, which deliver the best-known solutions, called *Hermitian–Gaussian modes.* But there are also other solutions, for example the *Laguerre–Gaussian modes,* which are found when applying cylindrical coordinates. The paraxial Helmholtz equation (2.30) is

$$\left(\frac{\partial^2}{\partial x^2} + \frac{2ikx}{q(z)}\frac{\partial}{\partial x} + \frac{\partial^2}{\partial y^2} + \frac{2iky}{q(z)}\frac{\partial}{\partial y} + 2ik\frac{\partial}{\partial z}\right)\mathcal{A}(x, y, z) = 0. \tag{2.31}$$

As for the principal mode, we look for amplitudes that depend symmetrically on x and y and cause only a small correction of the phase evolution along the longitudinal direction:

$$\mathcal{A}(x, y, z) = \mathcal{F}(x)\mathcal{G}(y)\exp\left[-i\mathcal{H}(z)\right].$$

We substitute this form in Eq. (2.31) and take into account that $1/(z - iz_0) = 2(1 - iz/z_0)/ikw^2(z)$. By claiming exclusively real solutions for $\mathcal{F}$, $\mathcal{G}$ and $\mathcal{H}$, imaginary contributions cancel and we get

$$\frac{1}{\mathcal{F}(x)}\left[\frac{\partial^2}{\partial x^2}\mathcal{F}(x) - \frac{4x}{w^2(z)}\frac{\partial}{\partial x}\mathcal{F}(x)\right]$$

$$+ \frac{1}{\mathcal{G}(y)}\left[\frac{\partial^2}{\partial y^2}\mathcal{G}(y) - \frac{4y}{w^2(z)}\frac{\partial}{\partial y}\mathcal{G}(y)\right] + 2k\frac{\partial}{\partial z}\mathcal{H}(z) \;=\; 0.$$

Expecting that the distribution of transverse amplitudes does not change along the z axis, we execute the variable transformation

$$u = \sqrt{2}\, x/w(z) \quad \text{and} \quad v = \sqrt{2}\, y/w(z)$$

(the factor $\sqrt{2}$ is necessary to normalize the new equations):

$$\frac{1}{\mathcal{F}(u)}\left[\mathcal{F}''(u) - 2u\mathcal{F}'(u)\right] + \frac{1}{\mathcal{G}(v)}\left[\mathcal{G}''(v) - 2v\mathcal{G}'(v)\right] + kw^2(z)\mathcal{H}'(z) = 0.$$

By this transformation we achieved a separation of the coordinates, and the equation can be solved via eigenvalue problems:

$$
\begin{aligned}
\mathcal{F}''(u) - 2u\mathcal{F}'(u) + 2m\mathcal{F}(u) &= 0, \\
\mathcal{G}''(v) - 2v\mathcal{G}'(v) + 2n\mathcal{G}(v) &= 0, \\
kw^2(z)\mathcal{H}'(z) - 2(m+n) &= 0.
\end{aligned}
\tag{2.32}
$$

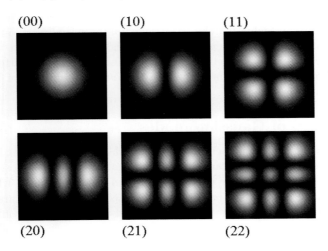

Fig. 2.9: *Transverse intensity distribution of Hermitian–Gaussian modes of low order* $(|\mathcal{A}_{mn}(x,y)|^2 = |H_m(x)H_n(y)|^2)$.

The equation for the (u, v) coordinates is known as Hermite's differential equation. Its solutions are called *Hermitian polynomials* $H_j(x)$, which are easy to determine according to the recursion relations

$$
\begin{aligned}
H_{j+1}(x) &= 2x H_j(x) - 2j H_{j-1}(x), \\
H_j(x) &= (-)^j\, e^{x^2}\, \frac{d^j}{dx^j}\left(e^{-x^2}\right).
\end{aligned}
\tag{2.33}
$$

The Hermitian polynomials of lowest order are

$$H_0(x) = 1, \quad H_1(x) = 2x, \quad H_2(x) = 4x^2 - 2, \quad H_3(x) = 8x^3 - 12x.$$

The modulus squared specifies the transverse intensity distribution and is illustrated in Fig. 2.9 for the mode of lowest order. They form a system of orthonormal functions with the orthogonality condition

$$\int_{-\infty}^{\infty} H_j(x)H_{j'}(x)\, e^{-x^2}\, dx = \frac{\delta_{jj'}}{2^j\, j!\sqrt{\pi}}.
\tag{2.34}$$

The third equation from (2.32) is solved by

$$\mathcal{H}(z) = (n + m)\eta(z) \tag{2.35}$$

with $\eta(z) = \tan^{-1}(z/z_0)$ (eq. (2.21)). It enhances the phase shift of the Gouy phase and plays an important role in the calculation of the resonance frequencies of optical resonators (see Chapter 5.6).

Thus the result for the modulation factor of the amplitude distribution for higher-order Gaussian or TEM_{mn} modes is

$$\mathcal{A}_{mn} = \mathrm{H}_m \left(\sqrt{2}\, x/w(z) \right) \mathrm{H}_n \left(\sqrt{2}\, y/w(z) \right) e^{-i(m+n)\eta(z)}, \tag{2.36}$$

and particularly the result for the TEM_{00} mode is reproduced, of course. All modes are described by a Gaussian envelope, modulated by Hermitian polynomials. Therefore they are called *Hermitian–Gaussian modes*.

A question might remain: Why have we chosen the Cartesian form of the paraxial Helmholtz equation, and why do cylindrical coordinates actually seldom appear? The reason is of technical nature, because at the interior of mirrors and windows small deviations from cylindrical symmetry are always present, and thus Cartesian Gaussian modes are preferred to Laguerre modes, which are found as solutions of equations with cylindrical symmetry.

2.3.4 Creation of Gaussian modes

In most experiments interest is focused on the TEM_{00} principal mode. By nature it is preferred in a laser resonator, because it has the smallest diffraction losses. According to Fig. 2.9 it is obvious that the effective face of a mode increases with the orders (m, n), so that the openings of a resonator (mirror edges, apertures) are of increasing importance. On the other hand, since the spatial amplification profile also has to match the desired mode, modes of very high order can be excited by intentionally misaligning a resonator (Fig. 2.10).

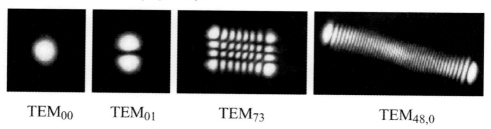

TEM$_{00}$ TEM$_{01}$ TEM$_{73}$ TEM$_{48,0}$

Fig. 2.10: *Gaussian modes of higher order from a simple titanium–sapphire laser. The TEM$_{48,0}$ mode has been reduced only a little bit. The asymmetry of the higher-order modes is caused by technical inaccuracies of the resonator elements (mirrors, laser crystal) [104].*

Controlled shaping of light fields can also be achieved by a filter; thereby the notion of *spatial filter* is used. Such a spatial filter is shown in Fig. 2.11, which in its most simple form consists of a convex lens (e.g. a microscope objective) and a so-called *pin-hole*, with a diameter adjusted to the TEM_{00} principal mode.

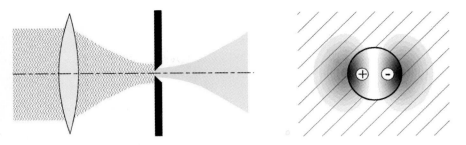

Fig. 2.11: *Spatial filter. In front of the aperture the beam consists of a superposition of many Gaussian modes. It is shown, for the example of a* TEM$_{01}$ *mode, how higher-order modes can be suppressed by the aperture. The fields in both 'ears' of the mode oscillate with opposite phase.*

Transmission of higher-order Gaussian modes is not only inhibited by the aperture, because the diameter increases rapidly with the order, but is also suppressed by the spatially alternating phase distribution. Therefore the aperture is not excited dipole-like, as is the case for the TEM$_{00}$ principal mode, but with a higher order, which, as everybody knows, radiates with lower intensity.

At the output a 'cleaned' Gaussian beam propagates, which has lost intensity, of course. Excellent suppression of higher-order modes is achieved when a single-mode optical waveguide is used instead of a pin-hole (see Chapter 3.3).

2.4   Polarization

We have already noticed in the previous section that electromagnetic waves are vector waves with direction, which can be described in terms of two orthogonal polarization vectors $\boldsymbol{\epsilon}$ and $\boldsymbol{\epsilon}'$ in free space.[7] We consider a transverse wave propagating in the $\mathbf{e}_z$ direction. The polarization must lie within the xy plane (unit vectors $\mathbf{e}_x$ and $\mathbf{e}_y$), and we consider two components, which may have different time-variant phases,

$$\mathbf{E}(z,t) = \mathcal{E}_x \mathbf{e}_x \cos(kz - \omega t) + \mathcal{E}_y \mathbf{e}_y \cos(kz - \omega t + \phi). \tag{2.37}$$

For $\phi = 0$, 2π, 4π, ..., these components have equal phases and the wave is linearly polarized,

$$\mathbf{E}(z,t) = (\mathcal{E}_x \mathbf{e}_x + \mathcal{E}_y \mathbf{e}_y) \cos(kz - \omega t).$$

For $\phi = \pi$, 3π, ..., they oscillate out of phase and in general yield an elliptically, or for $\mathcal{E}_x = \mathcal{E}_y$ circularly, polarized wave:

$$\mathbf{E}(z,t) = \mathcal{E}_x \mathbf{e}_x \cos(kz - \omega t) + \mathcal{E}_y \mathbf{e}_y \sin(kz - \omega t).$$

---

[7]The notion of 'polarization' is also used as *dielectric polarization* elsewhere. The kind of application for which it is used is always clear from the context.

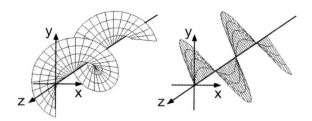

Fig. 2.12: *The field of a circularly polarized wave (left-hand side) rotates around the propagation axis every where with the same amplitude. The linearly polarized wave (right-hand side) is a common sine wave.*

Instead of Eq. (2.37) the field amplitude may also be written in the form

$$\mathbf{E}(z,t) = E_{\cos}(a\mathbf{e}_x + b\mathbf{e}_y)\cos(kz - \omega t + \alpha)$$
$$+ E_{\sin}(-b\mathbf{e}_x + a\mathbf{e}_y)\sin(kz - \omega t + \alpha),$$

with $a^2 + b^2 = 1$, which corresponds to the ellipse in Fig. 2.13 rotated by the angle α. By comparison of the coefficients at $(kz - \omega t) = 0$, $\pi/2$, one may calculate the angle α,

$$\tan(2\alpha) = \frac{2\mathcal{E}_x\mathcal{E}_y \cos\phi}{\mathcal{E}_x^2 - \mathcal{E}_y^2}.$$

Furthermore in Fig. 2.13 the decomposition of a linear and elliptical polarization into two circular waves is illustrated.

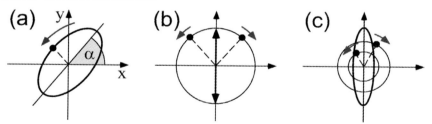

Fig. 2.13: *(a) Elliptically polarized wave. The linearly polarized wave (b) and the elliptical wave (c) can be decomposed into two counter-running circular waves.*

2.4.1   Jones vectors

In general any transverse polarized light wave can be decomposed into either two orthogonal and linear or two counter-running and circularly polarized waves. For example we find for the field of Eq. (2.37):

$$\mathbf{E}(z,t) = \mathfrak{Re}\{\mathcal{E}_x\mathbf{e}_x + \mathcal{E}_y\,e^{-i\phi}\mathbf{e}_y\}\,e^{-i(\omega t - kz)}$$
$$= \mathfrak{Re}\{(\mathcal{E}_x + i\,e^{-i\phi}\mathcal{E}_y)\mathbf{e}_+ + (\mathcal{E}_x - i\,e^{-i\phi}\mathcal{E}_y)\mathbf{e}_-\}\,e^{-i(\omega t - kz)}.$$

Jones suggested the orthogonal unit vectors

$$\left\{ \begin{array}{l} \mathbf{e}_+ = (\mathbf{e}_x + i\mathbf{e}_y)/\sqrt{2} \\ \mathbf{e}_- = (\mathbf{e}_x - i\mathbf{e}_y)/\sqrt{2} \end{array} \right\} \quad \text{and} \quad \left\{ \begin{array}{l} \mathbf{e}_x = (\mathbf{e}_+ + \mathbf{e}_-)/\sqrt{2} \\ \mathbf{e}_y = -i(\mathbf{e}_+ - \mathbf{e}_-)/\sqrt{2} \end{array} \right\} \quad (2.38)$$

for the characterization of a polarization: $\mathbf{e}_{x,y}$ for linear, and $\mathbf{e}_\pm$ for circular, components. Writing this for the individual components we find

$$\mathbf{e}_x = \begin{pmatrix} 1 \\ 0 \end{pmatrix}, \quad \mathbf{e}_y = \begin{pmatrix} 0 \\ 1 \end{pmatrix}, \quad \mathbf{e}_\pm = \frac{1}{\sqrt{2}} \begin{pmatrix} 1 \\ \pm i \end{pmatrix}.$$

It is obvious from Eq. (2.38) that any linearly polarized wave may be decomposed into two counter-running circularly polarized waves and vice versa. Optical elements affecting the polarization, like for example retardation plates, can be described very simply with this formalism (see Chapter 3.5.4).

2.4.2   Stokes parameters

For the characterization of a polarization state of a wave by Jones vectors, we need the amplitudes and directions for two orthogonal components ($\mathbf{e}_{x,y}$ or $\mathbf{e}_\pm$) at any given time. Polarizations may fluctuate in time also. Hence for characterization G. G. Stokes suggested the use of the time averaged quantities

$$\begin{array}{llll} S_0 & = & \langle \mathcal{E}_x^2 \rangle + \langle \mathcal{E}_y^2 \rangle, & S_2 & = & \langle 2\mathcal{E}_x \mathcal{E}_y \cos \phi \rangle, \\ S_1 & = & \langle \mathcal{E}_x^2 \rangle - \langle \mathcal{E}_y^2 \rangle, & S_3 & = & \langle 2\mathcal{E}_x \mathcal{E}_y \sin \phi \rangle. \end{array}$$

The first parameter S_0 is obviously proportional to the intensity, and since one direction is already fixed, only three parameters are independent of each other. Normalizing

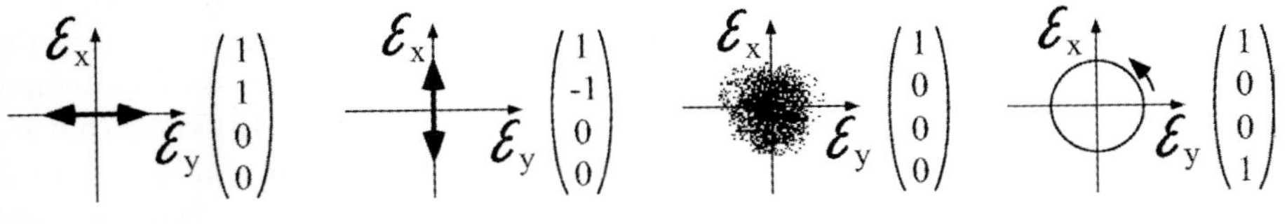

Fig. 2.14: *Stokes parameters and vectors for distinct polarization states. From left to right: linearly x polarized, linearly y polarized, unpolarized, right circularly polarized.*

the S parameters to $s_i = S_i/S_0$, then $s_0 = 1$ is always valid and furthermore[8]

$$V = s_1^2 + s_2^2 + s_3^2 \leq 1 \quad (= 1 \text{ for perfectly polarized light}).$$

According to the superposition principle for the superposition of two waves $S'' = S + S'$ holds for the Stokes parameters. Stokes parameters also describe unpolarized light as shown on Fig. 2.14.

---

[8]We will find in the chapter on light and matter (Chapter 6) that this structure appears again in the Bloch vectors of the analogous atomic two-state systems.

2.4.3   Polarization and projection

A quite astonishing property of the polarization may be demonstrated impressively with a polarization foil. A polarization foil generates polarized light from unpolarized light through absorption of the component that oscillates in parallel with the aligned organic molecules of the foil. More polarization components will be treated within the chapter on wave propagation in matter (Chapter 3).

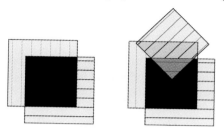

Fig. 2.15: *Transmission of crossed polarizers. The hatching indicates the direction of polarization. In the lower figure the third polarizer is inserted at 45° in between the other two polarizers.*

In Fig. 2.15, the left-hand side illustrates that two crossed polarizers result in the cancellation of the transmission. But it is quite astonishing that, when one more polarizer is inserted with polarization direction at 45° in between the two others, a quarter of the light transmitted by the first polarizer (neglecting losses) passes through the orthogonal polarizer! The polarization of the electromagnetic field is 'projected' onto the transmission direction of the polarizer, the polarizer affecting the field not the intensity.

2.5   Diffraction

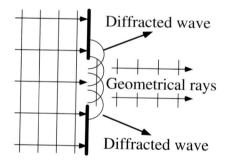

Fig. 2.16: *Huygens' principle: diffraction at an aperture.*

Light diffraction has played an important role in the development of the wave theory of light. Even famous physicists doubted for a long time that 'light comes around a corner like sound', but already Leonardo da Vinci (1452–1519) knew that some light falls into the shadow of an illuminated object – against the predictions of geometrical optics.

C. Huygens gave a first illustrative idea to wave theory by interpreting every point in space as an excitation source of a new wave, a concept called today ´Huygens' principle'. It makes indeed many diffraction phenomena accessible to our intuition, but it assumes ad hoc that waves are always transmitted into the forward direction only.

The general mathematical formalism of Huygens' principle is extremely elaborate, because the electric and magnetic radiation fields are vector fields, $\mathbf{E} = \mathbf{E}(x, y, z, t)$ and $\mathbf{B} = \mathbf{B}(x, y, z, t)$. Up to now there exist only a few general solved examples; the problem of planar wave propagation at an infinite thin edge solved in 1896 by A. Sommerfeld (1868–1951) counts as an exception.

An enormous simplification is achieved when substituting the vectorial field by scalar ones, whereby we have to determine the range of validity of the approximation. It is advantageous that light beams often propagate with only small changes of direction. Then the polarization changes only slightly and the scalar approximation describes the behaviour excellently.

2.5.1 Scalar diffraction theory

Here it is our objective to understand Huygens' principle by means of mathematics in scalar approximation by applying the superposition principle to the combined radiation field of multiple sources.[9]

Within this chapter we will for simplicity exclusively discuss the propagation of monochromatic waves:[10]

$$E(\mathbf{r}, t) = \mathcal{E}(\mathbf{r})\, e^{-i\omega t}.$$

The total field $\mathcal{E}(\mathbf{r_P})$ at a point P (Fig. 2.17) is composed of the sum of all contributions of the individual sources Q, Q', ... We know already that spherical waves emerging from a point-like source Q have the scalar form of Eq. (2.18),

$$\mathcal{E} = \mathcal{E}_Q\, e^{ikr}/kr.$$

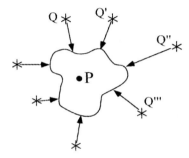

Fig. 2.17: *The light field at P is fed by the sources* Q, Q', Q'', *....*

To cover all fields incident on a point P, we look at the sources on a surface S and the effect of these on a very small volume with surface S' around P (Fig. 2.18). We can make use of the Green's integral theorem, well known from mathematics, which for two solutions ψ and ϕ of Helmholtz's equation (2.11) reads

$$\oint_S [\psi \boldsymbol{\nabla} \phi - \phi \boldsymbol{\nabla} \psi]\, d\mathbf{S} = \int_V [\psi \boldsymbol{\nabla}^2 \phi - \phi \boldsymbol{\nabla}^2 \psi]\, d^3 r = 0.$$

We let e^{ikr}/kr and $\mathcal{E}(\mathbf{r_P})$ be used for ψ and ϕ, and in Fig. 2.18(a) we cut out a sphere with very short radius r' and surface element $d\mathbf{S'} = r^2\, d\Omega'\, \mathbf{e}_r$ about point P to be contracted to this point,

$$\left(\oint_S d\mathbf{S} + \oint_{S'} d\mathbf{S'} \right) \left[\frac{e^{ikr}}{r} \boldsymbol{\nabla}\mathcal{E} - \mathcal{E}\boldsymbol{\nabla}\frac{e^{ikr}}{r} \right] = 0.$$

On the surface of the small sphere around P we have $d\mathbf{S'} \parallel \mathbf{e}_r$ and thus $\boldsymbol{\nabla}\mathcal{E} \cdot d\mathbf{S'} = (\partial \mathcal{E}/\partial r)r^2\, d\Omega'$. We furthermore use $-\boldsymbol{\nabla}\, e^{ikr}/r = (1/r^2 - ik/r)\, e^{ikr}\mathbf{e}_r$ and find

$$\oint_S \left[\frac{e^{ikr}}{r} \boldsymbol{\nabla}\mathcal{E} - \mathcal{E}\boldsymbol{\nabla}\frac{e^{ikr}}{r} \right] d\mathbf{S} = \oint_{S'} \left[\mathcal{E}(1 - ikr) + r\frac{\partial \mathcal{E}}{\partial r} \right] e^{ikr}\, d\Omega'. \tag{2.39}$$

---

[9]This section is mathematically more tedious. The reader may skip it and simply use the results in eqs. (2.44) and (2.45).

[10]This treatment requires spatial and temporal coherence of the light waves, which will be discussed in more detail in the chapter on interferometry (Chapter 5).

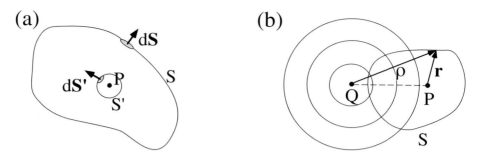

Fig. 2.18: *Kirchhoff's theorem. (a) Choice of the surfaces according to Eq. (2.39). (b) The surface S is excited by a source Q and radiates towards the point P.*

Now we let the radius of the volume around P decrease more and more ($r \to 0$) and with

$$\oint_{S'} \left(\mathcal{E} - ikr\mathcal{E} + r\frac{\partial \mathcal{E}}{\partial r} \right) e^{ikr} \, d\Omega' \xrightarrow{r \to 0} 4\pi\mathcal{E}|_{r=0} = 4\pi\mathcal{E}_P$$

we can prove Kirchhoff's integral theorem:

$$\mathcal{E}(\mathbf{r}_P) = \frac{1}{4\pi} \oint_S \left[\frac{e^{ikr}}{r} \boldsymbol{\nabla}\mathcal{E} - \mathcal{E} \boldsymbol{\nabla}\frac{e^{ikr}}{r} \right] d\mathbf{S}. \tag{2.40}$$

In principle we can now predict the field at point P if we know the distribution of fields on the surface S. Owing to its relatively wide generality, however, the Kirchhoff theorem does not give the impression that it might be very useful. Therefore we want to study further approximations and apply them to a point source Q illuminating the surface S (Fig. 2.18(b)). Let us assume that a scalar spherical wave of the form

$$\mathcal{E}(\boldsymbol{\rho}, t) = \frac{\mathcal{E}_Q}{k\rho} e^{i(\mathbf{k}\boldsymbol{\rho} - \omega t)}$$

propagates from there. We use spherical coordinates and just insert the spherical wave into Eq. (2.40),

$$\mathcal{E}(\mathbf{r}_P) = \frac{\mathcal{E}_Q}{4\pi k} \oint_S \left[\frac{e^{ikr}}{r} \left(\frac{\partial}{\partial \rho} \left(\frac{e^{ik\rho}}{\rho} \right) \right) \mathbf{e}_\rho - \frac{e^{ik\rho}}{\rho} \left(\frac{\partial}{\partial r} \left(\frac{e^{ikr}}{r} \right) \right) \mathbf{e}_r \right] d\mathbf{S}.$$

Then we make use of the approximation

$$\frac{\partial}{\partial \rho} \frac{e^{ik\rho}}{\rho} = k^2 e^{ik\rho} \left(\frac{i}{k\rho} - \frac{1}{(k\rho)^2} \right) \simeq e^{ik\rho} \frac{ik}{\rho}, \tag{2.41}$$

for ρ and r, which is excellent already for distances of only a few wavelengths, since $k\rho \gg 1$. Then also the Kirchhoff integral (2.40) can be simplified again crucially,

$$\mathcal{E}(\mathbf{r}_P) = -\frac{i\mathcal{E}_Q}{2\pi} \oint_S \frac{e^{ik(r+\rho)}}{r\rho} N(\mathbf{r}, \boldsymbol{\rho}) \, dS, \tag{2.42}$$

whereby we have introduced the Stokes' factor $N(\mathbf{r}, \boldsymbol{\rho})$:

$$N(\mathbf{r}, \boldsymbol{\rho}) = -\frac{\mathbf{e}_r \mathbf{e}_s - \mathbf{e}_\rho \mathbf{e}_s}{2} = -\tfrac{1}{2}[\cos(\mathbf{r}, \mathbf{e}_s) - \cos(\boldsymbol{\rho}, \mathbf{e}_s)]. \tag{2.43}$$

To understand the Stokes' factor (also obliquity factor) and its meaning (respectively to substitute it by the value '1' in most cases), we look at Fig. 2.19. Thereby we make use of a more realistic example, in which the rays are near the axis, which means that they propagate in the vicinity of the connecting line between Q and P. We can specify the 'excitation' originating from the surface element dS with

$$d\mathcal{E}_S = (\mathcal{E}_Q/k\rho) \exp(ik\rho) \cos(\boldsymbol{\rho}, \mathbf{e}_s)\, dS,$$

the 'modulus' at P with

$$d\mathcal{E}_P = d\mathcal{E}_S \cos(\mathbf{e}_r, \mathbf{e}_s) \exp(ikr)/r,$$

and thus find exactly the factors from Eq. (2.42).

A remarkable property of the Stokes' factor is the suppression of the radiation in the backward direction, because according to Eq. (2.43) $N \to 0$ holds for $\mathbf{e}_\rho \to \mathbf{e}_r$! In contrast to that, we find for near-axis rays in the forward direction $N \to 1$, and we want to restrict ourselves to this frequent case in the following. The right part of Fig. 2.19 shows the total angle distribution of the Stokes' factor for a planar incident wave with $\boldsymbol{\rho} = \mathbf{e}_s$.

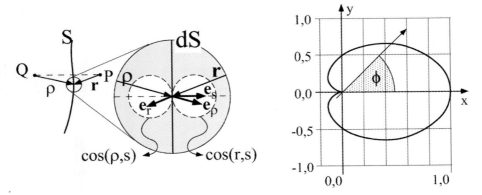

Fig. 2.19: *For the interpretation of the Stokes' factor. Left: geometric relations. Right: angle dependence of the Stokes' factor, $N(\mathbf{r}, \boldsymbol{\rho} \| \mathbf{e}_s) = [1 + \cos(\phi)]/2$.*

We finally consider the propagation of near-axis rays for $N \simeq 1$ in the geometry and with the notations of Fig. 2.20. Besides we assume that the surface S is illuminated with a planar wave. Then the field strength $\mathcal{E}_S \simeq \mathcal{E}_Q/k\rho$ is constant, but the intensity distribution may be characterized by a transmission function $\tau(\xi, \eta)$ (which in principle can be imaginary, if phase shifts are caused). According to Eq. (2.42) we can calculate the field strength at the point P

$$\mathcal{E}(\mathbf{r}_P) = -\frac{i\mathcal{E}_S}{\lambda} \oint_S \tau(\xi, \eta)\, \frac{e^{ikr}}{r}\, d\xi\, d\eta. \tag{2.44}$$

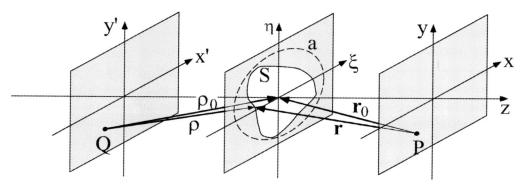

Fig. 2.20: *Fraunhofer diffraction for $N \simeq 1$.*

This result is also still too difficult for a general treatment. But further approximations are made easier by the circumstance that the distance between the diffracting object and the area of observation is in general large compared with the wavelength and the transverse dimensions, which are marked in Fig. 2.20 by a circle with radius a within the plane of the diffracting object. We express the distances r and r_0 by the coordinates of the respective planes,

$$r^2 = (x - \xi)^2 + (y - \eta)^2 + z^2 \qquad \text{and} \qquad r_0^2 = x^2 + y^2 + z^2.$$

We consider r as a function of r_0,

$$r^2 = r_0^2 \left(1 - \frac{2(x\xi + y\eta)}{r_0^2} + \frac{\xi^2 + \eta^2}{r_0^2} \right),$$

and expand r with $\kappa_x = -kx/r_0$ and $\kappa_y = -ky/r_0$,

$$r = r_0 \sqrt{1 + \frac{2(\kappa_x \xi + \kappa_y \eta)}{kr_0} + \frac{\xi^2 + \eta^2}{r_0^2}} \simeq r_0 \left(1 + \frac{\kappa_x \xi + \kappa_y \eta}{kr_0} + \frac{\xi^2 + \eta^2}{2r_0^2} \right).$$

Then the phase factor in Eq. (2.44) decomposes into three contributions,

$$\exp(ikr) \rightarrow \exp(ikr_0) \, \exp[i(\kappa_x \xi + \kappa_y \eta)] \, \exp\left(\frac{ik(\xi^2 + \eta^2)}{2r_0} \right).$$

The first factor just yields a general phase factor, the second depends linearly on the transverse coordinates of the diffracting plane and the plane of observation, the last one depends only on the coordinates of the diffracting plane (we have met the last factor already as the 'Fresnel factor', when discussing Gaussian beams (see p. 38)). In many experiments the Fresnel factor differs only a little from 1, because $ka^2/r_0 \ll 1$. Therefore it delivers the distinguishing property for the two important basic diffraction types, Fraunhofer and Fresnel diffraction ($r_0 \simeq z$):

(i)   Fraunhofer diffraction $a^2 \ll \lambda z$,

(ii) Fresnel diffraction $a^2 \geq \lambda z$ but $a \ll z$.   (2.45)

Since the 19th century diffraction phenomena have played an important role in the development of the wave theory of light, and up to now they are closely correlated with the names of Joseph von Fraunhofer (1787–1826) and Augustine Jean Fresnel (1788–1827). The radius $a = \sqrt{\lambda z}$ defines the region of validity of the Fraunhofer approximation within the diffracting plane. The usual condition is that in this case the object lies completely within the first *Fresnel zone* (see also p. 63). Besides, when the distance z to the diffracting object is just chosen large enough, one always reaches the far field limit, where Fraunhofer diffraction is valid.

2.5.2 Fraunhofer diffraction

The Fraunhofer approximation is applied in the far field of a diffracting object (e.g. a slit with typical dimension a), if the condition

$$a^2 \ll \lambda z$$

(2.45(i)) is fulfilled. For near-axis beams we can substitute the factor $1/r \simeq 1/r_0 \simeq 1/z$, and we find after inserting the approximations into Eq. (2.44) the expression

$$\mathcal{E}(\mathbf{r}_P) = -\frac{i\mathcal{E}_S\, e^{ikr_0}}{\lambda z} \oint_S \tau(\xi, \eta)\, e^{i(\kappa_x \xi + \kappa_y \eta)}\, d\xi\, d\eta. \tag{2.46}$$

But in the phase factor we keep r_0,

$$\exp\left(ikr_0\right) \simeq \exp\left(ikz\right) \exp\left(\frac{ik(x^2 + y^2)}{2z}\right), \tag{2.47}$$

because here even small deviations may lead to a fast phase rotation, which then plays an important role in interference phenomena.

After that the field amplitude at point P has the form of a spherical wave, which is modulated with the Fourier integral $T(\kappa_x, \kappa_y)$ of the transmission function $\tau(\xi, \eta)$,

$$T(\kappa_x, \kappa_y) = \int_\infty^\infty \int_\infty^\infty d\xi\, d\eta\, \tau(\xi, \eta)\, e^{i(\kappa_x \xi + \kappa_y \eta)}.$$

Finally the great impact of Fourier transformation in many areas of physics has been significantly supported by its relevance for the treatment of optical diffraction problems. Now we want to discuss some important examples.

Examples: Fraunhofer diffraction
1. Fraunhofer diffraction at a long single slit

We consider a long, quasi-one-dimensional slit (Fig. 2.21, width d) and assume again that the illumination may be inhomogeneous. Because we have introduced several approximations concerning ray propagation (e.g. Stokes' factor $N = 1$), we may not solve the one-dimensional case just by simple integration of the η coordinate in eq. (2.46) from $-\infty$ to ∞. Instead we have to work out the concept of Kirchhoff's integral theorem for a line-like (instead of a point-like) source. From a line-like source a cylindrical wave originates, the intensity of which does not decline like $1/z^2$ any more, but only with $1/z$. It turns out that the result has a very similar structure.

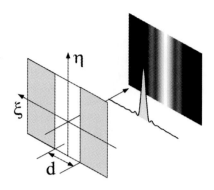

Fig. 2.21: *Diffraction at a long slit.*

The amplitude of the cylindrical wave must decline with $1/\sqrt{z}$ and the one-dimensional variant of Eq. (2.46) has the form

$$\mathcal{E}(\mathbf{r}_P) = -\frac{i\mathcal{E}_S\,e^{ikr_0}}{\lambda\sqrt{kz}} \oint_S \tau(\xi)\,e^{i\kappa_x\xi}\,d\xi.$$

In the case of a linear, infinitely long slit, the transmission function has the simple form $\tau(\xi) = 1$ for $|\xi| \le d/2$ and else $\tau(\xi) = 0$. One calculates

$$\mathcal{E}(x) = -\frac{i\mathcal{E}_S\,e^{ikr_0}}{\lambda\sqrt{kz}} \int_{-d/2}^{d/2} d\xi\,e^{i\kappa_x\xi} = \mathcal{E}_S\frac{d\,e^{ikr_0}}{\lambda\sqrt{kz}}\frac{\sin{(kxd/2z)}}{kxd/2z}.$$

The intensity distribution $I(x) \propto |\mathcal{E}(x)|^2$ is shown in Fig. 2.21 and distorted slightly in the grey colour scale for clarification.

2. Fraunhofer diffraction at a 'Gaussian transmitter'

We consider a Gaussian amplitude distribution, which one may create for example out of a planar wave by a filter with a Gaussian transmission profile. On the other hand we may just use the Gaussian beam from section 2.3 and insert an aperture only in thought – the physical result would be the same.

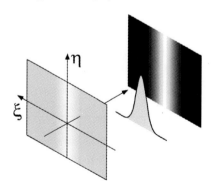

Fig. 2.22: *Diffraction at a 'Gaussian transmitter'.*

On a screen behind the aperture the intensity distribution must have been induced only by the diffraction at this fictitious aperture! We use the form and notation from section 2.3 with the fictitious transmission

$$\tau(\xi) = e^{-(\xi/w_0)^2}/\sqrt{\pi}.$$

The diffraction integral

$$\mathcal{E}(x) = i\mathcal{E}_0\frac{e^{ikr_0}}{\lambda\sqrt{kz}} \int_{-\infty}^{\infty} d\xi\,e^{i\kappa_x\xi}\frac{e^{-(\xi/w_0)^2}}{\sqrt{\pi}}$$

can be evaluated with

$$\int_{-\infty}^{\infty} d\xi\,\exp{[-(\xi/w_0)^2]}\,\exp{(i\kappa_x\xi)} = \sqrt{\pi}\,w_0\,\exp{[-(\kappa_x w_0/2)^2]}$$

and we find, using the notion on p. 40 (beam waist w_0, length of the Rayleigh zone z_0, etc.):

$$\mathcal{E}(x) = i\mathcal{E}_0 \frac{w_0\, e^{ikr_0}}{\lambda\sqrt{kz}}\, e^{-(xz_0/w_0z)^2} \simeq \mathcal{E}_0 \frac{w_0\, e^{ikz}}{\lambda\sqrt{kz}}\, e^{ikx^2/2z}\, e^{-(x/w(z))^2}.$$

The last approximation is valid in the far field ($z \gg z_0$) and we find after some conversions that it corresponds there exactly to the Gaussian TEM_{00} mode from Section 2.3. Indeed one could have started the search for stable modes in a mirror or lens system also from the viewpoint of diffraction. The amplitude distribution must be a self-reproducing solution (or eigenfunction) of the diffraction integral, which is 'diffraction-limited'. Indeed integral equations are not very popular in teaching, which is why usually the complementary path of differential equations according to Maxwell is struck.

In our discussion we have treated the x and y coordinates completely independently from each other. That is why wave propagation according to Gaussian optics occurs independently in x and y directions, an important condition for optical systems, the axial symmetry of which is broken, e.g. in ring resonators.

3. Fraunhofer diffraction at a circular aperture

One more element of diffraction, exceptionally important for optics, is the circular aperture, because diffraction occurs at all circle-like optical elements, among which lenses are counted for example. We will see that the resolution of optical instruments is limited by diffraction at these apertures, and that diffraction causes a fundamental limit for efficiency, the so-called *diffraction limit*.

We introduce polar coordinates (ρ, ψ) within the (η, ξ) plane and (r, ϕ) within the (x, y) plane of the screen. With these coordinates the diffraction integral from Eq. (2.46) reads as

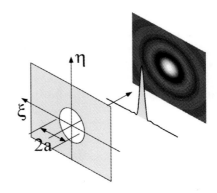

Fig. 2.23: *Diffraction at a circular aperture.*

$$\mathcal{E}(r) = -i\mathcal{E}_S \frac{e^{ikr_0}}{\lambda z} \int_0^a \rho\, d\rho \int_0^{2\pi} d\psi\, e^{-i(kr\rho/z)\cos(\phi-\psi)}.$$

This can be evaluated with the mathematical relations for Bessel functions,

$$J_0(x) = \frac{1}{2\pi} \int_0^{2\pi} \exp\left[ix\cos(\psi)\right] d\psi \qquad \text{and} \qquad \int_0^x dx'\, x'\, J_0(x') = x J_1(x).$$

The result is

$$\mathcal{E}(r) = -i\mathcal{E}_S\, e^{ikr_0}\, \frac{ka^2}{z}\, \frac{J_1(kar/z)}{(kar/z)}.$$

The central diffraction maximum is also called the 'Airy disc' (do not confuse this with the Airy function!).

The intensity distribution is determined by forming the modulus,

$$I(r) = I(r{=}0) \left(\frac{2J_1(kar/z)}{kar/z} \right)^2 .$$

The radius r_{Airy} of the Airy disc is defined by the first zero of the Bessel function $J_1(x{=}3.83) = 0$. From $kar_{\text{Airy}}/r_0 = 3.83 = 2\pi 1.22$ we find the radius

$$r_{\text{Airy}} = 1.22 \frac{z\lambda}{2a} .$$

With these specifications we may already determine the Rayleigh criterion for a lens of diameter $2a \to D$ and with focal length $z \to f$,

$$r_{\text{Airy}} = 1.22 \frac{f\lambda}{D}, \tag{2.48}$$

which matches the result of the treatment of Gaussian beams except for small constant factors (see p. 43).

2.5.3   Optical Fourier transformation, Fourier optics

According to Eq. (2.46) in the far field a diffracting object creates an amplitude distribution that corresponds to the complex amplitude distribution in the object plane and is a function of the spatial frequencies $\kappa_\eta = -k\eta/z$ and $\kappa_\xi = -k\xi/z$. A convex lens focuses incident beams and moves the Fourier transform of the amplitude distribution into the focal plane at the focal length f (Fig. 2.24):

$$\begin{aligned}
\mathcal{E}(\kappa_\eta, \kappa_\xi) &= \mathcal{A}(\eta, \xi) \oint_S \tau(x,y)\, e^{i(\kappa_\eta x + \kappa_\xi y)}\, dx\, dy \\
&= \mathcal{A}(\eta, \xi)\mathcal{F}\{\mathcal{E}(x,y)\}.
\end{aligned}$$

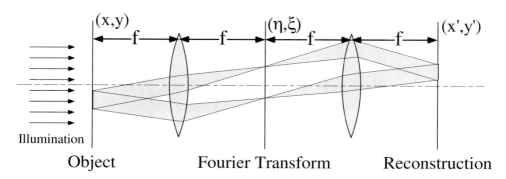

Fig. 2.24: *A lens as an optical Fourier transformer. The image can be reconstructed by a second lens. The properties of the image can be manipulated in Fourier space, i.e. the Fourier plane.*

Therefore when observing a Fraunhofer diffraction image, one uses conveniently a lens (directly after the diffracting object), to keep the working distance short. It can be shown that the factor $\mathcal{A}(\eta, \xi)$ is independent of (η, ξ) if the diffracting object is located at the front focal plane. Under these circumstances, studying the intensity distribution $I(\eta, \xi) \propto |\mathcal{E}(\kappa_\eta, \kappa_\xi)|^2 \propto |\mathcal{F}\{\mathcal{E}(x, y)\}|^2$, obviously one finds a power spectrum in the space frequencies of the diffracting object.

But the Fourier transformation of a diffracting object by a lens would not be that exciting if it were not for the fact that it forms the basis for the Abbe theory of imaging in a microscope (see p. 113) or more generally of *Fourier optics* [62]. The treatment of this goes beyond the scope of the present book, but, referring to Fig. 2.24 and without going into details, we want to point out that a second lens compensates or reverses again the Fourier transformation of the first lens. Within the focal plane of the first lens, the Fourier plane, the image can now be manipulated. Just by use of simple diaphragms (amplitude modulation) certain Fourier components can be suppressed and one obtains a smoothing of the images. On the other hand one can also apply phase modulation, e.g. by inserting glass retardation plates, which affect only selected diffraction orders. This procedure is also the basis for the phase contrast method in microscopy. Imaging can also include a magnification by application of lenses with different focal lengths.

2.5.4   Fresnel diffraction

For Fraunhofer diffraction, not only must the screen lie in the far field, but also the size a of the radiation source must fit into the first Fresnel zone with radius $r_0 = \sqrt{z\lambda}$, which means that $a \leq \sqrt{z\lambda}$ must be fulfilled. If this condition is not met, one may apply the Fresnel approximation, which for

$$a^2 \geq \lambda z / \pi \quad \text{but} \quad a \ll z$$

uses the full quadratic approximation in (x, y, η, ξ):

$$r^2 \;=\; (x - \eta)^2 + (y - \xi)^2 + z^2,$$

$$r \;=\; z\left(1 + \frac{(x - \eta)^2}{z^2} + \frac{(y - \xi)^2}{z^2}\right)^{1/2}$$

$$\;=\; z + \frac{(x - \eta)^2}{2z} + \frac{(y - \xi)^2}{2z} + \cdots.$$

Then according to Eq. (2.44) the diffraction integral reads as

$$\mathcal{E}(\mathbf{r_P}) = i\mathcal{E}_0 \frac{e^{ikz}}{\lambda z} \oint_S \tau(\eta, \xi) \exp\left(\frac{ik}{2z}[(x - \eta)^2 + (y - \xi)^2]\right) d\eta \, d\xi. \qquad (2.49)$$

Mathematically this is much more elaborate than the Fourier transformation in the Fraunhofer approximation (Eq. 2.46), but easy to treat with numerical methods.

Examples: Fresnel diffraction

1. Fresnel diffraction at a straight edge

First we introduce the normalized variable u,

$$\frac{k}{2z}(x - \eta)^2 := \frac{\pi}{2}u^2, \quad u_0 = u(\eta{=}0) = \sqrt{\frac{k}{\pi z}}\, x, \quad d\eta = -\sqrt{\frac{\pi z}{k}}\, du,$$

into the diffraction integral and substitute (K is constant)

$$\mathcal{E}(x) = K \int_0^\infty \exp\left[\frac{ik}{2z}(x - \eta)^2\right] d\eta \quad \xrightarrow{x \to u} \quad K\sqrt{\frac{\pi z}{k}} \int_{-\infty}^{u_0} \exp\left[i\frac{\pi}{2}u^2\right] du.$$

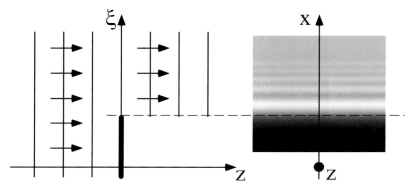

Fig. 2.25: *Fresnel diffraction at a straight edge.*

At a large distance $(x, u_0 \to \infty)$ from the edge we expect a homogeneous field and homogeneous intensity, which we can use for normalization:

$$I_0 = \frac{c\epsilon_0}{2}\mathcal{E}^2(x{\to}\infty) = \frac{c\epsilon_0}{2}\left|K\sqrt{\frac{\pi z}{k}}(1 + i)\right|^2 = \frac{c\epsilon_0}{2}K^2 z\lambda.$$

With that we can calculate the intensity, which can be expressed with the aid of the Fresnel integrals,

$$C(u) := \int_0^u du' \cos\left(\frac{\pi}{2}u'^2\right) \quad \text{and} \quad S(u) := \int_0^u du' \sin\left(\frac{\pi}{2}u'^2\right),$$

in a clear form:

$$I\left(x = \sqrt{\frac{\pi z}{k}}\, u_0\right) = \frac{c\epsilon_0}{2}|\mathcal{E}(x)|^2 = \frac{1}{2}I_0 \left|\int_{-\infty}^{u_0} \exp\left[i\frac{\pi}{2}u^2\right] du\right|^2$$

$$= \frac{I_0}{2}\left\{\left[C(u_0) + \frac{1}{2}\right]^2 + \left[S(u_0) + \frac{1}{2}\right]^2\right\}.$$

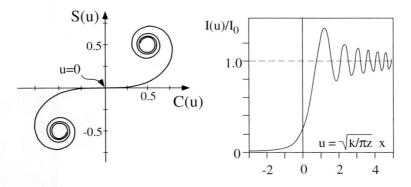

Fig. 2.26: *Cornu spiral and diffraction intensity behind a straight edge.*

As the result we gain the Cornu spiral and the intensity distribution behind a straight edge, which are both illustrated in Fig. 2.26.

2. Fresnel diffraction at a circular aperture

In order to evaluate the diffraction integral (2.49) for the case of near field diffraction at a circular aperture with radius a, we use $x = r\cos\phi'$, $y = r\sin\phi'$, $\eta = \rho\cos\phi$ and $\xi = \rho\sin\phi$:

$$\mathcal{E}(r,\phi) = i\mathcal{E}_S \frac{e^{ikz}\,e^{ikr^2/2z}}{\lambda z}$$

$$\times \int_0^a \int_0^{2\pi} e^{-ik\rho^2/2z}\, e^{-ir\rho\cos(\phi'-\phi)} \rho\,d\rho\,d\phi.$$

The angle integration can be carried out and substituting $\kappa := ka^2/z$ it yields the expected radially symmetric result

$$\mathcal{E}(r) = i\mathcal{E}_S\, e^{ikz}\, e^{i\kappa(r/a)^2/2} \kappa \int_0^1 e^{-i\kappa x^2/2} J_0(\kappa xr/a)x\,dx.$$

Now the integral can be evaluated numerically and then yields the diffraction images from Fig. 2.27. On the optical axis ($r = 0$) the integral can also be solved analytically with the result

$$\mathcal{E}(r{=}0) = i\mathcal{E}_S\, e^{ikz}\, 2\sin(\kappa/4)\, e^{i\kappa/4},$$

$$I(r{=}0) = 4 \times \tfrac{c\epsilon_0}{2}|\mathcal{E}_S|^2 \sin^2(ka^2/4z).$$

$$(2.50)$$

Accordingly along the axis one finds up to four-fold intensity of the incident planar wave! For $\kappa \ll 1$ the Fraunhofer approximation is reached and there $\sin(\kappa/4) \simeq \kappa/4 \propto 1/z$ is valid. On p. 63 we will interpret this result again with the aid of the Fresnel zones. Furthermore we will deal with the complementary problem, the circular obstacle, on p. 62.

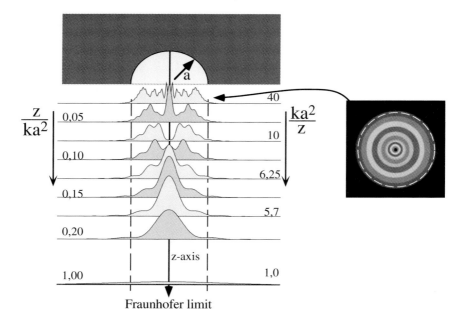

Fig. 2.27: *Example for Fresnel diffraction at a circular aperture from the Fresnel up to the Fraunhofer limit case. The right-hand figure indicates the intensity distribution at $ka^2/z = 40$.*

2.5.5   Babinet's principle

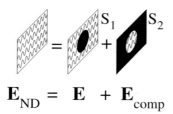

$$\mathbf{E}_{\mathrm{ND}} = \mathbf{E} + \mathbf{E}_{\mathrm{comp}}$$

Fig. 2.28: *Diffraction at a circular obstacle: Babinet's principle.*

Babinet's principle is nothing other than an application of the superposition principle (Section 2.1.6). It often allows a clever formulation for the analysis of diffraction phenomena, because it is particularly also linear within the diffracting plane. If we consider the light field, which is created by the two geometries S_1 and S_2, then the total field, which propagates without these objects, is just the sum of the two individual diffracting fields. According to Fig. 2.28, we can compose the non-diffracted field (index ND) out of the diffracted field and the corresponding complementary field:

$$\mathbf{E}_{\mathrm{ND}}(\mathbf{r}_{\mathrm{P}}) = \mathbf{E}(\mathbf{r}_{\mathrm{P}}) + \mathbf{E}_{\mathrm{comp}}(\mathbf{r}_{\mathrm{P}}).$$

This statement, Babinet's principle, seems fairly banal at first sight, but it allows a clever treatment of complementary geometries.

Example: Circular obstacle

We can construct the light field diffracted by a disc with Babinet's principle and the result from a circular aperture. It consists of just the difference between the non-diffracted field, in the most simple case a planar wave, and the complementary field, which originates from a circular aperture:

$$\mathcal{E}(r) = \mathcal{E}_S\, e^{ikz}\left(1 + i\, e^{i\kappa(r/a)^2/2}\kappa \int_0^1 e^{-i\kappa x^2/2} J_0(\kappa xr/a)x\, dx\right).$$

The diffraction image at a circular obstacle consists of the superposition of a planar wave and a diffraction wave of the circular aperture. In the centre a bright spot can *always* be seen, which has become famous as the 'hot spot':

$$\mathcal{E}(r{=}0) = \mathcal{E}_S\, e^{ikz}[1 + 2i\sin(\kappa/4)\, e^{i\kappa/4}] \quad \text{and} \quad I(r{=}0) = \frac{c\epsilon_0}{2}|\mathcal{E}_S|^2.$$

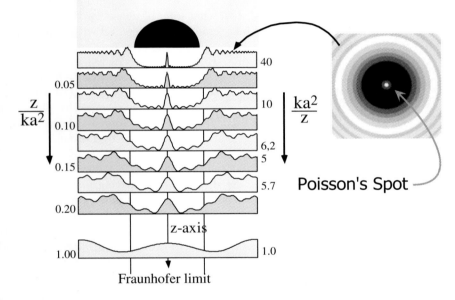

Fig. 2.29: *Fresnel diffraction at a circular obstacle. In the centre the hot spot can be recognized. Compare the complementary situation in Fig. 2.27.*

According to an anecdote, Poisson opposed Fresnel's diffraction theory on the grounds that the just achieved results were absurd; behind an aperture in the centre of the diffraction image a constant hot spot could not be observed. He was disproved by experiment – this observation is not simple, because the rims of the diffracting disc must be manufactured with optical precision (i.e. with only slight deviations in the micrometre range).

2.5.6 Fresnel zones and Fresnel lenses

In the case of Fraunhofer diffraction we can equate the Fresnel factor from eq. (2.5.4), $\exp[-ik(x^2 + y^2)/2z]$, with 1 according to $ka^2/z \ll 1$, but this is not the case for

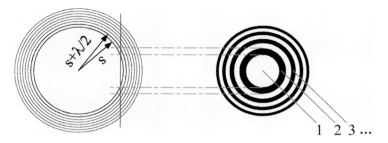

Fig. 2.30: *Fresnel zones and zone plate.*

Fresnel diffraction. This factor specifies with what kind of phase Φ_F the partial waves of the diffracting area contribute to the interference image, e.g. all with approximately $\Phi_F = 0$ in the Fraunhofer limit case.

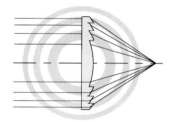

However, if we slowly increase the radius r of the diffracting object at a fixed distance z, then starting at $a_1 = \sqrt{z\lambda}$ the partial waves contribute with opposite phase, since $ka_1^2/z = \pi$. Therefore we can refer to the criterion established by Fresnel

$$a_N^2 = Nz\lambda, \qquad (2.51)$$

Fig. 2.31: *Fresnel's stepped lens.*

to divide the diffracting plane according to the character of its phase position.

In Fig. 2.30 the division with white and black zones is introduced, the outer radii of which increase according to Eq. (2.51). For clarification we look again at the diffraction at a circular aperture from the example on p. 61. According to Eq. (2.50) the brightness reaches a maximum on the axis at $a^2/z\lambda = 1,\ 3,\ \ldots$, while at $a^2/z\lambda = 2,\ 4,\ \ldots$ a minimum appears.

In a radially symmetric aperture every Fresnel zone contributes with the same area and intensity to the total field on the axis. Partial waves stemming from the odd Fresnel zones accumulate a path difference of $(N-1)\lambda/2 = 0,\ 2,\ 4,\ \ldots,\ \lambda$ on the axis, which results in constructive interference. On the other hand, a contribution with opposite phase is created from the even zones $(N\lambda/2)$, which results in cancellation of the light field for equivalent numbers of even and odd zones.

The suggestion to make use of this circumstance and use a diaphragm for every second zone dates back to Fresnel. The division into zones from Fig. 2.30 therefore stands exactly for the idea of a Fresnel zone plate. Alternatively one may also use a corresponding phase plate, which is better known under the name 'Fresnel lens' or 'Fresnel step lens' (Fig. 2.31). These lenses are often used in combination with large apertures, for example in overhead projectors.

3 Light propagation in matter

We have seen that we can describe diffraction at dielectric interfaces, such as glass plates, with the help of refractive indices introduced phenomenologically. We may also consider diffraction as the response of the glass plate to the incident electromagnetic light wave. The electric field shifts the charged constituents of the glass and thus causes a dynamic polarization. This in turn radiates an electromagnetic wave and acts back on the incident light wave through interference. Here we will discuss the properties of matter with macroscopic phenomenological indices of refraction. Some fundamental relations with the microscopic theory will be introduced in Chapter 6.

In the preceding chapter we discussed wave propagation in homogeneous matter and noticed that it differs from that in vacuum only by the phase velocity (Eq. (2.12)). Now we want to explore how interfaces or dielectrics with inhomogeneous refractive index affect the propagation of electromagnetic waves.

3.1 Dielectric interfaces

In order to discuss dielectric interfaces, we have to know how they affect electromagnetic fields. We will only cite the relations important for optics, and leave it to the reader to consult textbooks on electrodynamics for proofs of the rules (of mathematical boundary conditions) with the help of Maxwell's equations (2.8).

Suppose that an interface divides two media with refractive indices n_1 and n_2, and with normal unit vector $\mathbf{e}_N$. Then the electromagnetic radiation fields are fully characterized by

$$
\begin{aligned}
(\mathbf{E}_2 - \mathbf{E}_1) \times \mathbf{e}_N = 0 \quad &\text{and} \quad (n_2^2\mathbf{E}_2 - n_1^2\mathbf{E}_1) \cdot \mathbf{e}_N = 0, \\
(\mathbf{H}_2 - \mathbf{H}_1) \times \mathbf{e}_N = 0 \quad &\text{and} \quad (\mathbf{H}_2 - \mathbf{H}_1) \cdot \mathbf{e}_N = 0,
\end{aligned}
\tag{3.1}
$$

where $\mathbf{E}_{1,2}$ and $\mathbf{H}_{1,2}$ are to be taken in direct proximity to, but on different sides of, the interface. We further note that in optics we may often restrict ourselves to the application of vector products from (3.1), so that the scalar products are accounted for by Snell's law.

3.1.1 Diffraction and reflection at glass surfaces

In the case of a transverse electromagnetic wave incident on a dielectric interface, we can distinguish two polarization configurations: the polarization may be either linearly perpendicular (s) or parallel (p) to the plane of incidence (Fig. 3.1).

Optics, Light and Laser. Dieter Meschede
Copyright © 2004 Wiley-VCH Verlag GmbH & Co. KGaA
ISBN: 3-527-40364-7

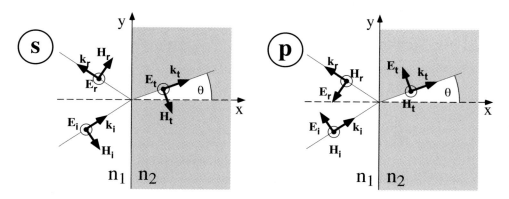

Fig. 3.1: *Electromagnetic fields at a dielectric interface for (left) s and (right) p polarization. The symbol $\odot$ indicates the field vectors perpendicular to the plane of the drawing.*

Waves with s (resp. p) polarization of the electric field are called s (resp. p) waves. Alternatively the notions σ and π polarizations (resp. TE and TM waves) are also used. We have to treat the two cases for each component individually. Then elliptical polarizations can be reduced to superpositions of these cases according to the superposition principle.

(a) s polarization

We consider the $\{\mathbf{E}, \mathbf{H}, \mathbf{k}\}_\alpha$ triads of the incident (i), reflected (r) and transmitted (t) waves and use the notation from Fig. 3.1 with

$$
\begin{aligned}
\mathbf{E}_\alpha &= \mathcal{E}_{0\alpha}\mathbf{e}_z\, e^{-i(\omega_\alpha t - \mathbf{k}_\alpha \mathbf{r})}, \\
\mathbf{H}_\alpha &= \frac{\mathcal{E}_{0\alpha}}{\mu_0 c \omega_\alpha}\mathbf{k}_\alpha \times \mathbf{e}_z\, e^{-i(\omega_\alpha t - \mathbf{k}_\alpha \mathbf{r})}.
\end{aligned}
\tag{3.2}
$$

The s-polarized electric field is perpendicular to the surface normal, which is why

$$
\mathbf{E}_t = \mathbf{E}_i + \mathbf{E}_r
\tag{3.3}
$$

is valid. If this relation is fulfilled everywhere and at all times at the interface, then obviously all waves must have the same frequency, and we can consider the time $t = 0$. Besides, according to (3.1), for arbitrary y the relation

$$
\mathcal{E}_{0t}\, e^{ik_{yt}y} = \mathcal{E}_{0i}\, e^{ik_{yi}y} + \mathcal{E}_{0r}\, e^{ik_{yr}y}
$$

must hold, and thus all y components of the $\mathbf{k}$ vectors must be equal:

$$
k_{yt} = k_{yi} = k_{yr}.
\tag{3.4}
$$

Next we consider the components individually for the reflected and transmitted parts. Since the reflected wave propagates within the same medium as the incident wave, according to $n_1^2 k_\alpha^2 = n_1^2(k_{x\alpha}^2 + k_{y\alpha}^2)$ the relations

$$
k_{xr}^2 = k_{xi}^2 \qquad \text{and} \qquad k_{xr} = -k_{xi}
$$

must be satisfied, because the positive sign creates one more incident wave, which is not physically meaningful. Thus the law of reflection is again established. For the transmitted wave, $k_t/n_2 = k_i/n_1$ must hold. From geometry one finds directly $k_i = k_{yi}/\sin\theta_t$ and thus also Snell's law (1.2) again,

$$n_1 \sin\theta_i = n_2 \sin\theta_t.$$

This is valid only for real refractive indices, but it can be generalized by the application of

$$k_{xt}^2 = k_t^2 - k_{yt}^2 = \frac{n_2^2}{n_1^2}k_i^2 - k_{yi}^2. \tag{3.5}$$

All results up to now have just confirmed the outcomes we knew already from ray optics. But by means of ray optics we could not make statements on the amplitude distribution, which is now possible by means of wave optics. According to (3.2) the tangential components of the **H** field are related to the **E** components,

$$H_{y\alpha} = -\frac{\varepsilon_0}{\mu_0 c\omega}k_{x\alpha}.$$

These must be continuous due to (3.1) and therefore fulfil the equations

$$\begin{aligned} k_{xt}E_{0t} &= k_{xi}E_{0i} + k_{xr}E_{0r} = k_{xi}(E_{0i} - E_{0r}), \\ E_{0t} &= E_{0i} + E_{0r}, \end{aligned} \tag{3.6}$$

which we have extended by the condition (3.3) to gain an equation system. It has the solutions

$$E_{0r} = \frac{k_{xi} - k_{xt}}{k_{xi} + k_{xt}}E_{0i} \quad \text{and} \quad E_{0t} = \frac{2k_{xi}}{k_{xi} + k_{xt}}E_{0i}.$$

With the amplitudes, the corresponding intensities can be calculated without any problems. The *reflection coefficient r* and the *transmission coefficient t* may also be described according to

$$r = \frac{E_{0r}}{E_{0i}} = \frac{n_1 \cos\theta_i - n_2 \cos\theta_t}{n_1 \cos\theta_i + n_2 \cos\theta_t},$$

$$t = \frac{E_{0t}}{E_{0i}} = \frac{2n_1 \cos\theta_i}{n_1 \cos\theta_i + n_2 \cos\theta_t},$$

and by the use of $n_1/n_2 = \sin\theta_t/\sin\theta_i$ according to Snell's law, these can be modified to yield

$$r = \frac{E_{0r}}{E_{0i}} = -\frac{\sin(\theta_i - \theta_t)}{\sin(\theta_i + \theta_t)} \quad \text{and} \quad t = \frac{2\sin(\theta_i - \theta_t)}{\sin(\theta_i + \theta_t)}.$$

The dependence of the reflection coefficient and the reflectivity on the angle of incidence θ_i is illustrated in Fig. 3.2. Among other things the figure shows the change of sign of the reflectivity coefficient for the reflection at a more dense medium; there a phase jump of 180° occurs.

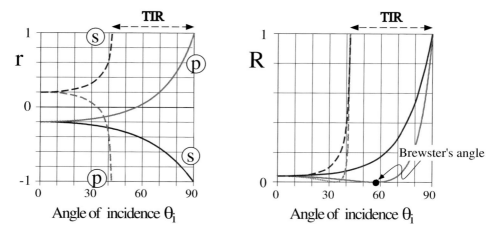

Fig. 3.2: *Reflection coefficient and reflectivity at a glass plate with refractive index $n = 1.5$ for s and p polarization. Full line: from vacuum into glass. Dashed line: from glass into vacuum.*

A very important special case occurs when light enters perpendicularly, i.e. at an angle $\theta_i = 0°$. Then for the *reflectivity R* and *transmission T*, the Fresnel formulae are valid

$$R = \frac{|\mathbf{E}_r|^2}{|\mathbf{E}_i|^2} = \left(\frac{n_1 - n_2}{n_1 + n_2}\right)^2 \qquad \text{and} \qquad T = \frac{|\mathbf{E}_t|^2}{|\mathbf{E}_i|^2} = \left(\frac{4n_1 n_2}{n_1 + n_2}\right)^2. \tag{3.7}$$

It is straightforward to calculate that, at a glass–air interface ($n_1 = 1$, $n_2 = 1.5$), 4% of the intensity is reflected.

(b) p polarization

The discussion of a p-polarized electric field oscillating within the plane of incidence follows the procedure just outlined, and may therefore be confined to the results. Snell's law is reproduced again, and for the amplitudes one finds the system of equations

$$\begin{aligned} k_t E_{0t} &= k_i E_{0i} + k_r E_{0r}, \\ k_i E_{0t} &= k_t (E_{0i} - E_{0r}), \end{aligned}$$

with the solutions

$$E_{0r} = \frac{k_t^2 - k_i^2}{k_t^2 + k_i^2} E_{0i} \qquad \text{and} \qquad E_{0t} = \frac{2k_i k_t}{k_t^2 + k_i^2} E_{0i}.$$

The reflection coefficient of the p wave obeys

$$r = \frac{E_{0r}}{E_{0i}} = -\frac{\tan(\theta_i - \theta_t)}{\tan(\theta_i + \theta_t)}$$

and is shown together with the reflectivity (dashed lines) in Fig. 3.2. It vanishes for

$$\theta_i - \theta_t = 0 \qquad \text{and} \qquad \theta_i + \theta_t = \pi/2.$$

The first condition is only fulfilled trivially for $n_1 = n_2$. The second one leads to the *Brewster condition*

$$\frac{n_2}{n_1} = \frac{\sin \theta_B}{\sin \theta_t} = \frac{\sin \theta_B}{\sin (\pi/2 - \theta_B)} = \tan \theta_B,$$

which yields the *Brewster angle* $\theta_B = 57°$ for the glass–air transition ($n = 1.5$). The Brewster condition may be interpreted physically with the angular distribution of the dipole radiation (see Section 2.2.3 and Fig. 3.3): the linear dielectric polarization in the refracting medium is transverse to the refracted beam and cannot radiate into the direction of the reflected wave, if the former makes a right angle with the refracted wave.

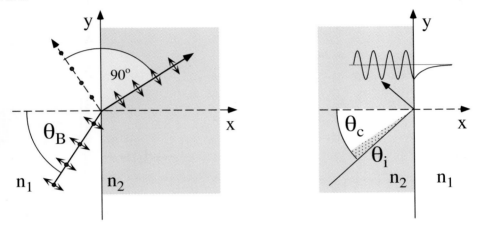

Fig. 3.3: *Left: At the Brewster angle θ_B only s-polarized light is reflected. Right: For total reflection at the denser medium ($n_1 > n_2$), an evanescent wave field is generated in the less dense medium.*

3.1.2 Total internal reflection (TIR)

We want to analyse total internal reflection (TIR) in more detail, the influence of which on the transition from a denser medium into a less dense one may be recognized already in Fig. 3.2. We consider the component $k_{xt} = k_2 \cos \theta_t$ describing the penetration of the wave into the less dense medium. We adopt the solutions for propagating waves below the critical angle $\theta_c = \sin^{-1}(n_2/n_1)$, $n_1 > n_2$, which we know already from Eq. (1.3), by generalization of Snell's condition for $\theta_i > \theta_c$ to imaginary values. With $W = \sin \theta_t = \sin \theta_i/\sin \theta_c > 1$, one may write

$$\cos \theta_t = (1 - \sin^2 \theta_t)^{1/2} = (1 - W^2)^{1/2} = iQ,$$

where Q is again a real number. Now we write the electric field for angles of incidence beyond the critical angle as a propagating wave,

$$\mathbf{E}(\mathbf{r}, t) = \mathbf{E}_{20} \exp[-i(\omega t - \mathbf{k}_2 \mathbf{r})].$$

With $\mathbf{k} = k_2(\cos\theta_t\,\mathbf{e}_x + \sin\theta_t\,\mathbf{e}_y)$, one gets

$$\mathbf{E} = \mathbf{E}_{20}\exp(-k_2Qx)\exp[-i(\omega t - k_2 Wy)].$$

For $\theta_i > \theta_c$ the wave propagates along the interface. Furthermore, it penetrates into the denser medium, but is attenuated exponentially with penetration depth $\delta_e = 2\pi/(k_2Q)$ (Fig. 3.3). The wave within the less dense medium often is called the *evanescent wave field* or the *laterally attenuated wave*.

Example: Penetration depth and energy transport for total reflection

According to the preceding section, the penetration depth of a totally reflected wave is

$$\delta_e = \frac{1}{k_2Q} = \frac{\lambda/2\pi}{\sqrt{n_2^2 - n_1^2\sin^2\theta_i}}.$$

For the case of a 90° prism from Fig. 1.7 (angle of incidence 45°, refractive index $n_1 = 1.5$), one calculates $Q = 0.25$ and $\delta_e = 0.27\,\mu$m at 600 nm.

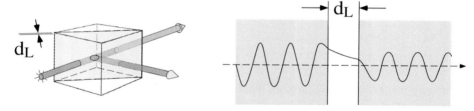

Fig. 3.4: *Frustrated total internal reflection. The width of the air gap must be less than the penetration depth of the evanescent wave.*

It is instructive to consider the energy transport according to Eq. (2.15) through the interface into the evanescent wave. It turns out that the normal component of the Poynting vector is purely imaginary,

$$\begin{aligned}
\langle\mathbf{S}\rangle\cdot\mathbf{e}_N &= \langle\mathbf{E}\times\mathbf{H}\rangle\cdot\mathbf{e}_N\\
&= \mathfrak{Re}\{c\epsilon_0/2|\mathbf{E}|^2 iQ\} = 0,
\end{aligned}$$

and therefore no energy transport occurs through the interface. Actually this situation changes if we position a second interface nearby, as indicated in Fig. 3.4. Then the so-called *frustrated total internal reflection* (FTIR) occurs. This applies not only when building optical beam splitters, but also to coupling light in different manners (by varying the air gap) into waveguides (see Fig. 3.6) or monolithic optical resonators, or for example to performing spectroscopy in the immediate vicinity of a surface.

3.2   Complex refractive index

So far we have considered real indices of refraction, which are a good approximation for absorption-free media. Absorption may be taken into account phenomenologically by the generalization of the refractive index to a complex quantity,

$$n = n' + in''.$$

Then in a homogeneous medium wave propagation may be described according to

$$\mathbf{E}(\mathbf{r}, t) = \mathbf{E}_0\, e^{-i(\omega t - n'\mathbf{kr})}\, e^{-n''\mathbf{kr}},$$

where obviously $\alpha = 2n''k_z$ specifies the attenuation of the intensity ($I \propto |\mathbf{E}|^2$), here for propagation along the z direction:

$$I(\mathbf{r}) = I(0)\exp(-\alpha z) = I(0)\exp(-2n''k_z z). \tag{3.8}$$

3.2.1   Refractive index of conducting materials

Nowadays dielectric multilayer mirrors are usually used for laser applications (see Chapter 5.7). But conventional mirrors made of evaporated metal layers also still play an important role in 'everyday optical technology' due to their low price and broadband effect. Metals are characterized by enormous conductivity, which also causes their high reflectivity. We consider a classical, phenomenological model for the conductivity σ, which goes back to Paul Drude (1863–1906). It has been shown to be extremely powerful, and more recently a microscopic proof has been found by solid-state quantum theory. In the Drude model the motion of a free electron is attenuated by friction forces with an attenuation rate τ^{-1},

$$m\left(\frac{dv}{dt} + \frac{v}{\tau}\right) = \mathfrak{Re}\{q\mathcal{E}_0\, e^{-i\omega t}\},$$

which takes into account all internal losses within a crystal in a lumped sum. In equilibrium the ansatz $v = v_0\, e^{-i\omega t}$ yields an average velocity amplitude

$$v_0 = \frac{q\mathcal{E}_0}{m}\frac{1}{-i\omega + 1/\tau} = -\frac{q\mathcal{E}_0\tau}{m}\frac{1}{1 - i\omega\tau}. \tag{3.9}$$

With charge carrier density $\mathcal{N}$ and current density $j = \sigma\mathcal{E} = \mathcal{N}qv$, one may determine the frequency-dependent conductivity of a metal,

$$\sigma(\omega) = \frac{\mathcal{N}q^2}{m}\frac{\tau}{1 - i\omega\tau} = \epsilon_0\omega_{\mathrm{p}}\frac{\omega_{\mathrm{p}}\tau}{1 - i\omega\tau}, \tag{3.10}$$

where we introduce the plasma frequency $\omega_{\mathrm{p}}^2 = \mathcal{N}q^2/m\epsilon_0$. The plasma frequencies of typical metals with large charge carrier densities ($\mathcal{N} = 10^{19}\,\mathrm{cm}^{-3}$) have values $\omega_{\mathrm{p}} \approx 10^{16}\,\mathrm{s}^{-1}$, which is beyond the frequencies of visible light. In semiconductors the conductivity may be adjusted by doping, and this frequency can be easily shifted into the visible or infrared spectral range.

To analyse the influence of conductivity on wave propagation, we refer to the third Maxwell equation (2.7) and introduce the current density we have just determined,

$$\nabla\times\mathbf{H} = \mu_0\sigma\mathbf{E} + \epsilon_0\frac{\partial}{\partial t}\mathbf{E}.$$

This gives rise to a modification of the wave equation (2.10),

$$\left(\nabla^2 - \frac{1}{c^2}\frac{\partial^2}{\partial t^2}\right)\mathbf{E}(\mathbf{r}, t) - \frac{\sigma}{\epsilon_0 c^2}\frac{\partial}{\partial t}\mathbf{E} = 0. \tag{3.11}$$

According to $k^2 = n^2(\omega)(\omega/c)^2$ the solution $\mathbf{E} = \mathcal{E}_0\epsilon\, e^{-i[\omega t - n(\omega)\mathbf{k r}]}$ yields a complex refractive index depending on the conductivity of the medium, which has to be determined phenomenologically,

$$n^2(\omega) = 1 + i\frac{\sigma(\omega)}{\epsilon_0\omega}. \tag{3.12}$$

It pays to distinguish the extreme cases of low and high frequencies.

(i) High frequencies: $\omega_{\mathrm{p}}\tau \gg \omega\tau \gg 1$

We expect this case for optical frequencies; according to (3.10) it holds that

$$\sigma \simeq i\epsilon_0\omega_{\mathrm{p}}^2/\omega \qquad \text{and} \qquad n^2(\omega) \simeq 1 - (\omega_{\mathrm{p}}/\omega)^2.$$

For $\omega < \omega_{\mathrm{p}}$, the refractive index is purely imaginary,

$$n = i\frac{(\omega_{\mathrm{p}}^2 - \omega^2)^{1/2}}{\omega} = in'', \tag{3.13}$$

and the wave no longer propagates in this medium. Instead for $\omega > \omega_{\mathrm{p}}$, the wave penetrates into the medium, as in the case of total internal reflection, to a depth of

$$\delta = (n''k)^{-1} = \frac{c}{\sqrt{\omega_{\mathrm{p}}^2 - \omega^2}}.$$

For $\tau^{-1} \ll \omega \ll \omega_{\mathrm{p}}$, we get that $n'' \approx \omega_{\mathrm{p}}/\omega$ is valid, and the penetration is called the 'anomalous skin effect' with an approximately constant penetration depth δ_{as}, which corresponds exactly to the plasma wavelength $\lambda = \omega_{\mathrm{p}}/2\pi c$,

$$\delta_{\mathrm{as}} = c/\omega_{\mathrm{p}} = \lambda_{\mathrm{p}}/2\pi.$$

(ii) Low frequencies: $\omega\tau \ll 1 \ll \omega_{\mathrm{p}}\tau$

At the lower end of the frequency spectrum, the conductivity is independent of frequency to good approximation,

$$\sigma(\omega) \simeq \epsilon_0\omega_{\mathrm{p}}^2\tau,$$

and in this case the imaginary part of the index of refraction is from Eq. (3.12)

$$n'' \simeq \omega_{\mathrm{p}}/2\sqrt{\tau/\omega}.$$

Now the refractive index determines the penetration depth, which is called the 'normal skin effect' for lower frequencies:

$$\delta_{\mathrm{ns}} = \frac{\lambda_{\mathrm{p}}}{\pi\sqrt{\omega\tau}}.$$

This case is less important in optics, but plays an important role in applications at radio frequencies.

3.2.2   Metallic reflection

noindent Now we can use the results of the previous chapter to discuss metallic reflection. However, we confine ourselves to perpendicular incidence. Oblique incidence has many interesting properties but requires elaborate mathematical treatment, which can be reviewed in the specialist literature.

For optical frequencies the limiting case of high frequencies ($\omega\tau \gg 1$) from the previous section applies, and we can use the purely imaginary refractive index (3.13)

$$n = in'' = i\sqrt{\omega_p^2 - \omega^2}/\omega.$$

We can take the boundary conditions from Eq. (3.6) and use them directly for the air–metal interface, $k_t/k_i = in''$,

$$in'' E_{0t} = E_{0i} + E_{0r},$$
$$E_{0t} = E_{0i} + E_{0r}.$$

Without any problems one finds

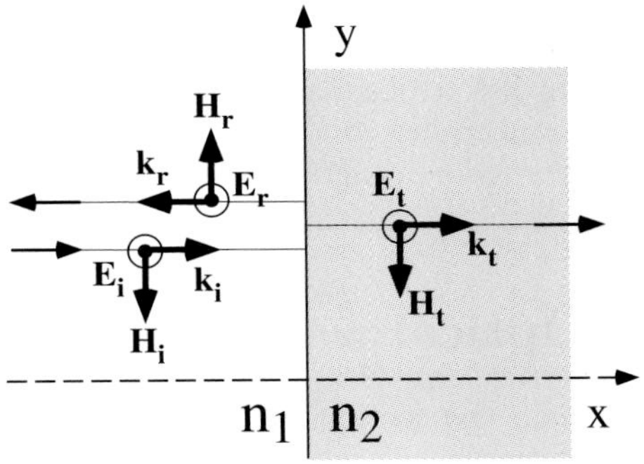

Fig. 3.5: *Electromagnetic fields reflected at perpendicular incidence.*

$$\mathcal{E}_{0r} = \frac{1 - in''}{1 + in''}\mathcal{E}_{0i} \quad \text{and} \quad \mathcal{E}_{0t} = \frac{2}{1 + in''}\mathcal{E}_{0i}.$$

An interesting result shows up, when calculating the reflectivity,

$$R = \frac{|\mathbf{E}_r|^2}{|\mathbf{E}_i|^2} = \frac{|1 - in''|^2}{|1 + in''|^2} = 1.$$

We have neglected ohmic losses (relaxation rate τ^{-1}!), which are, of course, always present in real metals. Indeed, one finds that within the visible spectral region important metals like Al, Au and Ag have reflectivities of the order of 90–98%. Normally this value is reduced by oxidized surfaces, so that metallic mirror surfaces either have to be deposited on the backside of a glass plate or are covered with a transparent and thin protective layer.

Example: Hagen–Rubens relation
To take into account the ohmic losses in Eq. (3.10), we use the approximation

$$\sigma(\omega) = \frac{\sigma_0}{1 - i\omega\tau} \simeq \frac{i\sigma_0}{\omega\tau}\left(1 - \frac{i}{\omega\tau}\right),$$

which gives for the refractive index at optical frequencies ($\omega_p \gg \omega \gg \tau^{-1}$) the approximation

$$n^2(\omega) \simeq -\frac{\omega_p^2}{\omega^2}\left(1 - \frac{i}{\omega\tau}\right).$$

Furthermore, we use $\sqrt{1 - i/\omega\tau} \simeq -i(1 - i/2\omega\tau)$ and find, with n'' according to (3.13) and with $n'/n'' = 1/2\omega\tau$ from

$$n^2(\omega) \simeq \frac{\omega_{\mathrm{p}}^2}{\omega^2}\left(i + \frac{1}{2\omega\tau}\right)^2 = n''^2\left(i + \frac{n'}{n''}\right)^2$$

for the reflectivity, the Hagen–Rubens relation

$$R = 1 - 4n'/n''^2 \simeq 1 - 2/\omega_{\mathrm{p}}\tau.$$

Aluminium has a plasma frequency $\omega_{\mathrm{p}} \approx 1.5 \times 10^{16}\,\mathrm{s}^{-1}$, which suggests that for an optimum reflectivity of a fresh layer of 95% for visible wavelengths the average time between scattering events is $\tau \approx 2 \times 10^{-15}\,\mathrm{s}$.

3.3 Optical waveguides and fibres

Following up the section on beam propagation in waveguides (Section 1.7), we now want to explore the characteristics of wave propagation and solve the corresponding Helmholtz equation. Again we concentrate on waveguides that have a cylindrical cross-section (commonly also called 'optical fibres'), and, as mentioned before in Section 1.7, constitute the backbone of optical networks – from short-range interconnections for local cross-linking of devices up to overseas cables for optical telecommunications.

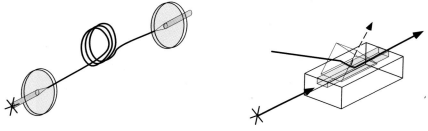

Fig. 3.6: *Types of waveguides. Left: Cylindrical, mechanically very flexible, fibres (light is coupled in and out by lenses) are used for transmission over long distances. Right: Waveguides with rectangular cross-section on the surface of suitable substrates (e.g. LiNbO₃) play an important role in integrated optics. Coupling can be performed via an edge or by frustrated total internal reflection (FTIR) with a prism on top.*

Waveguides are also an important basic element of integrated optics. Here, planar structures are preferred, onto which transverse structures can be fabricated by well-known techniques of semiconductor technology. In LiNbO$_3$, for example, the index of refraction may be varied within approximately 1% by in-diffusion of protons, creating waveguides on the surface of planar crystals that have a nearly rectangular profile of the index of refraction.

A mathematical investigation of the waveforms of an optical fibre is rather tedious and involved. As an example, let us sketch the treatment of the cylindrical fibre, the most important type for applications.

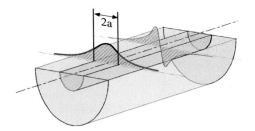

Fig. 3.7: *Step-index fibre with transverse field distribution. The curvature of the modes must be different in the core and the cladding.*

3.3.1 Step-index fibres

The index of refraction in a step-index fibre (Fig. 3.7) is cylindrically symmetric and homogeneous within the core and the cladding, respectively. Its value declines from n_1 within the core at $r = a$ step-like to the value n_2 of the cladding. According to the geometry we look for solutions of the form $\mathbf{E} = \mathbf{E}(r, \phi)\, e^{-i(\omega t - \beta z)}$. The wave equation for cylindrical (r, ϕ) components is complicated, since the $\mathbf{e}_r$ and $\mathbf{e}_\phi$ unit vectors are not constant.

For the $\mathcal{E}_z$ and $\mathcal{H}_z$ components, a scalar wave equation still holds, where $\boldsymbol{\nabla}_\perp(r, \phi)$ stands for the transverse part of the nabla operator,

$$(\boldsymbol{\nabla}_\perp^2 + \mathbf{k}^2 - \beta^2)\left\{ \begin{array}{c} \mathcal{E}_z \\ \mathcal{H}_z \end{array} \right\} = 0.$$

Fortunately, one gets a complete system of solutions if one first evaluates the components $\{\mathcal{E}_z, \mathcal{H}_z\}$ and then constructs $\{\mathcal{E}_r, \mathcal{E}_\phi, \mathcal{H}_r, \mathcal{H}_\phi\}$ by means of Maxwell's equations,

$$\boldsymbol{\nabla} \times \mathbf{H} = -i\omega\epsilon_0 n_i^2 \mathbf{E} \qquad \text{and} \qquad \boldsymbol{\nabla} \times \mathbf{E} = i\omega\mu_0 \mathbf{H}, \tag{3.14}$$

the result of which is given in Eqs. (3.18) and (3.17).

The propagation constant β must still be determined, and the Helmholtz equation for $\{\mathcal{E}_z, \mathcal{H}_z\}$ in cylindrical coordinates with $k_{1,2} = n_{1,2}\omega/c$ is

$$\left(\frac{\partial^2}{\partial r^2} + \frac{1}{r}\frac{\partial}{\partial r} + \frac{1}{r^2}\frac{\partial^2}{\partial \phi^2} + (k_i^2 - \beta^2) \right)\left\{ \begin{array}{c} E_z(r, \phi) \\ H_z(r, \phi) \end{array} \right\} = 0.$$

With the help of the trial solutions $\{\mathcal{E}_z, \mathcal{H}_z\} = \{e(r), h(r)\}e^{\pm i\ell\phi}$, this is reduced to a Bessel equation for the radial distribution of the amplitudes,

$$\left(\frac{\partial^2}{\partial r^2} + \frac{1}{r}\frac{\partial}{\partial r} + k_i^2 - \beta^2 - \frac{\ell^2}{r^2} \right)\left\{ \begin{array}{c} e(r) \\ h(r) \end{array} \right\} = 0.$$

The curvature of the radial amplitudes $\{e(r), h(r)\}$ depends on the sign of $k_i^2 - \beta^2$. Within the core we can permit positive, convex curvatures corresponding to oscillating solutions; but within the cladding the amplitude must decline rapidly and therefore must have a negative curvature – otherwise radiation results in an unwanted loss of energy (see Fig. 3.7):

within the core
within the cladding

$$\begin{aligned} 0 < k_\perp^2 &= k_1^2 - \beta^2, \\ 0 > -\kappa^2 &= k_2^2 - \beta^2. \end{aligned} \tag{3.15}$$

In other words, the propagation constant must have a value between the wavenumbers $k_i = n_i \omega / c$ of the homogeneous core and of the cladding material,

$$n_1 \omega / c \leq \beta \leq n_2 \omega / c,$$

and differ only a little from $k_{1,2}$ for small differences in the index of refraction $\Delta = (n_1 - n_2)/n_1$ (Eq. (1.7)). Such waveguides are called *weakly guiding*.

By definition it holds that $k_\perp^2 + \kappa^2 = k_1^2 - k_2^2$. Since $k_1 \simeq k_2$, the transverse wavevectors $k_\perp$ and κ are small compared with the propagation constant β,

$$k_\perp^2 + \kappa^2 \simeq 2\Delta(n_1 \omega / c)^2 \ll \beta^2.$$

The transverse solution must be finite, thereby keeping only the Bessel functions J_ℓ (modified Bessel functions K_ℓ) of the first kind within the core (cladding):

$$e(r) = \begin{cases} AJ_\ell(k_\perp r)/J_\ell(k_\perp a) & \overset{r \to 0}{\propto} \ (k_\perp r)^\ell & \text{core}, \\ AK_\ell(\kappa r)/K_\ell(\kappa a) & \overset{r \to \infty}{\propto} \ e^{-\kappa r}/\sqrt{\kappa r} & \text{cladding}, \end{cases}$$

$$h(r) = \begin{cases} BJ_\ell(k_\perp r)/J_\ell(k_\perp a) & \overset{r \to 0}{\propto} \ (k_\perp r)^\ell & \text{core}, \\ BK_\ell(\kappa r)/K_\ell(\kappa a) & \overset{r \to \infty}{\propto} \ e^{-\kappa r}/\sqrt{\kappa r} & \text{cladding}. \end{cases} \tag{3.16}$$

By defining the coefficients, we have already taken care that the components $\{\mathcal{E}_z, \mathcal{H}_z\}$ are continuous at $r = a$. For the $\{\mathcal{E}_\phi, \mathcal{H}_\phi\}$ contributions we obtain conditions from Eqs. (3.14),

$$\begin{aligned} \mathcal{E}_\phi(r, \phi) &= \frac{-i\beta}{k_i^2 - \beta^2} \left[\frac{i\ell}{r} e(r) + \frac{\omega \mu_0}{\beta} \frac{\partial}{\partial r} h(r) \right] e^{i\ell\phi}, \\ \mathcal{H}_\phi(r, \phi) &= \frac{i\beta}{k_i^2 - \beta^2} \left[\frac{i\ell}{r} h(r) - \frac{\omega \epsilon_0 n_i^2}{\beta} \frac{\partial}{\partial r} e(r) \right] e^{i\ell\phi}. \end{aligned} \tag{3.17}$$

Here we use appropriate wavenumbers k_i for core ($i = 1$) and cladding ($i = 2$). In weakly guiding fibres we usually have $\beta/(k_i^2 - \beta^2)r \simeq 1/2\Delta k_i a \gg 1$; hence the (r, ϕ) components are much stronger than the z components with which we started our solutions; these waves are nearly transverse.

Once the radial contributions are calculated, all six field components are known:

$$\begin{aligned} \mathcal{E}_r(r, \phi) &= \frac{-1}{\omega \epsilon_0 n_i^2} \left[\frac{\ell}{r} h(r) e^{i\ell\phi} + \beta \mathcal{H}_\phi(r, \phi) \right], \\ \mathcal{H}_r(r, \phi) &= \frac{1}{\omega \mu_0 n_i^2} \left[\frac{\ell}{r} e(r) e^{i\ell\phi} + \beta \mathcal{E}_\phi(r, \phi) \right]. \end{aligned} \tag{3.18}$$

To determine the propagation constant β, we substitute $\{e(r), h(r)\}$ from eq. (3.16), use boundary conditions (3.1) at $r = a$, and obtain after some algebra a system of linear equations in A and B. It is cast in transparent form with symbols $X = k_\perp a$, $Y = \kappa a$ and $J'(x) = dJ(x)/dx$,

$$B\frac{\omega\mu_0}{\beta}\left(\frac{J'_\ell(X)}{XJ_\ell(X)} + \frac{K'_\ell(Y)}{YK_\ell(Y)}\right) + i\ell A\left(\frac{1}{X^2} + \frac{1}{Y^2}\right) = 0,$$

$$A\frac{\omega\epsilon_0}{\beta}\left(\frac{n_1^2 J'_\ell(X)}{XJ_\ell(X)} + \frac{n_2^2 K'_\ell(Y)}{YK_\ell(Y)}\right) - i\ell B\left(\frac{1}{X^2} + \frac{1}{Y^2}\right) = 0,$$

and yields a characteristic eigenvalue equation

$$\left(\frac{J'_\ell(X)}{XJ_\ell(X)} + \frac{K'_\ell(Y)}{YK_\ell(Y)}\right)\left(\frac{k_1^2 J'_\ell(X)}{XJ_\ell(X)} + \frac{k_2^2 K'_\ell(Y)}{YK_\ell(Y)}\right) = \ell^2\beta^2\left(\frac{1}{X^2} + \frac{1}{Y^2}\right)^2.$$

Numerical treatment of this transcendental equation, for $\ell = 0, 1, 2, \ldots$, gives solutions $(X_{\ell m}, Y_{\ell m})$ and a propagation constant $\beta_{\ell m}$; this treatment is elaborate and is covered extensively in the literature [89]. As we did in section 1.7 on ray optics, we restrict ourselves to the case of weakly guiding waves at small differences of the indices of refraction $n_1 \simeq n_2$ or $k_1 \simeq k_2 \simeq \beta$ and end up with

$$\left(\frac{J'_\ell(X)}{XJ_\ell(X)} + \frac{K'_\ell(Y)}{YK_\ell(Y)}\right) = \pm\ell\left(\frac{1}{X^2} + \frac{1}{Y^2}\right). \tag{3.19}$$

The derivatives can be replaced by the identities

$$J'_\ell(X) = \pm J_{\ell\mp1}(X) \mp \frac{\ell J_\ell(X)}{X} \quad \text{and} \quad K'_\ell(Y) = -K_{\ell\mp1}(Y) \mp \frac{\ell K_\ell(Y)}{Y},$$

and substitution delivers the conditions for each sign in $\ell\pm1$, which may be associated with two classes of modes,

$$\begin{aligned}
\text{HE}_{\ell m} \text{ modes:} \quad & \frac{J_{\ell-1}(X_{\ell m})}{X_{\ell m}J_\ell(X_{\ell m})} = \frac{K_{\ell-1}(Y_{\ell m})}{Y_{\ell m}K_\ell(Y_{\ell m})}, \\
\text{EH}_{\ell m} \text{ modes:} \quad & \frac{J_{\ell+1}(X_{\ell m})}{X_{\ell m}J_\ell(X_{\ell m})} = -\frac{K_{\ell+1}(Y_{\ell m})}{Y_{\ell m}K_\ell(Y_{\ell m})}.
\end{aligned} \tag{3.20}$$

Since $k_\perp^2 + \kappa^2 = k_1^2 - k_2^2$, a further condition is

$$X_{\ell m}^2 + Y_{\ell m}^2 = V^2, \tag{3.21}$$

where the V parameter is a measure for the number of modes. It increases with frequency ω because of $V = \omega^2(n_1^2 - n_2^2)a^2/c^2$. Graphical solutions for the propagation constant can now be obtained from Fig. 3.8.

Eq. (3.19) gives one more condition for the coefficients (A, B) from eq. (3.16), which fix the amplitudes. The '+' sign holds for the HE modes, and the '-' sign for the EH modes:

$$[A \pm i(\omega\mu_0/\beta)B]\ell = 0. \tag{3.22}$$

From this it can be seen that the electric and magnetic fields are 90° out of phase.

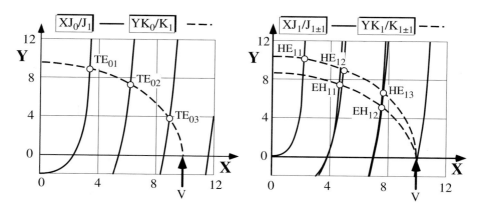

Fig. 3.8: *Graphical solution of the values for β in a step-index fibre for $V = 10$. Left: TE and TM modes. Right: HE and EH modes.*

Let us now have a look at some special cases.

(1) $\ell = 0$: TE and TM modes

For $\ell = 0$ we have $A = 0$ ($B = 0$), i.e. either the E or the H field is purely transverse. Hence for $\ell = 0$ the TE/TM denominations are sensibly used. We have indicated the graphical conditions for the (degenerate) TE_{0m} and TM_{0m} modes in Fig. 3.8. TE/TM modes will not be guided for $V \le X_{01} = k_\perp a = 2.405$ because $J_0(2.405) = 0$. There is no guiding at all below the corresponding cut-off frequency $\omega_{\mathrm{cut}} = 2.405c/a(n_1^2 - n_2^2)^{1/2}$, which is directly obtained from eq. (3.15). At $J_1(5.520) = 0$ the next mode appears, and the fibre is no longer monomode.

(2) $\ell > 0$: HE and EH modes

The lowest mode is the HE_{11} mode, which exists down to $X = 0$. Thus the core fixes the mode for arbitrarily weakly curved transverse amplitudes – the part of the energy propagating within the cladding increases more and more. On doing the mathematical treatment we assume the cladding to have an infinite extension – which involves technical limits. Owing to eqs. (3.20) and (3.22), the HE and EH modes differ. This difference not only is revealed by the unequal propagation constants, but also manifests itself in the domination of the z components in the corresponding H (HE), resp. E (EH), parts.

(3) $\ell > 0$: LP modes

By means of a recursion formula, one can show that the modes with indices $HE_{\ell m}$ and $EH_{\ell+2,m}$ are degenerate. Their linear superposition results in linearly polarized transverse E and H fields, which are orthogonal and have a weak longitudinal field. Modes constructed in that way are so-called *linearly polarized modes* (LP),

$$HE_{\ell m},\ EH_{\ell+2,m} \rightarrow LP_{\ell-1,m}.$$

3.3.2 Single-mode fibres

For X values below 2.405 the step-index fibre is counted among the so-called single-mode fibres. The lowest mode has a bell-shaped profile, which is very similar to the free TEM_{00} Gaussian mode. Coupling light into a multimode fibre in general excites a superposition of several modes, which propagate with different velocities ('mode dispersion'). At the exit of a waveguide, this light field is again converted to a free mode whose transverse and longitudinal profiles are distorted due to several dispersion contributions.

In a single-mode fibre this transverse distortion is suppressed, as is the effect of modal dispersion on the temporal structure of the pulses (see Section 3.4). A Gaussian beam may be very efficiently coupled into a bell-shaped fundamental mode of a single-mode fibre. Indeed, single-mode fibres are often used as very efficient spatial filters, because only the desired principal mode out of the launched amplitude distribution will propagate.

But even an ideal cylindrical step-index fibre is degenerate in respect of two orthogonal states of polarization. Therefore, in realistic fibres the state of polarization at the output cannot be predicted and fluctuates due to temperature changes and mechanical oscillations in the fibre curvature. These problems are solved by polarization-preserving single-mode fibres. They are realized, for example, with elliptical cross-section, which yields different propagation constants for the principal axes. The polarization of the coupled-in light must be parallel to the principal axis, to make use of the characteristics of conserving the polarization.

3.3.3 Graded-index fibre

The term 'quadratic index media' covers all the common systems like gradient-index fibres with parabolic index profile (see Fig. 3.9) that we have already dealt with in the section on ray optics and may be treated like the limiting case of an infinite thick lens. Realistic gradient fibres have a quadratic profile in the centre only, which then continues into a step-like form again. Instead, we look at a simplified, purely quadratic system, which reflects already the properties of a graded-index fibre. The index of refraction depends on the normalized radius r/a, and making use of the difference in the index of refraction, $\Delta = (n_1 - n_2)/n_1$ (see Section 1.7.3), we find

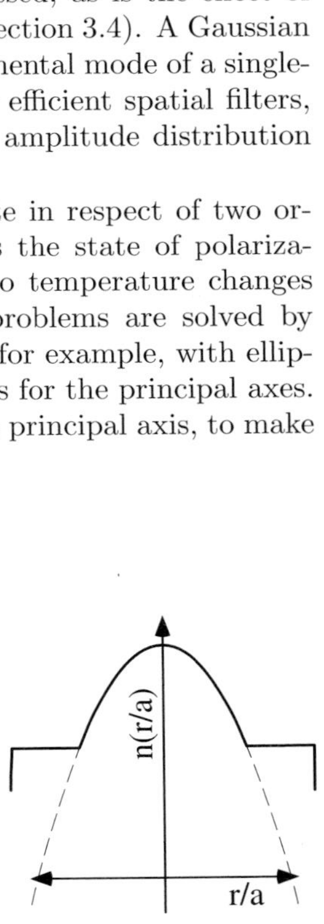

Fig. 3.9: *Simplified profile of the index of refraction of a GRIN fibre.*

$$n(\rho) = n_1[1 - \Delta(r/a)^2] \quad \text{and} \quad \Delta \ll 1.$$

We seek solutions of the Helmholtz equation (2.11), whose envelope does not change along the direction of propagation, i.e. of the form $\mathcal{E}(x, y, z) = \mathcal{A}(x, y) \exp(i\beta z)$, and get the modified equation

$$\{\nabla_\perp + n_1^2 k^2 - 2n_1^2 k^2 \Delta[(x/a)^2 + (y/y)^2] - \beta^2\}\mathcal{A}(x, y) = 0,$$

using $\{n_1 k[1 - \Delta(r/a)^2]\}^2 \simeq (n_1 k)^2[1 - 2\Delta(r/a) + \cdots]$. Let us assume now, as in the case of higher Gaussian modes (see Section 2.3.3), that the transverse distribution corresponds to modified Gaussian functions,

$$\mathcal{A}(x, y) = \mathcal{F}(x)\, e^{-(x^2/x_0^2)}\, \mathcal{G}(y)\, e^{-(y^2/y_0^2)}.$$

With this ansatz we find

$$\left(\mathcal{F}'' - \frac{4x}{x_0^2}\mathcal{F}' - \frac{2}{x_0^2}\mathcal{F} \right)\mathcal{G} + \left(\mathcal{G}'' - \frac{4y}{y_0^2}\mathcal{G}' - \frac{2}{y_0^2}\mathcal{G} \right)\mathcal{F} + n_1^2 k^2 \mathcal{F}\mathcal{G}$$
$$+ \left[\left(\frac{4}{x_0^4} - \frac{2n_1^2 k^2 \Delta}{a^2} \right)x^2 + \left(\frac{4}{y_0^4} - \frac{2n_1^2 k^2 \Delta}{a^2} \right)y^2 \right]\mathcal{F}\mathcal{G} - \beta^2 \mathcal{F}\mathcal{G} = 0,$$

where the unpleasant quadratic term in general can be eliminated by choosing

$$kx_0 = ky_0 = (ka)^{1/2}/(2n_1^2\Delta)^{1/4} \gg 1.$$

By substituting $\sqrt{2}\,x/x_0 \to u$ and $\sqrt{2}\,y/y_0 \to v$, we transform again to the Hermite differential equation that we already know from the higher Gaussian modes. With indices m and n we find

$$2(\mathcal{F}'' - 2u\mathcal{F}' + 2m\mathcal{F})\mathcal{G} + 2(\mathcal{G}'' - 2v\mathcal{G}' + 2n\mathcal{G})\mathcal{F}$$
$$+ [n_1^2 k^2 x_0^2 - \beta^2 x_0^2 - 4(m + n + 1)]\mathcal{F}\mathcal{G} = 0.$$

The terms of the upper row are constructed to vanish already on inserting the Hermite polynomials $\mathcal{H}_{m,n}$ (see p. 45, Eq. (2.32)). After a short calculation one gets for the propagation constant, which is the centre of interest,

$$\beta_{mn}(\omega) = n_{\text{eff}}\frac{\omega}{c} = \frac{n_1\omega}{c}\sqrt{1 - \frac{4\sqrt{2}\Delta(m + n + 1)}{n_1 ka}}.$$

The transverse distribution of the amplitudes also corresponds to the ones in Fig. 2.9. But in contrast to the Gaussian modes, the mode diameters (x_0, y_0) do not change. This example of a simplified GRIN fibre illustrates that multimode fibres characterized by a frequency-dependent index of refraction show 'mode dispersion' in addition to 'material dispersion'. This influences the form of pulses, because individual partial modes have different propagation velocities.

3.3.4   Fibre absorption

One could not imagine the success of optical fibres without their extraordinarily advantageous absorption properties (Fig. 3.10). On the short-wavelength side these are limited by Rayleigh scattering at small inhomogeneities ($\propto 1/\lambda^4$), and on the long-wavelength side by infrared absorption caused by the wings of the phonon spectrum. The wavelengths 1.3 and 1.55 μm, very important for telecommunications, coincide with very small absorption coefficients, and simultaneously the group velocity dispersion vanishes at 1.3 μm (see p. 84). In between we find resonances that are caused, for example, by OH contamination in the glass.

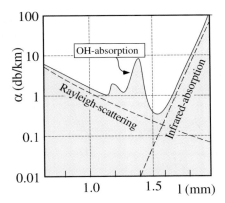

Fig. 3.10: *Absorption properties of optical fibres made of silicate glass.*

3.4   Light pulses in dispersive materials

Electromagnetic waves are used to transmit information. To make sure that there is enough power available at the other end of the transmission line for the message to be read by a receiver, the material (e.g. an optical fibre) in which transmission occurs must be sufficiently transparent. Of course, these conditions are valid for all kinds of electromagnetic waves used for transmission of information, for radio waves with ultrashort or long waves, and for microwave systems as well. For optical wavelengths, the properties of the transparent medium are generally described by two frequency-dependent indices: absorption is described by the absorption coefficient $\alpha(\omega)$ and dispersion by the refractive index $n(\omega)$.

Not only is a light pulse attenuated by the absorption of light energy, but also its shape is changed by the associated dispersion. Therefore, it is important to explore whether such a pulse is still detectable in its original shape at the end of a transmission line. We know that it is enough to describe a continuous, monochromatic field by an absorption coefficient $\alpha(\omega)$ and real index of refraction $n(\omega)$, the spectral properties of which are shown qualitatively in Fig. 3.11. The amplitude of the field at point z, taking the propagation coefficient $\beta(\omega) = n(\omega)\omega/c$ into account, then yields

at the start, $z = 0$: $E(0,t) = E_0\, e^{-i\omega t}$,

at the point z: $E(z,t) = E_0\, e^{-i[\omega t - \beta(\omega)z]}\, e^{-\alpha(\omega)z/2}$.

A light pulse can be described as a wavepacket, i.e. by the superposition of many partial waves. For that purpose we consider an electric field

$$E(0,t) = \mathsf{E}(0,t)\, e^{-i\omega_0 t},$$

with *carrier frequency* $\nu_0 = \omega_0/2\pi$ and time-variant envelope $\mathsf{E}(z,t)$, which describes the pulse shape, but in general varies slowly in comparison with the field oscillation itself,

$$\frac{\partial}{\partial t}\mathsf{E}(t) \ll \omega_0 \mathsf{E}(t). \tag{3.23}$$

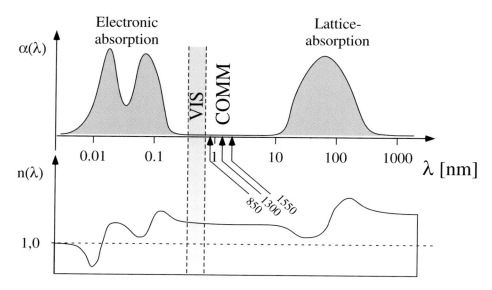

Fig. 3.11: *Qualitative trends of the absorption coefficient and refractive index as functions of wavelength for transparent optical materials. The narrow band of visible wavelengths (VIS, 400–700 nm) and the optical windows for telecommunications (COMM, 850, 1300, 1550 nm) are indicated.*

We determine the *field spectrum* $\mathcal{E}(z,\nu)$ of the light pulse by harmonic expansion:

$$\mathcal{E}(z,\nu) = \int_{-\infty}^{\infty} E(z,t)\,e^{i2\pi\nu t}\,dt = \int_{-\infty}^{\infty} \mathsf{E}(z,t)\,e^{i2\pi(\nu-\nu_0)t}\,dt,$$

$$E(z,t) = \int_{-\infty}^{\infty} \mathcal{E}(z,\nu)\,e^{-i2\pi\nu t}\,d\nu = \int_{-\infty}^{\infty} \mathcal{E}(z,\omega)\,e^{-i\omega t}\,d\omega\Big/2\pi.$$

(3.24)

Usually the spectrum of the wavepacket is located at $\nu = \nu_0$ because of Eq. (3.23) and its width is small compared to the oscillation frequency ν_0. In Fig. 3.12 we give two examples for important and common pulse shapes.

Characteristic quantities of pulsed laser radiation include the spectral bandwidth $\Delta\nu$ and the pulse length Δt, which are not easily defined and even more difficult to measure. We may for instance employ the conventional variance

$$\langle(\Delta\nu)^2\rangle = \langle\nu^2 - \nu_0^2\rangle = \int_{-\infty}^{\infty} (\nu-\nu_0)^2|\mathcal{E}(\nu)|^2\,d\nu\Big/\int_{-\infty}^{\infty} |\mathcal{E}(\nu)|^2\,d\nu,$$

and, accordingly, in the time domain,

$$\langle(\Delta t)^2\rangle = \int_{-\infty}^{\infty} (t-\langle t\rangle)^2|\mathsf{E}(t)|^2\,dt\Big/\int_{-\infty}^{\infty} |\mathsf{E}(t)|^2\,dt,$$

and show that the general relation

$$2\pi\Delta\nu t_{\mathrm{p}} \geq 1/2$$

(3.25)

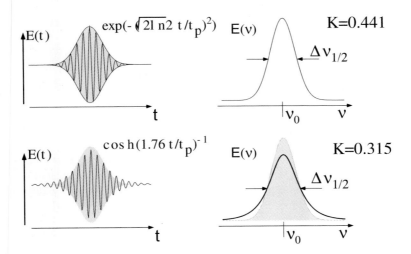

Fig. 3.12: *Two important pulse shapes in the time domain and in frequency or Fourier space. For illustration, the Gaussian pulse is superimposed on the sech or $\cosh^{-1}$ pulse. The amplitudes are chosen in such a way that the pulses have the same total energy ($\int |E(t)|^2 \, dt$). The K values specify the product of half-width times pulse length from Eq. (3.26).*

holds between these two quantities. The equals sign is valid only for pulses without frequency modulation, such pulses being called 'Fourier-limited'. From the experimental point of view it is easier to measure half-widths $\Delta\nu_{1/2}$ and $\Delta t_{1/2} = t_{\mathrm{p}}$ of the intensity. Then the pulse length times bandwidth product can be written as

$$2\pi\Delta\nu_{1/2}t_{\mathrm{p}} = K, \tag{3.26}$$

and this constant K is indicated for the two examples in Fig. 3.12. In general, its value is less than 0.5, because the half-width usually underestimates the variance. In Fig. 3.12 the much broader wings of the $\cosh^{-1}$ pulse can be seen as a reason for this.

For monochromatic waves the absorption coefficient α and the propagation constant β are often known for all partial waves of the wavepacket in the frequency domain. Then Eq. (3.24(i)) can also be described with the transfer function $\tau(z,\nu)$,

$$E(z,t) = E(0,t)\, e^{i\beta(\nu)z}\, e^{-\alpha(\nu)z/2} = \tau(z,\nu)E(0,t).$$

A pulse is composed of many partial waves, and the correlation between the pulse shapes at the start and the end of a transmission line is described by a linear, frequency-dependent transfer function $\tau(z,\nu)$ in Fourier space:

$$\mathcal{E}(z,\nu) = \tau(z,\nu)\mathcal{E}(0,\nu).$$

The temporal evolution of the field amplitude at the point z can now be determined according to

$$E(z,t) = \int_{-\infty}^{\infty} \tau(z,\nu)\mathcal{E}(0,\nu)\, e^{-i2\pi\nu t}\, d\nu.$$

Incidentally, according to the convolution theorem of Fourier transformation, a non-local correlation in the time domain is valid,

$$E(z,t) = \int_{-\infty}^{\infty} T(z, t - t') E(0, t') \, dt'$$

with

$$T(z,t) = \int_{-\infty}^{\infty} \tau(z, \nu) \, e^{-i2\pi\nu t} \, d\nu.$$

The optical bandwidth of common light pulses is generally narrow compared with the spectral properties of the transparent optical materials used in optical waveguides. Therefore, the following assumptions are reasonable. The frequency dependence of the absorption coefficient plays no role in pulse propagation. In good approximation it holds that

$$\alpha(\nu) \simeq \alpha(\nu_0) = \text{const.}$$

The pulse shape is changed very sensitively by the frequency-dependent dispersion, and the propagation constant $\beta(\nu) = 2\pi\nu n(\nu)/c$ can be described by the expansion

$$
\begin{aligned}
\beta(\nu) = \quad & \beta_0 + \frac{d\beta}{d\nu}(\nu - \nu_0) + \frac{1}{2}\frac{d^2\beta}{d\nu^2}(\nu - \nu_0)^2 + \cdots \\
= \quad & \beta_0 + \beta'(\nu - \nu_0) + \tfrac{1}{2}\beta''(\nu - \nu_0)^2.
\end{aligned}
\tag{3.27}
$$

Within this approximation the frequency dependence of the propagation constant $\beta(\nu)$ is described by the material-dependent parameters β_0, β' and β'', the interpretation of which we now want to introduce. With $\tau_0 = e^{-\alpha z/2}$ the corresponding transfer function reads as follows:

$$\tau(z,\nu) = \tau_0 \, e^{i\beta_0 z} \, e^{i\beta'(\nu - \nu_0)} \, e^{i\beta''(\nu - \nu_0)^2 z/2}.$$

3.4.1   Pulse distortion by dispersion

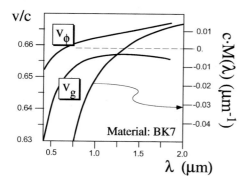

Fig. 3.13: *Example: Dispersion parameters of BK7 glass.*

Let us now discuss the influence of the dispersive contributions in more detail. If the dispersion is independent of frequency, then we obtain the wave equation (2.12) once more, in which the velocity of light in vacuum is substituted by the material-dependent phase velocity,

$$\beta_0 = 2\pi n(\nu_0)\nu_0/c = 2\pi\nu_0/v_\phi.$$

Let us first consider the case where $\beta'' = 0$. Indeed, this case occurs with glass, and one may realize qualitatively in Fig. 3.11 that somewhere between lattice absorption and electronic absorption the curvature of the refractive index must disappear. This happens at a wavelength of $\lambda = 1.3\,\mu\mathrm{m}$, which therefore offers an important window for transmission of information by optical communication. The pulse shape after a propagation length z is obtained from

$$E(z,t) = \tau_0\,e^{i\beta_0 z}\int_{-\infty}^{\infty} e^{i\beta'(\nu-\nu_0)z}\mathcal{E}(0,\nu)\,e^{-i2\pi\nu t}\,d\nu.$$

Substituting $\beta' z \to 2\pi t_\mathrm{g}$, after some algebra this yields the form

$$
\begin{aligned}
E(z,t) &= \tau_0\,e^{i\beta_0 z}\,e^{-i2\pi\nu_0 t}\int_{-\infty}^{\infty}\mathcal{E}(0,\nu)\,e^{-i2\pi(\nu-\nu_0)(t-t_\mathrm{g})}\,d\nu \\[4pt]
&= \tau_0\,e^{-i(2\pi\nu_0 t-\beta_0 z)}E(0,t-t_\mathrm{g}).
\end{aligned}
$$

The only effect of dispersion is a delay of the pulse transit time by $t_\mathrm{g} = z/v_\mathrm{g}$, which we interpret as a group delay time. This can be used for the definition of a group velocity v_g, which can be associated with a 'group index of refraction' n_g:

$$\frac{1}{v_\mathrm{g}} = \frac{1}{2\pi}\frac{d}{d\nu}\beta = \frac{1}{c}\left(n(\omega)+\omega\frac{d}{d\omega}n(\omega)\right) = \frac{n_\mathrm{g}(\omega)}{c}. \tag{3.28}$$

In most applications optical pulses propagate in a region of normal dispersion, i.e. at $dn/d\omega > 0$. Then according to Eq. (3.28) it holds that $v_\mathrm{g} < v_\phi = c/n(\omega)$. Red frequency contributions propagate faster in a medium than blue ones, but the pulse keeps its shape as long as the group velocity is constant ('dispersion-free'); this is a favourable condition for optical telecommunications, where a transmitter injects digital signals ('bit currents') in the form of pulses into optical waveguides, which have to be decoded by the receiver at the other end. In optical fibres this situation is similar to that in BK7 glass at $\lambda = 1.3\,\mu\mathrm{m}$, which can be seen in Fig. 3.13 for zero passage of the material parameter $M(\lambda)$ and will be discussed in the next section.

Example: Phase and group velocities in glasses

We can use the specifications from Tab. 3.1 to determine the index of refraction and the group refractive index as a measure of the phase velocity and group velocity in important optical glasses. The wavelength 850 nm is of substantial importance for working with short laser pulses, because, on the one hand, GaAs diode lasers with high modulation bandwidth exist in this range (up to pulse durations of 10 ps and less) and, on the other, the wavelength lies in the spectral centre of the Ti–sapphire laser, which is nowadays the most important primary oscillator for ultrashort laser pulses of 10–100 fs and below. There, with the Sellmeier formula (1.6) and the coefficients

Tab. 3.1: *Indices of refraction of selected glasses.*

Abbreviation	BK7	SF11	LaSF N9	BaK 1	F 2
Index of refraction at 850 nm					
n	1.5119	1.7621	1.8301	1.5642	1.6068
Group index of refraction					
n_g	1.5270	1.8034	1.8680	1.5810	1.6322
Material dispersion					
$cM(\lambda)$ (μm^{-1})	-0.032	-0.135	-0.118	-0.042	-0.075

from Tab. 1.1, we calculate the values for Tab. 3.1. The values for the group refractive index are always larger than the values of the (phase) refractive index by a few per cent.

For shorter and shorter pulses, the bandwidth increases according to eq. (3.25), and the frequency dependence of the group velocity influences the pulse propagation as well. This is specified as a function of frequency or wavelength by one of two parameters: the group velocity dispersion (GVD) $D_\nu(\nu)$ and the material dispersion parameter $M(\lambda)$:

$$D_\nu(\nu) = \frac{1}{(2\pi)^2}\frac{d^2}{d\nu^2}\beta = \frac{d}{d\omega}\left(\frac{1}{v_g}\right),$$

$$M(\lambda) = \frac{d}{d\lambda}\frac{1}{v_g} = -\frac{\omega^2}{2\pi c}D_\nu(\nu).$$

Like before, we gain the pulse shape from

$$E(z,t) = \tau_0\, e^{-i(\omega_0 t - \beta_0 z)}$$

$$\times \int_{-\infty}^{\infty} \mathcal{E}(0,\nu)\, e^{iD_\nu(\omega-\omega_0)^2 z/2}\, e^{-i(\omega-\omega_0)(t-t_g)}\,\frac{d\omega}{2\pi}. \tag{3.29}$$

This time the pulse is not only delayed, but also distorted in shape. We cannot specify this modification generally any more, but we have to look at instructive examples.

Example: Pulse distortion of a Gaussian pulse

At $z = 0$ the optical pulse $E(0,t) = E_0\, e^{-2\ln 2(t/t_p)^2}\, e^{-i\omega_0 t}$ with intensity half-width t_p has the spectrum

$$\mathcal{E}(0,\omega) = \mathcal{E}_0\, e^{-[(\omega-\omega_0)t_p]^2/8\ln 2}.$$

At the end of the propagation distance at $z = \ell$, the spectrum is deformed according to eq. (3.29). For the sake of simplicity we introduce $\ell_D = t_p^2/4\ln 2D_\nu$ and find

$$\mathcal{E}(\ell,\omega) = \mathcal{E}_0\, e^{-[(\omega-\omega_0)t_p]^2/8\ln 2}\, e^{i(\ell/\ell_D)[(\omega-\omega_0)t_p]^2/8\ln 2}.$$

Inverse Fourier transformation yields the time-dependent form

$$E(\ell,t) = \tau_0 E_0\, e^{-i(2\pi\nu_0 t - i\beta_0\ell)}$$

$$\times \exp\left(\frac{2\ln 2(t-t_g)^2}{t_p^2[1+(\ell/\ell_D)^2]}\right)\exp\left(i\frac{\ell}{\ell_D}\frac{2\ln 2(t-t_g)^2}{t_p^2[1+(\ell/\ell_D)^2]}\right).$$

Hence not only is the pulse delayed by t_g, but it is also stretched,

$$t'_\mathrm{p}(z{=}\ell) = t_\mathrm{p}\sqrt{1+(\ell/\ell_\mathrm{D})^2},\qquad\qquad(3.30)$$

and furthermore the spectrum exhibits the so-called 'frequency chirp', where the frequency changes during a pulse:

$$\nu(t) = \frac{1}{2\pi}\frac{d}{dt}\Phi(t) = \nu_0 + \frac{1}{\pi}\frac{\ell}{\ell_\mathrm{D}}\frac{t-z/v_\mathrm{g}}{t^2_\mathrm{p}[1+(\ell/\ell_\mathrm{D})^2]}.$$

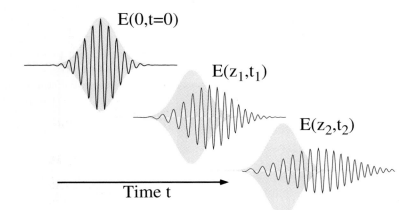

Fig. 3.14: *Pulse distortion manifests itself as pulse broadening and frequency chirp. The red frequency components run ahead (left-hand part of the pulse), whereas the blue ones lag behind (right-hand part). The neither distorted nor delayed pulse is also indicated for comparison.*

Now we can determine how far a pulse propagates within a material without significant change of shape. For example, according to Eq. (3.30) it holds that the pulse duration has increased at

$$\ell = \ell_\mathrm{D}\qquad\qquad(3.31)$$

by a factor of $\sqrt{2}$. This propagation length is also called the 'dispersion length' and plays a similar role in the transmission of pulses as the Rayleigh zone does for the propagation of Gaussian beams (see p. 40). For BK7 glass from Tab. 3.1 it holds that $D(\lambda{=}850\,\mathrm{nm}) = 0.04\,\mathrm{ps}^2/\mathrm{m}$. Then one finds for a GaAs diode laser and a conventional Ti–sapphire laser

GaAs diode laser:	$t_\mathrm{p} = 10\,\mathrm{ps}$	$\ell_\mathrm{D} = 200\,\mathrm{m}$,
Ti–sapphire laser:	$t_\mathrm{p} = 50\,\mathrm{fs}$	$\ell_\mathrm{D} = 5\,\mathrm{mm}$.

It turns out that a short pulse is heavily distorted even by a 5 mm BK7 glass window!

3.4.2   Solitons

All optical materials show dispersion, resulting in pulse distortion as introduced above, which is often undesirable in applications. However, in some materials one can use nonlinear properties, which will be discussed in more detail later in the chapter on nonlinear optics (Chapter 12), to compensate dynamically for the effects of dispersion. Here we are particularly interested in the optical Kerr effect, describing the intensity-dependent index of refraction,

$$n(I) = n_0 + n_2 I. \tag{3.32}$$

It is true that the values of the nonlinear index in glass are only in the range of $n_2 \approx 10^{-15}/(\mathrm{W\,cm^{-2}})$, but, since the power density in optical fibres is very high, this effect plays a role even at power levels of only a few milliwatts and enables the generation of so-called 'solitons' [28]. Under certain circumstances, these can propagate with a stable shape in an optical fibre for more than thousands of kilometres.

   We study the influence of nonlinearity in a one-dimensional wave equation, taking the linear contribution into account by the index of refraction, resp. the propagation constant β, as we did before,

$$\left(\frac{\partial^2}{\partial z^2} + \beta^2(\omega) \right) \mathsf{E}(z,t)\, e^{-i(\omega_0 t - \beta_0 z)} = \frac{1}{\epsilon_0 c^2} \frac{\partial^2}{\partial t^2} P_{\mathrm{NL}}(z,t), \tag{3.33}$$

and consider a harmonic field $E(z,t) = \mathsf{E}(z,t)\exp[-i(\omega_0 t - \beta_0 z)]$. In the wave equation we separate the linear and nonlinear contributions of the polarization,

$$P = \epsilon_0 (n^2 - 1)E \simeq \epsilon_0 (n_0^2 - 1 + 2 n_0 n_2 I + \cdots)E = \epsilon_0 (n_0^2 - 1)E + P_{\mathrm{NL}},$$

so that

$$P_{\mathrm{NL}}(z,t) = 2\epsilon_0 n_0 n_2 \frac{\epsilon_0 c^2}{2} |\mathsf{E}(z,t)|^2 \mathsf{E}(z,t)\, e^{-i(\omega_0 t - \beta_0 z)}.$$

To obtain approximate solutions, we use the so-called *slowly varying envelope approximation* (SVEA), where we neglect $\partial \mathsf{E}/\partial z \ll k\mathsf{E}$ second derivatives,

$$\frac{\partial^2}{\partial z^2} \mathsf{E}(z,t)\, e^{-i(\omega_0 t - \beta_0 z)} \simeq e^{-i\omega_0 t} \left(2 i \beta_0 \frac{\partial}{\partial z} - \beta_0^2 \right) \mathsf{E}(z,t).$$

We have already used this approximation when generating the paraxial Helmholtz equation (see Eq. (2.30)).

   The static dispersive properties of the materials are taken into account by $\Delta\omega = \omega - \omega_0$ and similarly to Eq. (3.27) by

$$\beta(\omega) \approx \beta_0 + \Delta\omega/v_{\mathrm{g}} + D_\nu (\Delta\omega)^2/2 + \cdots.$$

For bandwidths of the pulse that are not too large ($\Delta\omega \ll \omega_0$), we can use the equivalence $-i\Delta\omega\mathsf{E} \simeq \partial\mathsf{E}/\partial t$, etc. – thereby ignoring a more stringent mathematical transformation with the aid of a Fourier transformation – and write

$$\beta^2(\omega) \approx \beta_0^2 + \frac{2 i \beta_0}{v_{\mathrm{g}}} \frac{\partial}{\partial t} - \beta_0 D_\nu \frac{\partial^2}{\partial t^2} + \cdots.$$

Now inserting all contributions into Eq. (3.33), we get the equation of motion of a soliton as the final result after a few algebraic steps,

$$\left[\left(\frac{\partial}{\partial t} + \frac{1}{v_g}\frac{\partial}{\partial z}\right) + \frac{i}{2}D_\nu\frac{\partial^2}{\partial t^2} - i\gamma|E(z,t)|^2\right]E(z,t) = 0. \tag{3.34}$$

Obviously the propagation of a pulse with envelope $E(z,t)$ is described by a nonlinear coefficient

$$\gamma = \epsilon_0 c^2 n_2 \beta_0/n_0,$$

besides the two dispersion parameters, group velocity v_g and group velocity dispersion (GVD) D_ν.

Even with the considerable approximations that we have used so far, the solution of this equation still requires some mathematical effort. Therefore we want to restrict ourselves to the indication of the simplest soliton solution (*solitary envelope solution*). A pulse (pulse length τ_0) that has the shape

$$E(0,t) = E_0 \operatorname{sech}\left(\frac{t}{\tau_0}\right)$$

at the beginning of a fibre with dispersion length ℓ_D (see eq. (3.31)) can propagate keeping its shape

$$E(z,t) = E_0 \operatorname{sech}\left(\frac{t - z/v_g}{\tau_0}\right) e^{iz/4z_0}$$

if the conditions

$$\gamma \propto n_2 > 0 \qquad \text{and} \qquad D_\nu < 0$$

are fulfilled, and, besides, the amplitude has a value equal to [89]

$$E_0 = (|D_\nu|/\gamma)^{1/2}/\tau_0.$$

These conditions are found in optical fibres in the region of anomalous group velocity dispersion (GVD < 0), typically at $\lambda > 1.3\,\mu\text{m}$, with simultaneously moderate requirements for pulse power. Besides the fundamental solution, solitons of higher order exist, in analogy to the Gaussian modes, which are characterized by a periodic recurrence of their shape after a propagation length of ℓ_D, which we do not want to discuss here.

Linn Mollenauer, who, together with his colleagues [74], was the first to demonstrate long-distance transmission of optical solitons in optical fibres, introduced a very instructive model to illustrate the physical properties of a soliton (Fig. 3.15). He compares the differently coloured wavelength contributions of a pulse with a small field of runners of different speeds, which disperses very quickly without special influences. As shown in the lower part of the figure, however, the dispersion can be compensated by a soft, nonlinear floor.

Solitons play an important role in many other systems as well (one more example, spatial solitons, will be given in Section 13.2.1). The relationship of Eq. (3.34) with the nonlinear Schrödinger equation,

$$i\frac{\partial}{\partial x}\Psi + \frac{1}{2}\frac{\partial^2}{\partial t^2}\Psi + |\Psi|^2\Psi = 0,$$

may be demonstrated by the transformation into a moving frame of reference with $x = z - v_g t$ and the substitutions $\Psi = \tau_0\sqrt{\pi\gamma/|D_\nu|}\,E$ and $z/z_0 \rightarrow x = \pi|D_\nu|x/\tau_0^2$.

Fig. 3.15: *A soliton field of athletes (with kind permission from Linn Mollenauer).*

3.5   Anisotropic optical materials

When discussing the propagation of light in matter, we always assume the medium to be isotropic. Because of that isotropy, the induced dielectric displacement is always parallel to the inducing field and can be described for transparent materials by just one parameter, the index of refraction, $\mathbf{D} = \epsilon_0 n^2 \mathbf{E}$.

However, real crystals are very often anisotropic and the refractive index depends on the relative orientation of the electric field vectors with respect to the crystal axes.

3.5.1   Birefringence

Fig. 3.16: *The calcareous spar crystal (5 × 5 × 15 cm³) that Sir Michael Faraday gave to the German mathematician and physicist Julius Plücker as a present in about 1850.*

Birefringence in calcite (calcareous spar) has been fascinating physicists for a long time (see Fig. 3.16) and is one of the most prominent optical properties of anisotropic crystals. Birefringent elements play an important role in applications, as well, for example as retarder plates (p. 94), as a birefringent filter for frequency selection (p. 95) or as nonlinear crystals for frequency conversion (Section 12.4). Crystal anisotropies can be induced by external influences, like mechanical strain (strain birefringence) or electric fields (Pockels effect).

We restrict ourselves to the simplest case of uniaxial crystals, where the symmetry axis is called the 'optical axis' (O.A.), and thus the formal problem can be reduced from three to two dimensions. Light beams that are polarized parallel to the optical axis experience a different refractive index than beams with orthogonal polarization.

In a simple microscopic model we may illustrate that the charges of the crystal are bound to its axes by spring constants of different strengths (Fig. 3.17). Therefore, they

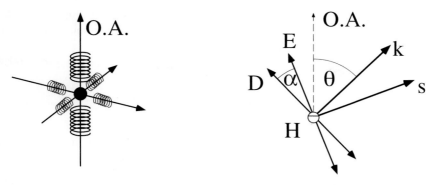

Fig. 3.17: *Left: Microscopic model of electromagnetic crystal anisotropy. Right: Electromagnetic field vectors and propagation vectors in an anisotropic crystal.*

are displaced by different amounts for identical excitation, and the relation between dielectric displacement $\mathbf{D}(\mathbf{r}, t)$ and the incident electric field $\mathbf{E}(\mathbf{r}, t)$ has to be described by a tensor, which has diagonal form, if the optical axis is used as one of the coordinate axes,

$$\mathbf{D} = \epsilon_0 \begin{pmatrix} n_o^2 & 0 & 0 \\ 0 & n_o^2 & 0 \\ 0 & 0 & n_e^2 \end{pmatrix} \mathbf{E}, \qquad \mathbf{E} = \begin{pmatrix} n_o^{-2} & 0 & 0 \\ 0 & n_o^{-2} & 0 \\ 0 & 0 & n_e^{-2} \end{pmatrix} \mathbf{D}/\epsilon_0.$$

In uniaxial crystals (unit vectors $\mathbf{e}_\perp \perp$ O.A., $\mathbf{e}_\parallel \parallel$ O.A.), there are two identical indices (ordinary index $n_\perp = n_o$) and one extraordinary refractive index ($n_\parallel = n_e \neq n_o$). Selected examples are collected in Tab. 3.2. The difference $\Delta n = n_o - n_e$ itself is often called birefringence, and may have positive or negative values.

Tab. 3.2: *Birefringence of important materials at $\lambda = 589\,nm$ and $T = 20°\,C$.*

Material	n_o	n_e	Δn	α_{max}
Quartz	1.5442	1.5533	0.0091	0.5°
Calcite	1.6584	1.4864	−0.1720	6.2°
LiNbO$_3$	2.304	2.215	−0.0890	2.3°

In Maxwell's equations (2.8) for optics we also have to use the correct tensor relation instead of $\mathbf{D} = n^2\mathbf{E}$ and write more exactly

$$\begin{aligned} i\mathbf{k} \cdot \mathbf{D} &= 0, & i\mathbf{k} \times \mathbf{E} &= i\mu_0\omega\mathbf{H}, \\ i\mathbf{k} \cdot \mathbf{H} &= 0, & i\mathbf{k} \times \mathbf{H} &= -i\omega\mathbf{D}. \end{aligned} \qquad (3.35)$$

From that we conclude directly

$$\mathbf{k} \times (\mathbf{k} \times \mathbf{E}) = -\omega^2 \mathbf{D}/\epsilon_0 c^2.$$

After some algebra ($\mathbf{k} \times (\mathbf{k} \times \mathbf{E}) = (\mathbf{k} \cdot \mathbf{E})\mathbf{k} - k^2\mathbf{E}$) we can write

$$\mathbf{D} = \epsilon_0 n^2 \left(\mathbf{E} - \frac{\mathbf{k}(\mathbf{k} \cdot \mathbf{E})}{k^2} \right),$$

introducing the index of refraction $n^2 = (ck/\omega)^2$, which describes the phase velocity $v_\theta = c/n$ of the wave. Its value has to be determined including the dependence on crystal parameters.

In the next step we decompose the propagation vector $\mathbf{k} = k_\perp \mathbf{e}_\perp + k_\parallel \mathbf{e}_\parallel$ and with $D_\perp = \epsilon_0 n_\perp^2 E_\perp$, etc., we may write the individual components as

$$k_\perp E_\perp = \frac{n^2 k_\perp^2 (\mathbf{k} \cdot \mathbf{E})}{(n^2 - n_{\mathrm{o}}^2) k^2} \qquad \text{and} \qquad k_\parallel E_\parallel = \frac{n^2 k_\parallel^2 (\mathbf{k} \cdot \mathbf{E})}{(n^2 - n_{\mathrm{e}}^2) k^2}.$$

The sum of these two components corresponds exactly to the scalar product $\mathbf{k} \cdot \mathbf{E}$, and with

$$\mathbf{k} \cdot \mathbf{E} = \left(\frac{n^2 k_\perp^2}{(n^2 - n_{\mathrm{o}}^2) k^2} + \frac{n^2 k_\parallel^2}{(n^2 - n_{\mathrm{e}}^2) k^2} \right) (\mathbf{k} \cdot \mathbf{E})$$

we obtain after short calculations a simplified form of the so-called Fresnel equation [11],

$$\frac{1}{n^2} = \frac{k_\perp^2 / k^2}{n^2 - n_{\mathrm{o}}^2} + \frac{k_\parallel^2 / k^2}{n^2 - n_{\mathrm{e}}^2},$$

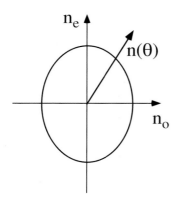

which yields an equation linear in n^2, because the n^4 contributions cancel after multiplying ($k^2 = k_\perp^2 + k_\parallel^2$). Finally, substituting the components of the propagation vector $\mathbf{k}$ by $k_\perp / k = \sin \theta$ and $k_\parallel / k = \cos \theta$, we reach the most important result for describing wave propagation in a uniaxial crystal:

$$\frac{1}{n^2(\theta)} = \frac{\cos^2 \theta}{n_{\mathrm{o}}^2} + \frac{\sin^2 \theta}{n_{\mathrm{e}}^2}. \qquad (3.36)$$

This equation describes the so-called 'index ellipsoid' of the refractive index in a uniaxial crystal, which we introduced in Fig. 3.18.

Fig. 3.18: *The index ellipsoid.*

3.5.2   Ordinary and extraordinary light rays

Now, we consider the incidence of a light ray onto a crystal, the crystal axis of which makes an angle θ with the propagation direction. If the light ray is polarized perpendicular to the optical axis (O.A., Fig. 3.19), then only the ordinary index of refraction plays a role. The ordinary light ray (E_{o}) obeys the ordinary Snell's law (Eq. (1.2)). If the polarization lies within the plane of propagation and optical axis, then different indices of refraction affect the components of the field parallel and perpendicular to the optical axis, and the light ray now propagates as an extraordinary light ray (E_{e}).

Since according to the boundary conditions in Eq. (3.1) the normal (z) component of the dielectric displacement is continuous, it must vanish for normal incidence.

Therefore, the dielectric displacement lies parallel to the polarization of the incident electric field. According to Eq. (3.35), the vector of propagation **k** is perpendicular to **D** and **H** and retains its direction in the extraordinary ray. The propagation direction of the ray continues to be determined by the Poynting vector **S**,

$$\mathbf{S} = \mathbf{E} \times \mathbf{H},$$

i.e. the direction of **S** makes the same angle with the wavevector **k** as it does with **E** and **D**. According to Fig. 3.19 it is sufficient to determine the angle

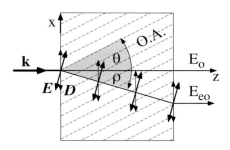

Fig. 3.19: *Ordinary and extraordinary rays in birefringence.*

$$\tan \rho = E_z / E_x$$

from the electric field components in the crystal, in order to specify the angle of deflection of the extraordinary ray.

The relation of **D** and **E** can be calculated without much effort, if we use the system of major axes with the optical axis,

$$\begin{pmatrix} D_z \\ D_x \end{pmatrix} = \begin{pmatrix} \cos\theta & -\sin\theta \\ \sin\theta & \cos\theta \end{pmatrix} \begin{pmatrix} n_e^2 & 0 \\ 0 & n_o^2 \end{pmatrix} \begin{pmatrix} \cos\theta & \sin\theta \\ -\sin\theta & \cos\theta \end{pmatrix} \begin{pmatrix} E_z \\ E_x \end{pmatrix}$$

$$= \begin{pmatrix} n_e^2 \cos^2\theta + n_o^2 \sin^2\theta & (n_e^2 - n_o^2)\sin\theta\cos\theta \\ (n_e^2 - n_o^2)\sin\theta\cos\theta & n_e^2\cos^2\theta + n_o^2\sin^2\theta \end{pmatrix} \begin{pmatrix} E_z \\ E_x \end{pmatrix}.$$

Because of the boundary conditions (3.1) the D_z component must vanish, and we may conclude directly that

$$\tan \rho = \frac{1}{2} \frac{(n_e^2 - n_o^2)\sin 2\theta}{n_e^2 \cos^2\theta + n_o^2 \sin^2\theta}.$$

The 'getting out of the way' of the extraordinary beam is called *beam walk-off* and must always be considered when using birefringent components. We can find an equivalent formulation of the *beam walk-off* angle using $n(\theta)$ from Eq. (3.36),

$$\tan \rho = \frac{n^2(\theta)}{2} \left(\frac{1}{n_o^2} - \frac{1}{n_e^2} \right) \sin 2\theta. \tag{3.37}$$

Example: Beam walk-off angle of quartz
We calculate the maximum deflection angle for birefringence in a quartz crystal with the common methods and find

$$\theta_{\max} = \arctan(n_e / n_o) = 45.2°.$$

For θ_{max} we calculate the beam walk-off angle according to eq. (3.37),

$$\rho = 0.53°.$$

One could say that in general the beam walk-off angle amounts to only a few degrees; even for a material such as calcareous spar (see Tab. 3.2) with strong birefringence it is only about 6°. In nonlinear optics, for example, in the case of the so-called angular phase matching (see Section 12.4) the efficiency of frequency conversion is limited by beam walk-off.

3.5.3 Retarder plates

An important application of birefringent materials is the so-called 'retarder plates', with which the states of polarization can be manipulated, by adjusting the optical axis perpendicular to the direction of propagation. Ordinary and extraordinary light rays then propagate collinearly through the crystal, and their components are given by the projection onto the optical axis; the angles of those are adjusted relative to the incident polarization by rotation (see Fig. 3.20).

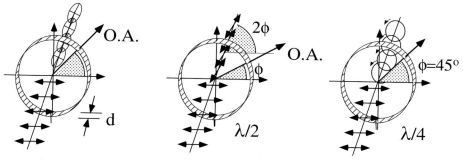

Fig. 3.20: *Retarder plates transform an incoming state of polarization into a new state, depending on their thickness and the angle of orientation of the optical axis O.A. (left). Important special cases: λ/2 plate (centre); and λ/4 plate (right).*

For the discussion we conveniently use Jones vectors, which have e.g. the form $\mathbf{E} = a\mathbf{e}_x + b\mathbf{e}_y$ for a basis of linearly polarized light (Section 2.4.1). The ordinary and extraordinary beams are delayed with respect to each other within a plate of thickness d with phase shifts $\exp(i\alpha_o) = \exp(in_o\omega d/c)$, resp. $\exp(i\alpha_e) = \exp(in_e\omega d/c)$. In rotating the electric field first towards the coordinate system of the optical axis and then back again, the general transformation matrix can be specified as

$$\mathbf{E}' = \begin{pmatrix} \cos\phi & -\sin\phi \\ \sin\phi & \cos\phi \end{pmatrix} \begin{pmatrix} e^{i\alpha_o} & 0 \\ 0 & e^{i\alpha_e} \end{pmatrix} \begin{pmatrix} \cos\phi & \sin\phi \\ -\sin\phi & \cos\phi \end{pmatrix} \mathbf{E}.$$

From that we gain after some manipulation

$$\mathbf{E}' = \begin{pmatrix} e^{i\alpha_o}\cos^2\phi + e^{i\alpha_e}\sin^2\phi & (e^{i\alpha_o} - e^{i\alpha_e})\sin\phi\cos\phi \\ -(e^{i\alpha_o} - e^{i\alpha_e})\sin\phi\cos\phi & e^{i\alpha_o}\cos^2\phi + e^{i\alpha_e}\sin^2\phi \end{pmatrix} \mathbf{E}. \qquad (3.38)$$

Let us now consider two important special cases, the λ/2 plate and the λ/4 plate.

(i) $\lambda/2$ plate

For the $\lambda/2$ plate, the special case $\exp(i\alpha_o) = -\exp(i\alpha_e)$ is chosen. Therefore, the optical pathlengths of the ordinary and extraordinary rays must differ by exactly half a wavelength. In this case, the Jones matrix $\mathbf{M}_{\lambda/2}$ reads as

$$\mathbf{M}_{\lambda/2} = e^{i\alpha_o} \begin{pmatrix} \cos 2\phi & \sin 2\phi \\ -\sin 2\phi & \cos 2\phi \end{pmatrix}$$

and shows a rotation of an arbitrary initial state by angle 2ϕ (Fig. 3.20 centre).

(ii) $\lambda/4$ plate

For the $\lambda/4$ plate, the special case $\exp(i\alpha_o) = i\exp(i\alpha_e)$ is chosen, which corresponds to a difference in the optical pathlengths of a quarter wavelength. The Jones matrix $\mathbf{M}_{\lambda/4}$ in this case reads as

$$\begin{aligned}
\mathbf{M}_{\lambda/4} &= \frac{e^{i\alpha_o}}{2} \begin{pmatrix} (1+i)+(1-i)\cos 2\phi & (1-i)\sin 2\phi \\ -(1-i)\sin 2\phi & (1+i)-(1-i)\cos 2\phi \end{pmatrix} \\
&= \frac{e^{i(\alpha_o+\pi/4)}}{\sqrt{2}} \begin{pmatrix} 1 & -i \\ i & 1 \end{pmatrix} \quad \text{for } \phi = \pi/4.
\end{aligned}$$

In particular, for the angle adjustment of $\phi = 45°$, the $\lambda/4$ plates transform linear polarizations into circular ones, and vice versa.

The differences in pathlengths of retarder plates are in general not exactly equal to $\lambda/2$ and $\lambda/4$, but to $\lambda(n+1)/2$ and $\lambda(n+1)/4$ instead, and the number of total waves n is called the order. They serve their purpose independently of their order, but due to the dispersion, which in addition has different temperature coefficients for n_o and n_e, retarder plates of higher order are much more sensitive to variations in frequency or temperature than retarder plates with lower order.

So-called retarder plates of zero order consist of two plates with nearly the same thickness but unequal differences in optical path $\lambda/2$ or $\lambda/4$. If two plates with crossed optical axes are mounted on top of one another,[1] then the influences of higher orders are compensated and there remains an effective plate of lower order, which is less sensitive to spectral and temperature changes. 'Real' plates of zero order would generally be too thin and therefore too fragile for manufacturing.

Lyot filters

The relative delay of the two partial waves in a birefringent plate of thickness d, the optical axis of which makes an angle ϕ with the x axis, is $\Delta = 2\pi(n_o - n_e)d/\lambda$ and is wavelength-dependent. Combining retarder plates with polarizers, one can achieve wavelength-dependent and frequency-dependent transmission. Such applications are called birefringent filters or Lyot filters.

---

[1] They are often 'optically contacted', i.e. they are connected via two very well polished surfaces (whose planarities must be much better than an optical wavelength) only by adhesive forces.

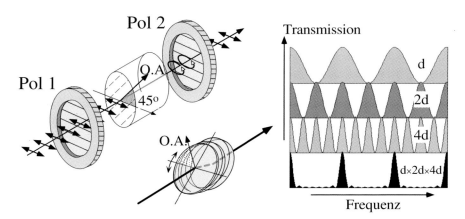

Fig. 3.21: *Top left: Lyot filter built from two parallel polarizers and a wavelength dependent retarder plate oriented at 45°. Optimum extinction occurs for wavelength where the retarder plate corresponds to a λ/2-plate. Right: Transmission curves of Lyot filters with single retarder plate of increasing thickness d, 2d, 4d, and of a filter composed of the three plates. Bottom left: Three-plate Lyot filter at Brewster angle for use within a laser resonator.*

In Fig. 3.21 a retarder plate (now of higher order) is positioned between two parallel polarizers. Only for distinct wavelengths does it serve as a $\lambda/2$ plate, for example, and cancel transmission.

The incident light is transformed in general into elliptically polarized light depending on the orientation of the optical axis. We can calculate the transmission of a light field polarized linearly in the x direction according to Eq. (3.38),

$$E'_x = \exp\left(i\frac{\alpha_o + \alpha_e}{2}\right)\left[\cos\left(\frac{\alpha_o - \alpha_e}{2}\right) + i\sin\left(\frac{\alpha_o - \alpha_e}{2}\right)\cos 2\phi\right]E_x,$$

and with $(\alpha_o - \alpha_e) = (n_o - n_e)2\pi\nu d/c$ we find the transmitted intensity correlated with the incident intensity I_x

$$I_T = I_x\left[\cos^2\left(\frac{(n_o - n_e)\pi\nu d}{c}\right) + \sin^2\left(\frac{(n_o - n_e)\pi\nu d}{c}\right)\cos^2 2\phi\right].$$

In particular, for $\phi = 45°$ one finds a transmission modulated by 100% with the period (or the 'free spectral range') $\Delta\nu = c/(n_o - n_e)d$. Positioning several Lyot filters with thicknesses $d_m = 2^m d$ one behind the other, the free spectral range is maintained, but the width of the transmission curve is reduced quickly.

Lyot filters, resp. birefringent filters, may be positioned in the ray path at the Brewster angle as well, to reduce losses substantially (Fig. 3.21). The optical axis lies within the plane of the plate, and the central wavelength of the filter with the lowest losses can be tuned by rotating the axis. Such elements are mainly used in broadband laser oscillators (e.g. Ti–sapphire lasers, dye lasers, Section 7.9.1) for rough wavelength selection.

3.5.4   Birefringent polarizers

One more important application of birefringent materials is their use as polarizers. From the many variants we introduce the Glan air polarizer. Its effect is based on the various critical angles of total internal reflection for the ordinary beam (which is reflected for devices made of calcareous spar) and the extraordinary beam (Fig. 3.22).

Applying a polarizer, both the extinction ratio and the acceptance angle are the most relevant numbers to determine the alignment sensitivity, depending on the difference of the refractive indices for ordinary and extraordinary beams. With Glan polarizers very high extinction ratios of $1 : 10^6$ and more can be achieved. One variant is the Glan–Thompson polarizer, where a glue is inserted between the two prisms with refractive index between n_o and n_e. Then total internal

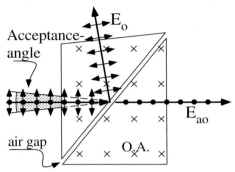

Fig. 3.22: *Glan air polarizer. The acceptance angle is defined by the critical angles for total internal reflection for the ordinary and the extraordinary ray.*

reflection occurs for the partial wave with the lower index of refraction only, the other one is always transmitted, in Fig. 3.22 the extraordinary wave.

3.6   Optical modulators

Materials in which the index of refraction can be controlled or switched by electric or magnetic fields offer numerous possibilities to influence the polarization or phase of light fields, thereby realizing mechanically inertia-free optical modulators for amplitude, frequency, phase or beam direction. We will pick out several important examples.

3.6.1   Pockels cell and electro-optical modulators

The electro-optical effect addresses the linear dependence of the refractive index on the electric field strength and is also called the Pockels effect. If the index of refraction depends quadratically on the field strength, resp. linearly on the intensity, then we talk about the optical Kerr effect, which will be discussed in more depth in the chapter on nonlinear optics (Chapter 13.2). We came across self-modulation of an optical wave by the Kerr effect in Section 3.4.2 on solitons already.

The electric field is created by electrodes mounted on the faces of the crystal. The changes in the index of refraction are in general determined by crystal symmetry. Here, we confine ourselves to a simple and important example, the uniaxial KDP crystal.

The KDP crystal is mounted between two crossed polarizers and its optical axis is adjusted parallel to the propagation direction. A longitudinal electric field is created with transparent electrodes (Fig. 3.23).

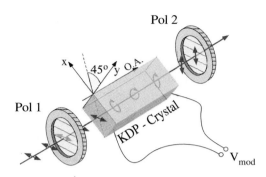

Fig. 3.23: *Electro-optical modulator with a KDP crystal, switched to blocking mode.*

In the field-free state, there is axial symmetry, which is cancelled by the external field and induces an optically marginal biaxial crystal. Thereby the indices of refraction in the x and y directions, which are tilted by 45° against the position of the polarizer, are changed by the same modulus of the angle, but in the opposite direction:

$$n_{ox} = n_o - rn_o^3 U/2d \quad \text{and}$$
$$n_{oy} = n_o + rn_o^3 U/2d.$$

In this arrangement the transmission is proportional to

$$I_T = I_0 \cos^2(2\pi rn_o^3 U/d).$$

When applying electro-optical modulators (EOMs), the half-wave voltage, where the difference in the indices of refraction creates a phase delay of the x and y components of $\lambda/2$, is among the most important technical criteria. The maximum modulation frequency is determined by the capacitive properties of the driver circuit. At very high frequencies (>200 MHz) transit time effects add, necessitating the use of travelling-wave modulators, in which the radio-frequency wave and the optical wave co-propagate.

Example: Half-wave voltage of KDP
The electro-optical coefficient of KDP is $r = 11\,\text{pm V}^{-1}$ at a refractive index of $n_o = 1.51$. For a crystal length of $d = 10\,\text{mm}$, the half-wave voltage at a wavelength of $\lambda = 633\,\text{nm}$ is calculated as ($E = U/d$)

$$U = 2 \times \frac{\lambda}{2} \frac{1}{rn_o^3} = 84\,\text{V cm}^{-1}.$$

In this case the half-wave voltage does not depend at all on the length of the crystal. Therefore it is more convenient to choose arrangements with transverse electro-optical coefficients.

Example: Phase modulation with an EOM
If one adjusts the linear polarization of a light beam parallel to the principal axis of a crystal and leaves out the polarizers in Fig. 3.23, then the beam experiences not an amplitude modulation but a phase modulation, resp. frequency modulation. The

index of refraction depends linearly on the driving voltage and causes a phase variation at the output of the EOM,

$$\begin{aligned}
\Phi(t) &= \omega t + m\sin(\Omega t),\\
E(t) &= \Re\left\{E_0\exp(-i\omega t)\exp[-im\sin(\Omega t)]\right\},
\end{aligned}\tag{3.39}$$

where the modulation index m specifies the amplitude and is correlated with the material parameters through

$$m = \omega r n_o^3 U/2c.$$

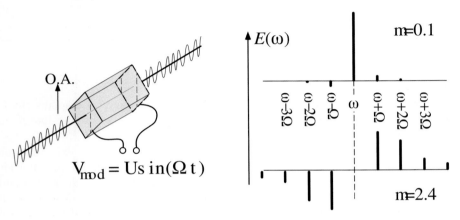

Fig. 3.24: *Phase modulation with an EOM. The spectra are illustrated for modulation indices $m = 0.1$ (top) and $m = 2.4$ (bottom). The lengths of the bars indicate the contribution of the sideband, and the direction indicates the phase position according to Eq. (3.40).*

The corresponding instantaneous frequency experiences a harmonic modulation as well,

$$\omega(t) = \frac{d}{dt}\Phi(t) = \omega + m\Omega\cos(\Omega t),$$

where the modulation shift $M = m\Omega$ appears. Actually, we cannot strictly distinguish between phase modulation (PM) and frequency modulation (FM). However, the modulation index allows a rough and common categorization into distinct regions:

$$\begin{aligned}
m &< 1 \qquad \text{phase modulation,}\\
m &\geq 1 \qquad \text{frequency modulation.}
\end{aligned}$$

The difference gets more pronounced if we decompose the electromagnetic wave (3.39) with intensity

$$\begin{aligned}
e^{-im\sin(\Omega t)} = {} & J_0(m) + 2[J_2(m)\cos(2\Omega t) + J_4(m)\cos(4\Omega t) + \cdots]\\
& - 2i[J_1(m)\sin(\Omega t) + J_3(m)\sin(3\Omega t) + \cdots]
\end{aligned}$$

into its Fourier frequencies:

$$\begin{aligned}
E(t) = {} & E_0\,e^{-i\omega t}[J_0(m) + J_1(m)(e^{-i\Omega t} - e^{i\Omega t})\\
& + J_2(m)(e^{-i2\Omega t} + e^{i2\Omega t}) + J_3(m)(e^{-i3\Omega t} - e^{i3\Omega t}) + \cdots].
\end{aligned}\tag{3.40}$$

We present these spectra for the cases $m = 0.1$ and $m = 2.4$ in Fig. 3.24. For a small modulation index (PM), the intensity at a carrier frequency ω is barely changed, but sidebands appear at a distance of the modulation frequency. The intensity of the sidebands is proportional to $J_\ell^2(m)$. For a big shift (FM), the intensity is distributed to many sidebands, and in our special case the carrier is even completely suppressed due to $J_0(2, 4) = 0$.

In contrast to harmonic amplitude modulation (AM), where exactly two sidebands are created, many sidebands appear for PM/FM modulation. Another important difference is that the AM variation can be shown ('demodulated') with a simple photodetector, but PM/FM information cannot.

3.6.2   Liquid crystal modulators

Liquid crystal (LC) modulators are well known from liquid crystal displays (LCDs). By 'liquid crystals' we mean a certain type of order of slab-like or disc-like organic molecules within a liquid (which appear quite often).

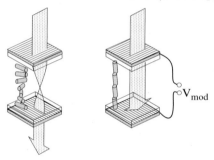

Fig. 3.25: *Liquid crystal modulator.*

In the nematic phase (there exist also smectic and cholesteric phases), all the molecular slabs point in one direction, without aligning their centres. If the molecules are exposed to a surface with a preferred direction (grooves, anisotropic plastics), then they become oriented in this direction. The enclosure of a liquid crystal between glass plates with crossed grooves causes the rotated nematic phase shown in Fig. 3.25, where the molecular axes are rotated continuously from one direction into the other.

The rotated nematic phase rotates the plane of polarization of an incident polarized light wave by 90°. But the molecular rods can be aligned parallel to the field lines of an electric field in the direction of propagation. Then the polarization is not changed during transmission. Thus, an electric field can be used to switch the transmitted amplitude. LC displays use the same principle, but work in general in a reflection mode.

3.6.3   Spatial light modulators

The digital revolution is more and more also entering the world of optical devices. It has led to the development of modulators allowing spatial control of the intensity or phase of an extended light field, so called spatial light modulators or SLMs. Conceptually it is straightforward to use the fabrication methods of microelectronics and divide the LC described in the last section into an array of small and individually addressable pixels, and LCDs are an ubiquitous component of electric and electronic tools. With

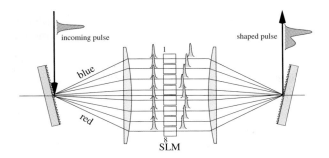

Fig. 3.26: *Puls shaping using a spatial light modulator (SLM). The incoming ultrashort pulse (typically in the femtosecond domain)) is dispersed into its spectral components. The SLM modifies the intensity of individual channels (here: 1-8) and generates small delays. With a second grating the pulse is recombined.*

improved optical quality LC arrays can be used to actively control a wavefront incident on such a device.

While applications for versatile digital display technology are fairly obvious (see below) we here introduce another example where SLMs are used to control the shape of ultrashort pulses (For the generation of femtosecond pulses see section 13.2.1.). In Fig. 3.26 a very short pulse is dispersed by a grating and in combination with a lens a parallel wave front is generated with spatially varying color. Without SLM the second grating would undo the dispersion and simply restore the original pulse. The SLM can now be configured, if necessary by inserting additional optical elements such as polarizers etc., to introduce attenuation or delays in each channel (typically 128 and more) individually. On recombination the pulse is now very different from the incoming pulse. This pulse shaping method is used to improve for instance the efficiency of chemical reactions induced by femtosecond laser pulses ('femto chemistry') [16].

In 1987 Larry Hornbeck of Texas Instruments invented the digital mirror device (DMD) which can realize more than 1.3 million hinge-mounted mirrors on a single silicon chip. Each individual mirror in Fig. 3.27 has a square length of about 15 μm and corresponds to a pixel of a digital image. It is separated from adjacent mirrors by 1 μm, and it tilts up to 12° in less than 1 ms by micro electromechanical actuators. White and black is generated by directing each mir-

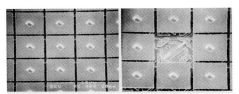

Fig. 3.27: *A sector of 3x3 mirrors out of an array of 1280x1024. On the right side one mirror is removed to expose the electro-mechanical actuators. With permission by Texas Instruments, from www.dlp.com/dlp_technology.*

ror in and out of the light beam from the projection lamp. Since each mirror can be switched on several thousand times per second, also gray scales can be realized by varying the 'on' versus the 'off' time of the mirror.

The DMD offers digital light processing (DLP^{TM}) with excellent quality and is currently revolutionizing display technologies from large scale cinemas to home entertainment.

3.6.4   Acousto-optical modulators

If a sound wave propagates within a crystal, it causes periodic density fluctuations, which induce a variation of the refractive index at the same frequency and wavelength. The periodic fluctuation of the refractive index has an effect like a propagating optical grating, at which the light ray is diffracted. Diffraction may be interpreted as a Bragg scattering or Bragg refraction off this grating.

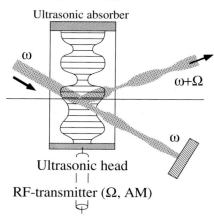

Fig. 3.28: *Acousto-optical modulator.*

An acousto-optical modulator (AOM) consists of a crystal, at the end of which is glued a piezo-element to excite ultrasonic waves (Fig. 3.28), the transducer. To avoid reflection and standing waves, a sound absorber is installed at its other end.

The ultrasonic head is set vibrating mechanically with a radio frequency (typically 10–1000 MHz) and radiates sound waves through the modulator crystal. Then the light ray transits a so-called extensive sound wave field and experiences diffraction in this 'Bragg region' in one order, only. If the light ray transits through a thin sound wave field, as is the case with an optical grating, then several, here undesirable, diffraction orders occur. This boundary case is called the 'Raman–Nath region'.

In order to discuss the influence of the sound wave on the propagation of the light ray in more detail, we consider the variation of the index of refraction in the x direction caused by a sound wave with frequency Ω and wavevector $\mathbf{q} = q\mathbf{e}_x$,

$$n(t) = n_0 + \delta n(t) = n_0 + \delta n_0 \cos(\Omega t - qx).$$

We use the wave equation in the form of Eq. (2.12) and take into account that $[n_0 + \delta n(t)]^2 \simeq n_0^2 + 2n_0\delta n(t) + \cdots$. Furthermore, we confine ourselves to the variations of the x components, because we do not expect any change through the sound wave in the other directions,

$$\left[\frac{\partial^2}{\partial x^2} - k_y^2 - k_z^2 - \left(\frac{n_0^2}{c^2} - \frac{2n_0\delta n(t)}{c^2} + \cdots\right)\frac{\partial^2}{\partial t^2}\right]\mathbf{E}(\mathbf{r}, t) = 0. \tag{3.41}$$

Now we shall consider how the amplitude of the incident wave evolves, which for simplification has only a linear polarization component,

$$E_i(\mathbf{r}, t) = E_{i0}(x, t)\, e^{-i(\omega t - \mathbf{k}\mathbf{r})}.$$

The modulated index of refraction leads to a time-dependent variation at frequencies $\omega \pm \Omega$; therefore, we can 'guess' an additional field $E_a 8\mathbf{r}, t)$, which we may interpret as a reflected field,

$$E_a(\mathbf{r}, t) = E_{a0}(x, t) e^{-i(\omega' t - \mathbf{k_r} \mathbf{r})},$$

with $\omega' = \omega + \Omega$ and $\mathbf{k}' = \mathbf{k} + \mathbf{q}$, arising from diffraction off the sound wave. The oscillating refractive index has no influence on the propagation vector; therefore, even at this point it must hold that

$$\mathbf{k}_r^2 = (\mathbf{k} + \mathbf{q})^2 = n_0^2(\omega + \Omega)^2/c^2 \simeq (n_0\omega/c)^2$$

(because $\Omega \ll \omega$). From that the Bragg condition immediately follows:

$$q = -2k_x.$$

Now, we study Eq. (3.41) with a total field $E = E_i + E_a$, and again assume that the change in amplitude is negligible on the scale of a wavelength, i.e.

$$\partial^2/\partial x^2 \left[E(x) e^{ikx}\right] \simeq \left[-k^2 + 2ikE'(x)\right] e^{ikx}.$$

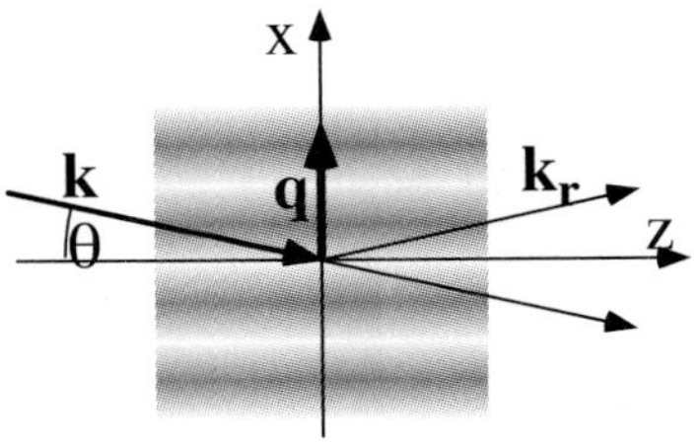

Fig. 3.29: *Bragg geometry.*

With $k_x^2 + k_y^2 + k_z^2 = (n_0\omega/c)^2$ and $(k_x + K)^2 + k_y^2 + k_z^2 = [n_0(\omega + \Omega)/c]^2$, after a short calculation we obtain the equation

$$\left[2ik_x \frac{\partial}{\partial x} + \frac{2\omega^2 n_0 \delta n_0}{c^2} \cos(\Omega t - qx)\right] E_{i0}(x) e^{-i(\omega t - k_x x)}$$

$$+ \left[-2ik_x \frac{\partial}{\partial x} + \frac{2\omega^2 n_0 \delta n_0}{c^2} \cos(\Omega t - qx)\right] E_{a0}(x) e^{-i[(\omega + \Omega)t + k_x x]} = 0.$$

To get a more simplified system for the two amplitudes E_{i0} and E_{a0}, we use the cos terms in their complex form, sort according to the oscillator frequencies and ignore oscillating terms, where the incident field does not participate:

$$2ik_x \frac{\partial}{\partial x} E_{i0} + \frac{\omega^2 n_0 \delta n_0}{c^2} E_{a0} = 0,$$

$$-2ik_x \frac{\partial}{\partial x} E_{a0} + \frac{\omega^2 n_0 \delta n_0}{c^2} E_{i0} = 0.$$

Finally, we substitute the x dependence by the dependence along the principal propagation direction z (thereby E_a propagates in the opposite direction to E_i). With $k_x = k \sin \theta = (n_0\omega/c) \sin \theta$ it holds that

$$i \frac{\partial}{\partial z} E_{i0} + \frac{k \delta n_0}{2n_0 \sin \theta} E_{a0} = 0,$$

$$i \frac{\partial}{\partial z} E_{a0} + \frac{k \delta n_0}{2n_0 \sin \theta} E_{i0} = 0.$$

(3.42)

The solutions of this system are well-known harmonic oscillations with frequencies

$$\gamma = \frac{k \delta n_0}{n_0 \sin \theta}.$$

In general, as $E_{a0} = 0$ is valid at the entrance of an AOM, we find the pendulum solution

$$
\begin{aligned}
E_i(z,t) &= E_{i0}\cos(\gamma z/2)\,e^{-i(\omega t - \mathbf{kr})}, \\
E_r(z,t) &= E_{a0}\sin(\gamma z/2)\,e^{-i[(\omega + \Omega)t - \mathbf{k_r r}]}.
\end{aligned}
$$

So the reflected beam is actually frequency-shifted, as guessed above. For small z the reflected intensity is proportional to $(\gamma z)^2$. The modulation amplitude of the refractive index at sound intensity I_S is

$$
\delta n_0 = \sqrt{\mathcal{M} I_S/2}.
$$

The $\mathcal{M}$ coefficient depends on the material parameters and is introduced here only phenomenologically. For small powers the reflected (in other words, diffracted) intensity is proportional to $|E_{a0}|^2$, and thus, according to this result, proportional to the applied sound power.

3.6.5   Faraday rotators

Certain materials show the Faraday effect, where the oscillation plane of linearly polarized light is rotated independently of the initial orientation proportional to a longitudinal magnetic field,

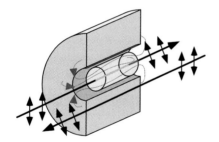

$$
\mathbf{E'} = \begin{pmatrix} \cos\alpha & -\sin\alpha \\ \sin\alpha & \cos\alpha \end{pmatrix} \mathbf{E}
$$

$$
\text{with} \qquad \alpha = VB\ell,
$$

Fig. 3.30: *Faraday rotation. Only those field lines are shown which pass through the whole crystal.*

where V (units $\deg\,\mathrm{m^{-1}\,T^{-1}}$) is the *Verdet constant* , B is the magnetic field strength and ℓ is the crystal length. The magnetization of a Faraday crystal affects right-hand and left-hand polarized refractive indices with different indices of refraction: $n_{\pm} = n_0 \pm VB\lambda/2\pi$.

In contrast to the retarder plates in Section 3.5.3, the polarization transformation of an electromagnetic wave is not reversed in a Faraday rotator, if the wave is returned into the same configuration. The Faraday rotator is 'non-reciprocal' and therefore is suited extremely well for the design of isolators and diodes. As a result of the typically very small Verdet constants, relatively high magnetic field strengths are necessary. They can be more conveniently realized with permanent magnets made of SmCo or NdFeB. [112]

Tab. 3.3: *Verdet constant of selected materials at 589 nm.*

Material	Quartz	Heavy flint	TGG*
V $(\deg\,\mathrm{m}^{-1}\,\mathrm{T}^{-1})$	209	528	-145

*TGG = terbium–gallium garnet.

3.6.6   Optical isolators and diodes

In most applications laser light is sent to the device under test via various optical components. Thereby back-reflections always occur, which even for very low intensities cause undesirable amplitude and frequency fluctuations of the laser light. Optical isolators offer the possibility to decouple experiment and light source from each other. The components introduced in the preceding sections play a central role in this.

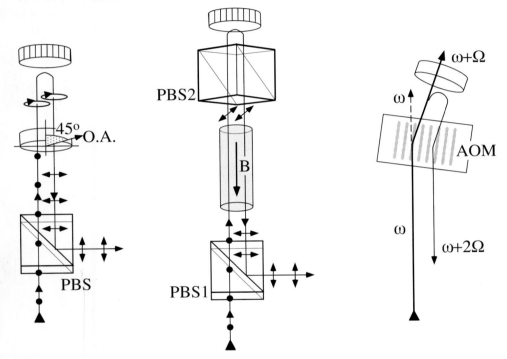

Fig. 3.31: *Optical isolators. (i) Left: $\lambda/4$ plate isolator. (ii) Centre: Faraday isolator. (iii) Right: The effect of an AOM isolator is based on frequency shifts. PBS = polarizing beam splitter.*

In Fig. 3.31 we present three concepts that may be applied to suppress reflections from the upper reflector:

(i) The isolator on the left uses a $\lambda/4$ plate, which transforms the linear polarization into circular polarization, e.g. a right-handed one. After reflection, the handedness is preserved but the wave propagates backwards. After the second passage through the retarder plate, the action of a $\lambda/2$ plate is realized. The polarization is thus rotated by $90°$ and the wave is deflected at the polarizing beam splitter (see section 3.5.4). This arrangement is only sensitive to the reflection of circularly polarized light and hence is of limited use.

(ii) In contrast to the previous case, the Faraday isolator allows the suppression of arbitrary reflections only in combination with a second polarizer between rotator and mirror. One disadvantage is the technically impracticable rotation by $45°$, which can be compensated with a $\lambda/2$ plate or a second rotator stage [112]. A two-stage isolator also offers typically $60\,\mathrm{dB}$ extinction of reflections, in contrast to the typical $30\,\mathrm{dB}$ of a single-stage unit.

(iii) From time to time the acousto-optical modulator is applied for isolation purposes. Its isolation effect is based on the frequency shift of the reflected light by twice the modulation frequency, which, for example, lies outside of the bandwidth of the laser light source.

4 Optical images

Images traditionally are among the most important applications in optics. The basic element in imaging is the convex lens, which for stigmatic imaging merges into one point again all the rays that originated from a single point. With the help of geometry (Fig. 4.1), we can understand the most important properties of a (real) optical image:

1. a beam parallel to the axis is guided through the focal point by a convex lens;

2. a beam that reaches the lens via the focal point leaves the lens parallel to the axis;

3. a beam passing through the centre of the lens is not diffracted.

From geometrical considerations, we can connect the distances g and b of object G and image B, respectively, with the focal length f, and deduce the lens equation,

$$\frac{x'}{f} = \frac{f}{x}.$$
(4.1)

We have already come across the form of the imaging equation ,

$$\frac{1}{f} = \frac{1}{g} + \frac{1}{b},$$
(4.2)

in Eq. (1.23), when discussing matrix optics. It evolves from the lens equation, when one uses the object and image distances, $g = f + x$ and $b = f + x'$.

We dedicate this chapter to lens imaging, as it is the basis for various optical instruments, which have substantially influenced the development of optics and have made possible – literally – our insights into the macro- and micro-cosmos. Besides

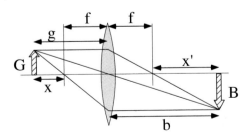

Fig. 4.1: *Conventional construction of a lens image with the common notations.*

Optics, Light and Laser. Dieter Meschede
Copyright © 2004 Wiley-VCH Verlag GmbH & Co. KGaA
ISBN: 3-527-40364-7

the basic principles, we first raise the question of the resolution capability of such instruments: what objects at very large distances can we make visible, and what are the smallest objects that we can observe with a microscope.

In the lens of our eye, imaging occurs as well, and we have to take this into account for all our vision processes; therefore, we begin by presenting some of the more important properties of our 'own vision instrument'.

4.1   The human eye

Unfortunately, it is not possible here to go into the physiological origin of the vision process, and for that we refer the reader to the relevant literature.

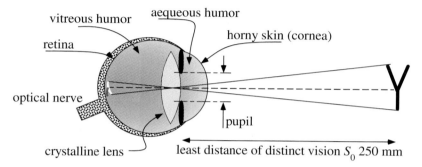

Fig. 4.2: *Human eye, reduced to the most important optical 'components'.*

For our purposes it is sufficient to construct a 'reduced artificial standard eye'. The eye body in general has a diameter of 25 mm, and several important optical properties are collected in Tab. 4.1. The refractive power of the total eye is achieved predominantly by the curvature of the cornea (typical radius 5.6 mm, difference in refractive index with respect to air $\Delta n \sim 0.37$), while the variable crystalline lens guarantees 'focusing' by contraction. Recently, laser ablation with femtosecond lasers has become established as a very promising method for reshaping the cornea, and thus a patient's ability to see can be improved.

By adaptation of the focal length of our eyes, we are generally able to recognize objects at a distance of 150 mm or more. As a standardized distance for optical instruments, often the conventional *least distance of distinct vision* of $S_0 = 250$ mm is chosen, where the best results are achieved with vision aids.

4.2   Magnifying glass and eyepiece

The simplest, and since ancient times very popular, optical instrument is the convex lens used as a magnifier. The effect can most quickly be understood by considering the angle α at which an object of height y is seen, since this angle determines our physiological impression of its size – a mountain 1000 m high at a distance of 10 km

Tab. 4.1: *Optical properties of the human eye.*

Vitreous humour, aqueous humour	$n = 1.336 \sim 4/3$
Cornea	$n = 1.368$
Crystal lens	$n = 1.37\text{--}1.42$
Focal length, front	$f = 14\text{--}17\,\text{mm}$
Focal length, back	$f = 19\text{--}23\,\text{mm}$
Clear vision distance	$150\,\text{mm--}\infty,\ S_0 = 250\,\text{mm}$
Pupil (diameter)	$d = 1\text{--}8\,\text{mm}$
Pupil (shutter time)	$\tau = 1\,\text{s}$
Resolving power at 250 mm	$\Delta x = 10\,\mu\text{m}$
Sensitivity (retina)	$1.5 \times 10^{-17}\,\text{W/vision cell} \sim 30\,\text{photon\,s}^{-1}$

seems to have the same size as a matchbox at a distance of 25 cm. Only our knowledge of their distance identifies objects according to their real sizes.

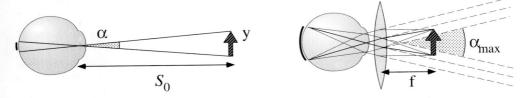

Fig. 4.3: *Vision without (left) and with magnifying glass. The magnification is caused by widening of the vision angle. The object lies in the focal length of the magnifier, and the position of the virtual image was chosen to be at the least distance of distinct vision S_0 in this example.*

Without technical aid, we view an object of size y with the eye (Fig. 4.3) at an angle $\alpha = \tan(y/S_0) \simeq y/S_0$, which is determined by the least distance of distinct vision S_0. Now, we hold the magnifier directly in front of the eye: the magnifier widens the angle at which we see the object. If we bring the object close to the focal length, $x \sim f$, then parallel rays reach the eye, so that the object appears to be removed to infinity. From geometrical relations, we can determine that

$$\alpha_{max} = \frac{y}{f}.$$

From that we can directly deduce the maximum magnification M of the magnifying glass,

$$M = \frac{\alpha_{max}}{\alpha} = \frac{S_0}{f}.$$

Thus, the smaller the focal length of a magnifying lens, the stronger is its magnification. However, since the least distance of distinct vision is defined as $S_0 \sim 250\,\text{mm}$, and because thicker and thicker, more curved, lenses are necessary at smaller focal lengths, the practicable magnification of magnifying lenses is limited to $M \leq 25$.

In contrast to the *real* image, discussed in the first section, the magnifying glass generates a *virtual* image. If the magnifying glass is not held directly in front of the eye like in Fig. 4.3, then the magnification is a little bit less, as one may straightforwardly find out from geometrical considerations. However, the difference is in general marginal, and, anyway, an individual user looks for a suitable working distance by manual variation of the distances of magnifier, eye and object.

In optical devices, such as microscopes and telescopes, real intermediate images are generated, which are then observed with a so-called *eyepiece*. The eyepiece in general consists of two lenses to correct for chromatic aberrations, which we will discuss in Section 4.5.3. In the Huygens eyepiece (Fig. 4.4), a real intermediate image is generated by the field lens, which is looked at with the eyepiece. The eyepiece fulfils exactly the task of a magnifying lens with an effective focal length f_{ocu} and a magnification $M_{ocu} = S_0/f_{ocu}$ for an eye that is adapted to infinite vision distance.

Example: Effective focal length and magnification of a Huygens eyepiece

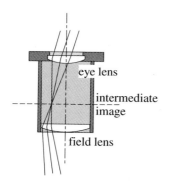

eye lens

intermediate image

field lens

Fig. 4.4: *Huygens eyepiece with path of rays.*

A Huygens eyepiece consists of two lenses at a separation of $d = (f_1 + f_2)/2$, because there the minimal chromatic aberrations occur (see Section 4.5.3). We determine the effective focal length and magnification of a system consisting of two lenses with $f_1 = 30\,\text{mm}$ and $f_2 = 15\,\text{mm}$, for instance by application of the matrix formalism of 1.9. The system has an effective focal length of

$$\frac{1}{f_{ocu}} = \frac{1}{f_1} + \frac{1}{f_2} - \frac{f_1 + f_2}{2 f_1 f_2}$$

$$= \frac{f_1 + f_2}{2 f_1 f_2} = (20\,\text{mm})^{-1}$$

and thus a magnification of $M_{ocu} = 250/20 = 12.5\times$.

4.3   Microscopes

To see small things 'big' is one of the oldest dreams of mankind and continues to constitute a driving force for our scientific curiosity. The magnifying glass alone is not sufficient, as we know already, to make visible the structure of very small objects, like e.g. the details of a biological cell. However, by adding one or two lenses, which generate a real image at first, it has been possible since the 19th century to achieve up to 2000-fold magnifications – the microscope [56] ´opens' our eyes.

We consider a microscope (Fig. 4.5) in which an objective 'obj' with focal length f_{obj} generates a real intermediate image. The intermediate image plane is a suitable position to install, for example, a graticule, the lines of which are seen simultaneously with the object under test. For that purpose, an eyepiece 'ocu' with focal length f_{ocu} or more simply a scaling factor $M_{ocu} = S_0/f_{ocu} = 250\,\text{mm}/f_{ocu}$ is used, typically with magnification factors $10\times$ or $20\times$. In practice, objectives and eyepieces are lens combinations, in order to correct for aberrations (see Section 4.5.3). The total focal length f_μ of the composite microscope is evaluated according to Eq. (1.25) as

$$\frac{1}{f_\mu} = \frac{1}{f_{obj}} + \frac{1}{f_{ocu}} - \frac{t}{f_{obj}f_{ocu}}.$$

In general, microscopes have tubes with well-defined lengths of $t = 160\,\text{mm}$, and since $t \gg f_{obj}, f_{ocu}$, one may approximately specify

$$f_\mu \simeq -\frac{f_{obj}f_{ocu}}{t} = -\frac{f_{obj}f_{ocu}}{160\,\text{mm}}.$$

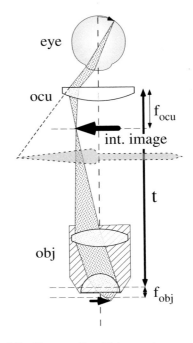

Fig. 4.5: *Beam path within a microscope: t is the length of tube; f_{obj} and f_{ocu} are the focal lengths of objective and eyepiece; the black arrow shows the position of the intermediate image.*

We can determine the image size in two steps: (i) The object lies approximately in the focal plane of the objective, whereas the distance between the real image and the objective differs only a little bit from the length of the tube t. According to Eq. (4.1) it then holds that $y/f_{obj} \simeq y'/t$ and the objective gives rise to a magnification $M_{obj} = y'/y \simeq t/f_{obj}$. (ii) The eyepiece further magnifies the image by the factor $M_{ocu} = y''/y' = S_0/f_{ocu}$, as explained in Section 4.2 on the magnifying glass. The total magnification M_μ of the microscope is then

$$M_\mu = M_{ocu}M_{obj} \simeq \frac{S_0}{f_{ocu}}\frac{t}{f_{obj}} = \frac{S_0}{f_\mu}.$$

This last result shows that the microscope in fact acts like an effective magnifying lens of extremely short focal length.

Example: Magnification of a microscope

We construct a microscope with an eyepiece, magnification $10\times$, and an objective with focal length $f_{obj} = 8\,\text{mm}$. The magnification of the objective amounts to $M_{obj} =$

$160\,\mathrm{mm}/8\,\mathrm{mm} = 20$. The total magnification can be calculated according to $M_\mu = 10 \times 20 = 200$.

Standard microscopes are designed for a quick exchange of the optical elements, to change the magnification easily. Both eyepiece and objective are usually specified with the magnification, e.g. $100\times$; the components of different manufacturers are in general interchangeable. The total magnification can be determined according to the procedure described above without difficulties. For precision measurements it is necessary to calibrate the magnification factor by means of a suitable length standard.

4.3.1   Resolving power of microscopes

So far, we have looked at the microscope only from the geometrical optics point of view and assumed that a point is imaged into an ideal point, again and again. However, as a result of diffraction at the apertures of the lenses, this is not possible, so the resolving power is limited by diffraction. A first measure for the resolving power can be gained from the result for the diameter of the Airy disc. We require that the separation Δx_{min} of the Airy discs of two distinct objects shall be at least as large as their diameter, see Eq. (2.48):

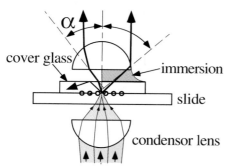

$$\Delta x_{\mathrm{min}} \geq 1.22 \frac{f_{\mathrm{obj}}\lambda}{D}.$$

The systematic approach is given by the numerical aperture NA (or the Abbean sine condition – see next subsection). It is defined as in the case of the acceptance angle of optical fibres (Eq. 1.9) as the sine of half of the aperture angle (Fig. 4.6), i.e. of the extreme rays that still contribute to the image:

$$\mathrm{NA} = n \sin\alpha.$$

Thereby n specifies the index of refraction in the object space. The spatial resolution of a microscope is usually defined by (Section 4.3.2)

Fig. 4.6: *Resolving power and numerical aperture. The right half shows the enhancement of the resolution with immersion oil. The resolution is influenced by illumination, as well. Here, a condenser lens is applied, which illuminates a maximum solid angle.*

$$\Delta x_{\mathrm{min}} \geq \frac{\lambda}{\mathrm{NA}}. \tag{4.3}$$

For smaller magnifications, longer focal lengths and therefore smaller angles occur. Then these two conditions are equivalent due to $\sin\alpha \simeq \tan\alpha \simeq D/2f_{\mathrm{obj}}$.

Since the object under test is always very close to the focal plane, the NA is a property of the objective used and is in general specified on standard components. The resolving power therefore is enhanced by short wavelengths (optical microscopes use blue or even ultraviolet light for high resolution) and a large NA. In air with short focal lengths, which means objectives with large magnification, NA values of about 0.7 are achieved. In order to achieve the theoretical values of resolution, when designing the objective the cover glass must be considered, which in general covers the objective as well (see Fig. 4.6). Thereby, for example, total internal reflection is troublesome, and limits the maximum angle within the cover glass to about 40°. By means of immersion fluids the NA value available can be enhanced significantly, using for example a liquid with an index of refraction that is adjusted to match that of the cover glass. Also, the exact form is of importance, if one is to achieve the total theoretical resolution; but the details are far beyond the scope of this book. The best optical microscopes achieve resolutions of about $0.2\,\mu m$ with blue illumination. Shorter wavelengths and hence further improved resolution can be achieved by using alternative 'light', like e.g. electrons in an electron microscope.

4.3.2   Abbe theory of resolution

To determine the resolving power of a microscope even more accurately, we want to consider a periodic structure (a grating with period Δx) that we observe with a microscope. Ernst Abbe (1840–1905), professor of physics and mathematics at the University of Jena, Germany, and close coworker of Carl Zeiss (1816–1888), provided crucial experimental and theoretical contributions to the development of modern microscopy.

The simplified situation for the grating is illustrated in Fig. 4.7, and the focal plane or Fourier plane of the objective now plays a very crucial role. There, bundles of parallel rays are focused, and one observes the Fraunhofer diffraction image of the object, which is simple only for the chosen example of a one-dimensional grating. However, the following point is crucial. Within the focal plane, the object generates the Fourier transform of the complex amplitude distribution in the object plane, as we have already seen in Section 2.5.3. A structure with a certain size a can only be reconstructed if, apart from the zero order, at least one more diffraction order enters the objective and contributes to the image. This is the Abbean sine condition:

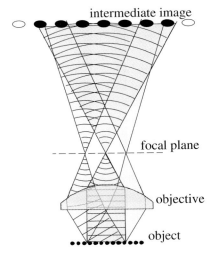

Fig. 4.7: *Fourier image of a microscope, after Ernst Abbe.*

$$a \geq \lambda / \sin \alpha.$$

In the optical microscope, the Fourier spectrum of a diffracting or scattering object is reconstructed by the eyepiece and eye or camera objective to yield a magnified image. In principle, the reconstruction can be gained by a calculation or a numerical procedure. In this sense, the scattering experiments of high-energy physics, where the far field of the diffraction of extremely short-wave matter waves off very small diffracting objects is measured, are nothing other than giant microscopes.

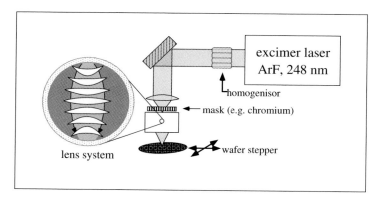

Fig. 4.8: *Optical lithography. Principle of the wafer stepper and UV illumination unit. The lens systems contain numerous components for the correction of aberrations.*

Excursion: Optical lithography

From many points of view, optical lithography is the reverse of microscopy, because lithography, which is nowadays one of the most powerful driving forces of the world economy, is primarily concerned with the miniaturization of electronic circuits to the smallest possible dimensions. The principle is introduced schematically in Fig. 4.8. With a 'wafer stepper', a mask ('reticule'), which contains the structure of the desired circuit, is projected at reduced scale step by step onto cm^2 large areas. Thereby, adequate film material ('resist') is chemically altered such that afterwards in eventually several processing stages transistors and transmission lines can be produced. Manufacturers of lithography objectives, which nowadays may consist of 60 and more individual lenses, succeeded impressively in guiding the resolution of their wafer steppers directly along the resolution limit according to Eq. (4.3). At present miniaturization of electronic circuits is limited by the wavelength in use, nowadays in general the wavelengths of the KrF* laser at 248 nm and the ArF* laser at 193 nm. Further progress will entail enormous costs, because at these short wavelengths tremendous problems arise in manufacturing and processing of suitable, i.e. transparent and homogeneous, optical materials.

4.3.3   Depth of focus and confocal microscopy

Every user of a microscope knows that he or she has to adjust the image to make it 'sharp', and that the range of adjustments for which sharp images are generated decreases with increasing magnification. The longitudinal distance towards the direction of the optical axis where two points can both be just imaged sharply is called the 'depth of focus'. A quantitative measure of the depth of focus may be obtained, for example, from the geometrical considerations in Fig. 4.9.

The movement of an object point by δg out of the 'true' object plane causes a spot with diameter Δx in the intermediate image plane. From Eq. (4.2) we can derive that for $\delta g/g \ll 1$ the image distance moves approximately by $\delta b/b \sim -\delta g/g$. Geometrical considerations then yield directly the result $\Delta x = |\delta b\, D/2b| =$

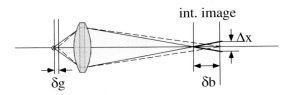

Fig. 4.9: *Geometry of the depth of focus.*

$|\delta g\, D/2g| \simeq |\delta g\, D/2f|$. If we require that this spot should stay smaller than the diffraction disc of the object point, then we find for a maximum tolerable movement Δz:

$$\Delta z \leq \Delta x_{\min} \frac{f}{D/2} \sim \frac{\lambda f^2}{(D/2)^2} \sim \frac{\lambda}{\mathrm{NA}^2}.$$

Then, for larger magnifications, the depth of focus becomes very small, as well; it reaches the order of a wavelength. The small depth of focus for the reverse process of microscopy, reduction in optical projection lithography, causes high demands on the mechanical tolerances of wafer steppers in optical lithography. We can draw an analogy to Gaussian ray optics (see p. 42): the length of the Rayleigh zone of the focused coherent light beam has the same ratio with the diameter of the focal spot as the depth of focus has!

Confocal microscopy uses the short depth of focus of an image with short focal length and large numerical aperture to gain – beyond 'planar' information – three-dimensional information of the device under test. In Fig. 4.10 the basic principle of confocal microscopy is shown: a coherent light beam creates a narrow spot with little depth of focus within the probe.   Only the light intensity reflected, resp. scattered, out of the spot is focused onto an aperture. Structures in other planes are projected into other planes and therefore they are largely suppressed by the aperture.

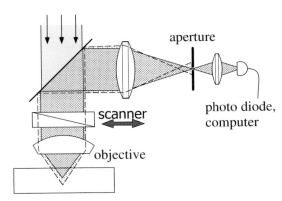

Fig. 4.10: *Principle of confocal microscopy.*

The intensity transmitted through the aperture is continuously monitored by a photodetector and evaluated by a computer to yield an image. Confocal microscopy is an example of scanning probe microscopy, since the focal spot has to scan the sample. In our picture this is achieved by a movable beam steerer ('scanner'). Confocal microscopy achieves resolutions of about $1\,\mu$m; the advantage is access to the third dimension, which is, of course, only possible in transparent samples.

4.3.4   Scanning near-field optical microscopy (SNOM)

The limited resolution of a microscope is a 'result' of Maxwell's equations. In free space the curvature of the electric field cannot occur on a scale much shorter than a wavelength. An ideal point light source is imaged by an optical imaging system into a small but finite spot at best, and this limits the resolution by diffraction to a value of about half a wavelength $\lambda/2$ of the light in use.

In the near field of a radiating system, this limit can be exceeded, since in the presence of polarizable material the propagation of electromagnetic waves is no longer restricted by the diffraction limit. A typical arrangement is introduced in Fig. 4.11. An optical fibre is pulled by a pipette pulling device to yield a tip, the radius of curvature of which is less than 100 nm. This receives a cladding, e.g. out of relatively low-loss aluminium, which leaves only a small aperture, which serves as radiation source or detector of the local light field (or both simultaneously).

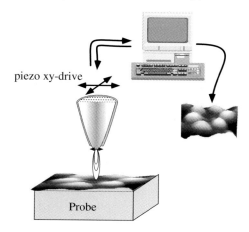

piezo xy-drive

Probe

Fig. 4.11: *Scanning near-field optical microscope. An aperture at the end of an optical fibre is used as a source or detector of radiation fields with a resolution of less than optical wavelengths.*

The end of the fibre, which is made to oscillate by a piezo-drive, experiences an attractive van der Waals force at a typical distance of micrometres and an attenuating force that may be used to adjust the distance as in force microscopy and therefore gives information about the surface topography of the sample. The optical information is recorded by detecting at the end of the fibre the light picked up or reflected at the tip. With smaller and smaller apertures, the spatial resolution increases, and can be pushed significantly below the wavelength in use (typically $\lambda/20$); note that it depends essentially on the diameter of the aperture and not on the wavelength. On the other hand, the detection sensitivity decreases more and more, because the sensitivity decreases with a high power of the diameter of the aperture, and even 1 mW of light power damages the apertures.

After the celebrated success of scanning microscopy, which was initiated by tun-
nelling microscopy and force microscopy in the 1980s, nowadays *scanning near-field
optical microscopy (SNOM)* has been established as yet another new method of scanning microscopy.

4.4   Telescopes

Binoculars and telescopes are used to make terrestrial or astronomical objects more
visible. In general, they are composed of two lenses or mirrors, the focal points of
which coincide exactly, i.e. their separation is $d = f_1 + f_2$. In the Galilean telescope in
Fig. 4.12, a concave (diverging) lens with negative focal length is used. Under these
circumstances, the image matrix of the system reads as follows according to Eq. (1.24):

$$M_{\text{tel}} = \begin{pmatrix} -f_2/f_1 & d \\ 0 & -f_1/f_2 \end{pmatrix}.$$

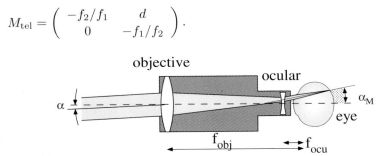

Fig. 4.12: *Angle magnification in a Galilean telescope.*

The total refractive power of this system vanishes, $\mathcal{D}_{\text{tel}} = 0$. Such systems are called
afocal [85]; their action is based on angle magnification only. The objects are located
effectively always at very large distances. From there, parallel bundles of rays originate,
which are transformed into parallel bundles of rays at other, larger angles.

4.4.1   Theoretical resolving power of a telescope

Before we determine the magnification, we want to consider the kinds of objects we
might be able to recognize. Therefore, we have to recall the resolving power of a convex
lens, which we have determined already in Eq. (2.45). There, we have already seen
that at a fixed wavelength the aperture of any imaging optics determines the smallest
angle at which two point-like objects can still be distinguished. We reformulate this
condition for telescopes:

$$\text{minimum structural dimension} \simeq \frac{\text{wavelength} \times \text{distance}}{\text{aperture}}.$$

The consequences for (1) the human eye (pupil 1 cm), (2) a telescope with 10 cm
mirror and (3) the 2.4 m mirror of the Hubble Space Telescope (HST) have been
illustrated in Fig. 4.13. The shapes of objects can be recognized above the limiting
lines 1–3, whereas below those lines the objects cannot be distinguished from points.

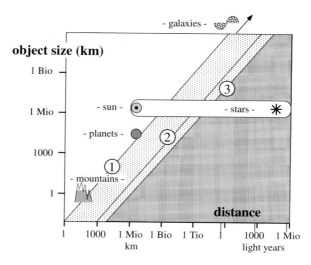

Fig. 4.13: *Pattern recognition of objects far away, with the eye (1), a telescope with 10 cm mirror (2) and the Hubble Space Telescope (3) [1 light year = 9.5 trillion km].*

4.4.2   Magnification of a telescope

In Fig. 4.12 we introduced the concept invented by Galileo Galilei, which is composed of a convex lens with focal length f_{obj} and a concave eyepiece with a focal length f_{ocu}. Geometrical considerations, such as the calculation of the system matrix M_{tel}, show easily that the magnification of the angle of binoculars is

$$\text{magnification } M = \frac{\alpha_M}{\alpha} = -\frac{f_{\text{obj}}}{f_{\text{ocu}}}.$$

A negative sign of M means that the image is inverted; therefore, the Galilean telescope of Fig. 4.12 offers a non-inverted image due to the concave lens with negative f_{ocu}. Telescopes are large-volume devices, because large apertures and lengths of focus are advantageous. The minimum device length is

$$\ell_{\text{telescope}} = f_{\text{obj}} + f_{\text{ocu}}.$$

4.4.3   Image distortions of telescopes

Like all optical instruments, telescopes are affected by several aberrations (see next section). We restrict ourselves to selected problems; the Schmidt mirror sets an example for the correction of spherical aberrations from p. 125.

Lens telescopes and reflector telescopes

Chromatic aberrations, which we will discuss in detail in Section 4.5.3, were identified very early as an obstacle to improving lens telescopes technically. Isaac Newton (1688) was one of the first to discover that refractive lens optics, suffering from strong dispersion, should be substituted by the reflective optics of reflector telescopes, which nowadays has become the standard device layout.

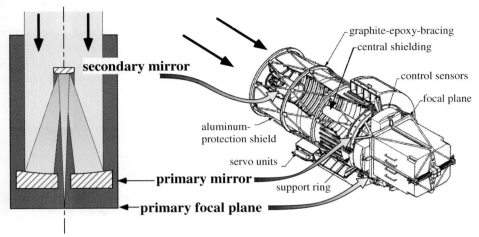

Fig. 4.14: *Reflector telescope (Cassegrain type) and Hubble Space Telescope.*

In Fig. 4.14 the Cassegrain concept is shown, which consists of a primary concave mirror and a secondary convex mirror. If the primary mirror has parabolic shape, then the secondary must have hyperbolic shape; however, other types (with other types of aberrations) are possible, as well. One of the newest instruments of this type is the Hubble Space Telescope, (HST) which since 1990 has delivered more and more new and fascinating pictures of stars and galaxies far away, not influenced by atmospheric fluctuations [48].

In the original configuration, a mistake in the calculation of the mirror properties had resulted in aberrations, which inhibited the total theoretically available resolution of the HST! However, there was real delight after the optics of the telescope had been corrected by an additional pair of mirrors – after, so to speak, fitting 'spectacles' to the HST.

To evaluate the quality of an imaging system, often the so-called *point-spread function* is used, by means of which the im-

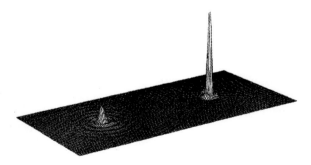

Fig. 4.15: *Point-spread function of the Hubble Space Telescope after installation of the COSTAR correction optics. After [22].*

age of a point according to wave theory is described, taking the exact form of the imaging system into account. In Fig. 4.15 the result for the calculation for the HST before and after the installation of the correction optics is shown.

Atmospheric turbulence

The HST, with a mirror diameter of 2.4 m, does not have an extraordinary diameter; with 10 m the new Keck Telescope at the Keck Observatory in Hawaii offers much more than that. However, the resolution of the HST is much superior to that of terrestrial telescopes, because the resolving power of the latter is limited to effectively 10 cm by turbulent motion of the atmosphere (like the optical telescope in Fig. 4.13)! However, owing to their collecting power, giant terrestrial telescopes offer the possibility to study objects with very low light power in more detail.

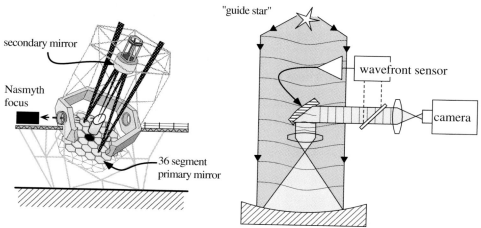

Fig. 4.16: *Left: Schematic of the new telescope at the Keck Observatory. Right: Artificial or reference stars for the application of adaptive optics.*

For the installation of huge mirror telescopes, one looks for environments with very favourable atmospheric conditions, e.g. in the Andes of Chile or on the Hawaiian Islands. The 10 m telescope at the Keck Observatory is one of the most modern facilities (construction year 1992). To use the total theoretical efficiency of a mirror, the optical shape – a sphere, hyperboloid or whatever – must be kept to within subwavelength precision. However, with increasing size, this requirement is more and more difficult to fulfil, because these heavy mirrors are even distorted by the influence of gravity, thus causing aberrations. Therefore, the Keck mirror was manufactured with 36 segments, the positions and shapes of which can be corrected by hydraulic positioning elements in order to achieve optimum imaging results.

Actively tuned optical components are used more and more and are summarized under the term 'adaptive optics'. With newer developments it is possible to compensate for atmospheric turbulence that changes on a time scale of ca. 100 ms. Typically, for that purpose the wavefront must be analysed and used to control a deformable mirror within a feedback loop. The wavefront in the upper atmosphere may be assessed, observing atmospherically, by analysis of the light from a very bright reference star or by the positioning of an 'artificial star' (Fig. 4.16) [33], such as a laser light source, for example, in the upper atmosphere.

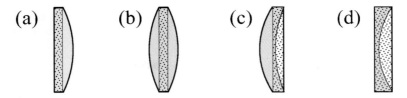

Fig. 4.17: *Important lens types: (a) planar convex lens; (b) biconvex lens; (c) convergent meniscus lens; (d) planar concave lens.*

4.5   Lenses: designs and aberrations

The spherical biconvex lens is, so to speak, the cardinal case of a convex lens and is the lens usually illustrated in figures. All spherical lenses cause aberrations, however, and the application of certain designs depends completely on the area of application. As a rule of thumb, we recall the paraxial approximation: the linearized form of Snell's law ($\sin\theta \rightarrow \theta$, eq. (1.14)) is the better fulfilled, the smaller are the angles of refraction! Therefore, it is convenient to distribute the refraction of a beam of rays, passing through a lens, as evenly as possible to the two refracting surfaces. At selected points aberrations can be compensated by adequate choice of the surfaces. In a multi-lens system (doublet, triplet, ...), several curved surfaces and thus degrees of freedom are available. However, the perfect lens system, correcting for several types of aberrations at the same time (see below), cannot be realized in this way, and thus all multi-lens systems ('objectives') are in general designed for specific applications.

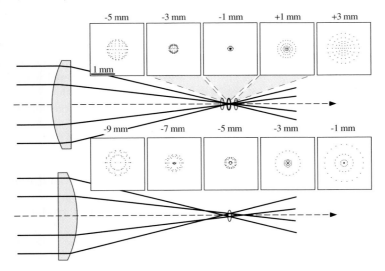

Fig. 4.18: *Spot diagrams of a planar convex lens for two different orientations (after commercial software for the analysis of aberrations). The specifications of the distances refer to the distances to the nominal focal point (here 66 mm).*

Before introducing the technical discussion of aberrations, we want to collect some intuitive arguments for dealing with one or two lenses. More complex systems must be analysed numerically.

4.5.1   Types of lenses

Planar convex lenses

This type of lens only has one curved surface and therefore may be manufactured quite cheaply. For typical indices of refraction of technical glasses of $n = 1.5$, one finds according to Eq. (1.19) $f = -1/\mathcal{D} \simeq 2R$. To focus a light ray, the planar convex lens may be used in two different orientations. Fig. 4.18 indicates how spherical aberrations primarily affect the ability to focus. The so-called 'spot diagrams' show the evolution of the size of a spot along the optical axis. Obviously, it is convenient to distribute the focal powers to several surfaces – indeed, in the orientation of the lower part in Fig. 4.18, refraction occurs only on one side of the lens, resulting in reduced focusing.

Biconvex lenses and doublets

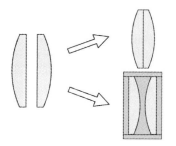

Fig. 4.19: *Biconvex lens and planar convex doublet.*

We may imagine a biconvex lens as composed of two planar convex lenses back to back, as indicated in Fig. 4.19. Therefore, the focal powers add and we find for common glasses again according to Eq. (1.19) with $n \sim 1.5$:
$$f \simeq R.$$
For $1:1$ imaging, the biconvex lens singlets have minimum spherical aberration, which is important, for example, for collimators. However, the refractive powers of planar convex lenses add in exactly the same way if they are mounted with their spherical surfaces opposing each other.

Thereby, in a $1:1$ image the refractive power is distributed to four surfaces and one achieves further reduction of aberrations.

Meniscus lenses

Meniscus lenses may minimize as singlets the aberrations for a given distance between object and image. Indeed, they are first of all part of multi-lens objectives and serve, for example, to change the length of focus of other lenses, without introducing additional aberrations or coma. Such systems are called *aplanatic* [64].

4.5.2   Aberrations: Seidel aberrations

Here, we want to describe briefly the fundamental formal method, going back to Seidel, to classify aberrations. Since it is now necessary to deal with non-axial contributions as well, the complex numbers $r_0 = x + iy$ are convenient for the discussion.

We use the notation from Fig. 4.20, following the discussion of matrix optics (Section 1.9). Then, for the correlation between the location of the object at r_0 with slope r_0' and an image point at $r(z)$:

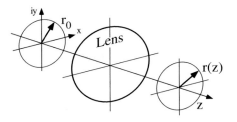

$$r(z) = g(z; x, x', y, y')$$
$$= f(z; r_0, r_0{}^*, r_0', r_0'{}^*).$$

One may use the Laurent expansion, known from the theory of complex numbers,

Fig. 4.20: *Notation of aberrations.*

$$r(z) = \sum_{\alpha\beta\gamma\delta \geq 0} C_{\alpha\beta\gamma\delta}\, r_0{}^\alpha r_0{}^{*\beta} r_0'{}^\gamma r_0'{}^{*\delta}. \tag{4.4}$$

A rotation by the angle Θ in the plane of the object, $r_0 \rightarrow r_0\, e^{i\Theta}$, must cause a rotation by the same angle in the image plane,

$$r(z)\, e^{i\Theta} = \sum_{\alpha\beta\gamma\delta} C_{\alpha\beta\gamma\delta}\, r_0{}^\alpha r_0{}^{*\beta} r_0'{}^\gamma r_0'{}^{*\delta}\, e^{i\Theta(\alpha - \beta + \delta - \gamma)}.$$

From that one finds directly the first condition

$$
\begin{aligned}
&\text{(i)} &&\alpha - \beta + \gamma - \delta = 1, \\
&\text{(ii)} &&\alpha + \beta + \gamma + \delta = 1, 3, 5, \ldots,
\end{aligned}
\tag{4.5}
$$

while the second follows from the special case $\Theta = \pi$, resp. $r(z) \rightarrow -r(z)$, from direct reflection at the optical axis. It determines that only odd orders 1, 3, ... may occur.

(a) Ray propagation in first order

In first order ($\alpha + \beta + \gamma + \delta = 1$ in Eq. (4.5)), one finds $\beta = \delta = 0$ and

$$r(z) = C_{1000}r_0 + C_{0010}r_0'.$$

This form corresponds exactly to the linear approximation, which we already used as the basis of matrix optics and discussed in detail in Section 1.9.

(b) Ray propagation in third order

In third order ($\alpha + \beta + \gamma + \delta = 3$), in total six contributions arise, the prefactors of which are known as 'Seidel coefficients'. We find the conditions $\alpha + \gamma = 2$ and $\beta + \delta = 1$, which can be fulfilled with six different coefficients $C_{\alpha\beta\gamma\delta}$ and are itemized in Tab. 4.2.

From the table we will now discuss several selected aberrations and the corrections of those in more detail. The coefficients are properties of the lens or the lens system, and in the past the theoretical determination of those has been possible only for certain applications due to the enormous numerical calculation expenditure. Nowadays, these tasks are done by suitable computer software.

Tab. 4.2: *Seidel coefficients of aberrations.*

Coefficient	α	β	γ	δ	$\propto$	Aberration
C_{0021}	0	0	2	1	r'^3	spherical aberration
C_{1011}	1	0	1	1	$r'^2 r$	coma I
C_{0120}	0	1	2	0	$r'^2 r$	coma II
C_{1110}	1	1	1	0	$r' r^2$	astigmatism
C_{2001}	2	0	0	1	$r' r^2$	curvature of the image field
C_{2100}	2	1	0	0	r^3	distortion

Aperture aberration or spherical aberration

We have already introduced the effect of spherical aberration in Fig. 4.18 with the example of a planar convex lens with spot diagrams. It depends only on the aperture angle (r'_0 in Eq. (4.4)), may be reduced by limiting the aperture, and is therefore called 'aperture aberration'. However, on doing this the imaging system very quickly loses light intensity. Therefore, for practical applications further corrections are necessary, which can be achieved by choice of a combination of convenient radii of curvature ('aplanatic systems') or by the use of a lens system, for example. In particular, spherical aberration is often corrected at the same time as chromatic aberration (see Section 4.5.3).

Example: Aperture aberration of a thin lens

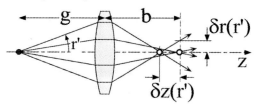

Fig. 4.21: *Spherical aberrations.*

Since spherical aberration is determined by the aperture angle only, we consider a point on the axis, $r_0 = 0$, at a distance g from the lens (Fig. 4.21). As we have already discussed in more detail on p. 21, the image point must also lie at $r(z) = 0$ and must be independent of r'_0. From the combination of the linear approximation with the Seidel approximation, one finds

$$r(z) = 0 = gz \left(\frac{1}{g} + \frac{1}{z} - \frac{1}{f} \right) r'_0 + C_{0021} r'^3_0.$$

Within the paraxial approximation Eq. (4.2) is fulfilled exactly for $z = b$. But here the intercept with the optical axis depends on r'_0. In linear approximation for small shifts it holds that $z = b + \delta z(r')$ and $r(z) = 0$ for r'_0 is valid for

$$\delta z = \frac{b}{g} C_{0021} r'^3_0$$

Here, we have determined the so-called *longitudinal spherical aberration*. In a similar way the transverse spherical aberration ($\delta r(r')$ in Fig. 4.21) may be calculated.

Example: Schmidt mirror

An interesting variant of the commonly used Cassegrain concept is the so-called Schmidt telescope, which is additionally equipped with a compensator plate made of glass. It corrects not only the aperture aberration, but also chromatic aberrations, coma and astigmatism. Thereby large image fields up to 6° are achieved that are very suitable for a celestial survey campaign. Standard telescopes do not achieve more than about 1.5°.

Schmidt's idea first takes into account that a parabolic mirror may generate perfect images very close to the axis, but on the other hand causes strong comatic distortions even at small distances, while a spherical mirror creates a much more regular image of a circular observation plane. In the vicinity of the axis the location of the spherical mirror may be described according to the expansion

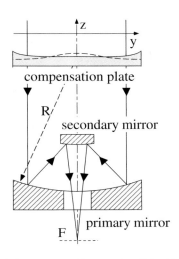

Fig. 4.22: *Cassegrain–Schmidt telescope.*

$$z = \frac{y^2}{4f} + \frac{y^4}{64f^3} + \cdots,$$

where the first term corresponds exactly to the paraboloidal form. The compensator plate with refractive index n compensates exactly for the difference in optical path-length between spherical and paraboloidal surface, if the variation of the thickness is

$$\Delta(y) = \frac{y^4}{(n-1)32f^3}$$

(the factor of 2 occurs due to reflection). This form – the solid variant in Fig. 4.22 – increases towards the aperture of the telescope, whereas the dashed variant in Fig. 4.22 minimizes chromatic aberrations as well [11].

When the compensator plate is mounted within the plane of the centre of curvature of the primary mirror, then the correction is valid also for larger angles of incidence within good approximation.

Astigmatism

If the object points do not lie on the optical axis, then the axial symmetry is violated and we have to discuss the 'sagittal' and the 'tangential' planes of beam propagation separately.[1] The effective length of focus of a lens depends on the angle of incidence, as

---

[1] Astigmatism of an optical lens also occurs for a component that is perfectly rotationally symmetric. It should be distinguished from astigmatism of the eye, which is caused by cylindrical asymmetry of the cornea and creates image points at different distances even for axial points.

can be recognized in Fig. 4.23, where the light rays of the sagittal and tangential planes are concentrated into two different focal lines. Between these two lines there exists a plane where one may identify an image point of 'least confusion' as a compromise.

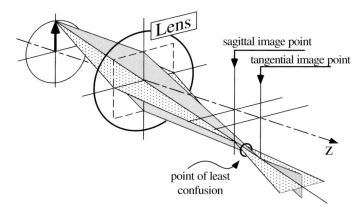

Fig. 4.23: *Astigmatism of a lens. Within the sagittal (dotted) and tangential planes (shaded), the image points lie at different distances.*

Example: Astigmatism of tilted planar plates

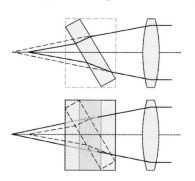

Fig. 4.24: *Astigmatism of a tilted planar plate.*

When light passes through a planar plate at an oblique angle, this leads to different effective focal lengths and thereby to astigmatism. We have illustrated this qualitatively in Fig. 4.24. Thus a planar plate may be used to compensate for the astigmatism of other components as well. Astigmatism occurs nowadays, for example, as a property of light beams emanating from diode lasers, which in edge-emitting configuration do not have axial symmetry in general (see section 9).

In laser resonators, optical components are often installed at the Brewster angle. If curved concave mirrors are used, then the astigmatism of those (see p. 17) may be used for compensation by suitable choice of the angle [60].

Coma and distortion

Among all image aberrations, the one called 'coma' (from the Greek word for *long hair*) or asymmetry aberration is the most annoying. Coma causes a comet-like tail (that is where the name comes from) for non-axial object points, which we have illustrated qualitatively in Fig. 4.25.

According to Tab. 4.2 the image field curvature has a similar form to astigmatism, but it is axially symmetric. Distortion has two variants, pin-cushion and barrel distortion, which are also indicated in Fig. 4.25. This contribution depends on the radius only.

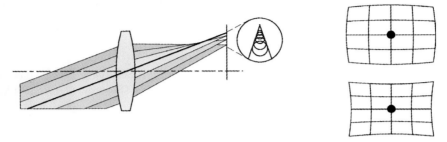

Fig. 4.25: *Coma (left), barrel distortion (top right) and pin-cushion distortion (bottom right).*

4.5.3   Chromatic aberration

Chromatic aberration is caused by *dispersion* of optical materials, since the index of refraction of the glasses used in lenses depends on the wavelength. The refractive power of a convex lens is in general higher for blue light than for red light. We discuss the effect of dispersion with the lens maker's equation (1.19) for a lens with refractive index $n(\lambda)$ and radii of curvature R and R':

$$\frac{1}{f} = \frac{1}{g} + \frac{1}{b} = -(n-1)\left(\frac{1}{R'} - \frac{1}{R}\right).$$

The object distance is fixed, of course, but the image distance changes with the index of refraction,

$$\Delta\frac{1}{b} = -\Delta n\left(\frac{1}{R'} - \frac{1}{R}\right) = \frac{\Delta n}{n-1}\frac{1}{f}$$

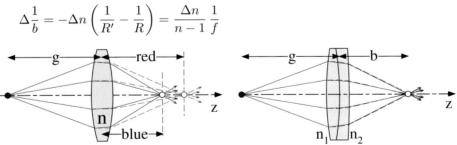

Fig. 4.26: *Chromatic aberration and correction with so-called achromats.*

We know (see p. 22) that the refractive powers $\mathcal{D}$ of two directly neighbouring lenses add, and due to $\mathcal{D} = 1/f$ it holds that $1/f_{\text{tot}} = 1/f_1 + 1/f_2$. If the focal length of the combined system is not to change with wavelength any more, then

$$\Delta\frac{1}{f_{\text{tot}}} = \frac{\Delta n_1}{n_1 - 1}\frac{1}{f_1} + \frac{\Delta n_2}{n_2 - 1}\frac{1}{f_2} = 0,$$

and we find the condition, to correct for chromatic aberration:

$$f_2 \frac{\Delta n_1}{n_1 - 1} = -f_1 \frac{\Delta n_2}{n_2 - 1}. \tag{4.6}$$

To treat this situation more precisely, we have to use the linear expansion of the refractive index,

$$\Delta n_i = \frac{dn_i}{d\lambda} \Delta\lambda + \frac{1}{2} \frac{d^2 n_i}{d\lambda^2} (\Delta\lambda)^2 + \cdots .$$

However, because certain standard wavelengths for $\Delta\lambda$ have been agreed (see Tab. 1.1), the above expression is sufficient. Since dispersion has the same sign for all kinds of known glasses, a lens without chromatic aberrations, which is called *achromatic*, must be composed of a convex and a concave lens (see Fig. 4.26). Lenses also play an important role in particle optics; there, it is much more difficult than in light optics to construct achromatic systems, since divergent lenses cannot be constructed so easily

Incidentally, the radii of curvature of the two lenses are not yet determined by the condition (4.6) for correction of chromatic aberration. This degree of freedom is often used to correct not only for chromatic aberration, but simultaneously for spherical aberration of a lens. Therefore, with an achromat one often gets a lens that is corrected spherically, as well.

5 Coherence and interferometry

The principle of superposition from Section 2.1.6 delivers all the requirements needed to deal with the interference of wave fields. So one could treat interferometry and coherence just as part of wave optics, or as an implementation of the principle of superposition. But interference in interferometry is decisively determined by the phase relations of the partial waves. In particular, we will consider the quite unwieldy concept of coherence, leading to quantitative measures for the role of phases, which are always subject to fluctuations in the real world.

Because of this enormous significance, we will devote this chapter to these aspects of wave optics. Nearly every field of physics dealing with wave and especially interference phenomena has taken up the concept of coherence, like e.g. quantum mechanics, which calls the interference of two states 'coherence'. With the help of quantum mechanics, interference experiments are described and interpreted with matter waves.

Tab. 5.1: *Basic interferometer types.*

Coherence type	Two-beam interferometer	Multiple-beam interferometer
Transverse	Young's double slit	Optical grating
Longitudinal	Michelson interferometer	Fabry–Perot interferometer

The wealth of literature dealing with interferometry is not easily comprehensible, not least due to its significance, e.g. for the technique of precise length measurement. In this book, we are focusing on the types shown in Tab. 5.1 underlying all variants.

5.1 Young's double slit

The double-slit experiment first carried out by Thomas Young (1773–1829), an early advocate of the wave theory of light, is certainly among the most famous experiments of physics because it is one of the most simple arrangements to achieve interference. The concept is emulated in numerous variants in order to prove the wave properties of different phenomena, e.g. of matter waves of electron beams [76] or atomic beams [18], which are discussed later in a short digression (see p. 135).

The fundamental effect of interference for light emanating from a double-slit arrangement instead of a single slit is shown in Fig. 5.1. The conditions for and properties of this interference phenomenon are discussed in detail in Section 5.3.

Optics, Light and Laser. Dieter Meschede
Copyright © 2004 Wiley-VCH Verlag GmbH & Co. KGaA
ISBN: 3-527-40364-7

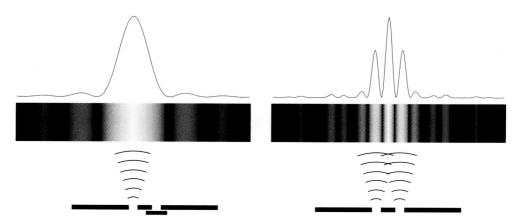

Fig. 5.1: *Young's double-slit experiment. On the right is shown the interference pattern from a double slit. On the left is shown the pattern of a single slit (one slit blocked) for comparison. A logarithmic grey scale has been chosen to make the interference patterns visible.*

5.2   Coherence and correlation

The concept of 'coherence' means the 'capability of interference' of wave fields. We shall see how we can also quantitatively describe it by 'coherence length' and 'coherence time'. These terms come from optical wave theory and state the distances or periods of time over which a fixed phase relation exists between (at least) two partial waves, so that Within this interval the principle of superposition can be applied without any trouble.

When one calculates the distribution of intensity from the superposition of two *coherent* partial waves $E_{1,2}(\mathbf{r}, t)$, first the amplitudes have to be added and then the square of the amount has to be set up:

$$
\begin{aligned}
I_{\text{coh}}(\mathbf{r}, t) &= \frac{c\epsilon_0}{2} |E_1(\mathbf{r}, t) + E_2(\mathbf{r}, t)|^2 \\
&= I_1(\mathbf{r}, t) + I_2(\mathbf{r}, t) + c\epsilon_0 \, \mathfrak{Re}\{E_1(\mathbf{r}, t) E_2^*(\mathbf{r}, t)\}.
\end{aligned}
\tag{5.1}
$$

In the *incoherent* case, however, the intensities are simply added,

$$
I_{\text{inc}}(\mathbf{r}, t) = \frac{c\epsilon_0}{2} \left[|E_1(\mathbf{r}, t)|^2 + |E_2(\mathbf{r}, t)|^2 \right] = I_1(\mathbf{r}, t) + I_2(\mathbf{r}, t),
$$

and we see immediately that the difference is determined by the superposition term.

This quantity I_{coh} though can only be observed if there is a fixed phase correlation between E_1 and E_2 at least during the time of the measurement, because every real detector carries out an average over a finite time and space interval. The times of fluctuation depend on the nature of the light source. For example thermal light sources exhibit fluctuations on the scale of pico- and femtoseconds, which is not reached by detectors working typically on the nanosecond scale.

5.2.1   Correlation functions

Quantitatively the relative time evolution of the phase of superposed fields can be understood by the concept of *correlation*. We define the general complex *correlation function*, also known as the *coherence function*, as

$$\Gamma_{12}(\mathbf{r}_1, \mathbf{r}_2, t, \tau) = \langle E_1(\mathbf{r}_1, t+\tau) E_2^*(\mathbf{r}_2, t) \rangle$$

$$= \frac{1}{T_D} \int_{t-T_D/2}^{t+T_D/2} E_1(\mathbf{r}_1, t'+\tau) E_2^*(\mathbf{r}_2, t') \, dt',$$

which in the average (brackets $\langle\ \rangle$) only takes into account the finite integration time T_D of the detector. It is obvious that the interference term in eq. (5.1) is a special case of this function. More exactly, this is the first-order correlation function. Fully developed theories of coherence make extensive use of correlation functions of higher orders, too. In the second order, for example, there are four field amplitudes related to each other [70].

In interferometry we will consider correlations that do not change with time, so that after averaging only the dependence of the delay is left. Additionally, we will generally determine the intensity of the superposition of waves, i.e. we will consider Γ_{12} at only *one* point $\mathbf{r} = \mathbf{r}_1 = \mathbf{r}_2$, so that the simplified form

$$\Gamma_{12}(\mathbf{r}, \tau) = \frac{c\epsilon_0}{2} \langle E_1(\mathbf{r}, t+\tau) E_2^*(\mathbf{r}, t) \rangle$$

$$= \lim_{T \to \infty} \frac{1}{T} \int_0^T E_1(\mathbf{r}, t+\tau) E_2^*(\mathbf{r}, t) \, dt \tag{5.2}$$

is sufficient. In the case of very large delay times τ, we expect in general the loss of phase relations between E_1 and E_2, so that Γ_{12} statistically fluctuates around 0 and vanishes on average:

$$\Gamma_{12}(\mathbf{r}, \tau \to \infty) \to 0.$$

To make the connection with Eq. (5.1), we have to take into consideration that, in a typical interferometry experiment, the partial waves are created with the help of beam splitters from the same light source. The delay τ then reflects the different optical pathlengths of the partial waves to the point of superposition. The function $\Gamma_{12}(\mathbf{r}, \tau)$ describes their capability of forming interference stripes.

It is very convenient to define the normalized correlation function $\gamma_{12}(\mathbf{r}, \tau)$ which is a quantitative measure for the interference contrast,

$$\gamma_{12}(\mathbf{r}, \tau) = \frac{(c\epsilon_0/2)\langle E_1(\mathbf{r}, \tau) E_2^*(\mathbf{r}, 0) \rangle}{\sqrt{\langle I_1(\mathbf{r}) \rangle \langle I_2(\mathbf{r}) \rangle}}. \tag{5.3}$$

The function γ_{12} is complex and takes values in the range

$$0 \le |\gamma_{12}(\mathbf{r}, \tau)| \le 1.$$

An important special case of Eq. (5.3) is the autocorrelation function,

$$\gamma_{11}(\mathbf{r}, \tau) = \frac{\langle E_1(\mathbf{r}, \tau) E_1^*(\mathbf{r}, 0) \rangle}{\langle I_1(\mathbf{r}) \rangle}, \tag{5.4}$$

which in this case relates the amplitude of an electromagnetic field to itself with delay τ. We shall see its important role in the quantitative analysis of coherence features.

Now we can summarize the calculation of intensity for coherent and incoherent superposition by

$$\langle I(\mathbf{r}) \rangle = \langle I_1(\mathbf{r}) \rangle + \langle I_2(\mathbf{r}) \rangle + 2\sqrt{\langle I_1(\mathbf{r}) \rangle \langle I_2(\mathbf{r}) \rangle}\,\mathfrak{Re}\{\gamma_{12}(\mathbf{r}, \tau)\}.$$

In interferometry, the different paths of light beams coming from the same source generally cause a delay $\tau = (s_1 - s_2)/c$. In order also to define a quantitative measure of coherence, we introduce the *visibility*

$$V = \frac{I_{\max} - I_{\min}}{I_{\max} + I_{\min}}, \tag{5.5}$$

with $I_{\max}$ and $I_{\min}$ describing the maxima and minima of an interference pattern, respectively. Obviously, $V(\tau)$ also takes values between 0 and 1. In an interferometric experiment, the degree of coherence can be measured by determination of the visibility.

The capability of interference could not have been taken for granted and has played an important part in the development of wave theory. The reason for the great significance of interferometry for wave theory is to be found in the fact that the physical features of a wave, i.e. *phase* and *amplitude*, can only be measured by superposition with another wave, i.e. by an interferometric experiment. Whether interference can be observed is crucially dependent on the coherence properties of the waves.

5.2.2   Beam splitter

The central element of an interferometric arrangement is the beam splitter. In the past, only by *separation* of an optical wave from a single light source[1] could one create two separated partial waves that were able to interfere.

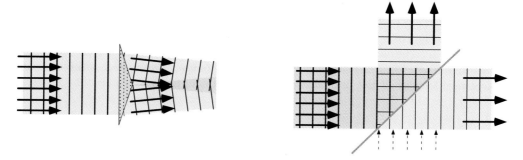

Fig. 5.2: *Wavefront (left) and amplitude (right) beam splitters. Beam splitters have a second entrance, which is not always as visible as for the right-hand type.*

One can differentiate between two different types of beam splitters, as shown in Fig. 5.2. The 'wavefront splitter' with the double slit is the classic form, and the

---

[1] Today, we are able to synchronize two individual laser light sources so well that we can carry out interference experiments using them.

'amplitude splitter' is usually in the form of a partially reflecting glass substrate. In the case of advanced applications, the existence of a second entrance gains importance. The second entrance can easily be seen in the right-hand interferometer in Fig. 5.2.

5.3   The double-slit experiment

Let us now consider in detail the incidence of a planar wave on a double slit (Fig. 5.3). Both slits act as new virtual and phase-synchronous ('coherent') light sources. To understand the interference pattern on the screen, we have to determine the difference between the two optical paths '1' and '2'. If the distance z between the double slits and the screen is very much larger than the distance d between the slits themselves and the extent x of the interference pattern, i.e. $d, x \ll z$, we can determine the path difference Δ_{12} between paths 1 and 2 in a geometrical way according to the construction from Fig. 5.1, and calculate the intensity distribution according to Eq. (5.1).

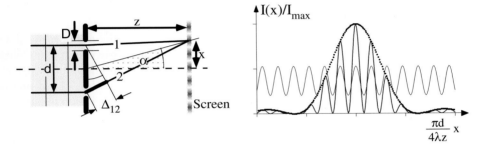

Fig. 5.3: *Analysis of the diffraction pattern from Fig. 5.1. Left: Notations and geometry of the double slit. Right: The interference pattern is understood as the product of the single-slit diffraction pattern (dotted line) and sinusoidal modulation (thin line). Here, width $D = distance/4 = d/4$.*

If the path difference is an integer multiple of the wavelength, $\Delta_{12} = n\lambda$, we expect constructive interference; in the case of half-integer multiples, we have destructive interference. The path difference Δ_{12} at angle α is

$$\Delta_{12} = d \sin \alpha,$$

and on the screen for $\alpha \simeq x/z$ we expect a periodic fringe pattern varying as

$$I(x) = \frac{I_0}{2} \left(1 + \cos \frac{2\pi d}{\lambda} \frac{x}{z} \right),$$

with maximum intensity $I(x{=}0) = I_0$.

During this analysis, we have assumed that the two slits are infinitesimally narrow. In a real experiment, of course, they have finite width, so we have to take into account single slit diffraction, also. The superposition of the two phenomena can be taken into account by means of Fraunhofer diffraction at a slit according to p. 55. The situation becomes very simple, if we displace the slits by $\xi = \pm d/2 = \pm \xi_0$ from the axis. Calling the box-shaped function for the slit (width D) again by $\tau(\xi)$, we get for the diffraction integral with $\kappa_x = 2\pi x/\lambda z$:

$$
\mathcal{E} \propto \oint_S [\tau(\xi - \xi_0) e^{i\kappa_x\xi} + \tau(\xi + \xi_0) e^{i\kappa_x\xi}]\, d\xi
$$

$$
= \oint_S \tau(\xi) e^{i\kappa_x\xi}\, d\xi \left(e^{i\kappa_x d/2} + e^{-i\kappa_x d/2} \right).
$$

The intensity distribution is calculated as usual from $I = c\epsilon_0|\mathcal{E}|^2/2$ for linear polarization,

$$
I = \frac{I_0}{2}\left(1 + \cos\frac{2\pi d}{\lambda}\frac{x}{z}\right)\frac{\sin^2(\pi x D/\lambda z)}{(\pi x D/\lambda z)^2},
$$

and we recognize immediately the complete interference pattern containing the product of the diffraction images of the single slit and of the double slit (Figs. 5.1 and 5.3).

5.3.1   Transverse coherence

If the light source has a finite extent, we can visualize it as consisting of point-like light sources that illuminate the double slit with the same colour or wavelength but with completely independent phases. In this case an additional phase difference appears that can be determined according to a similar construction as in Fig. 5.1. If one of these point sources S lies at an angle β to the axis, the whole phase difference is

$$
\Phi_{12} = k\Delta_{12} \simeq \frac{2\pi d}{\lambda}(\alpha - \beta)
$$

for small angles α and β.

According to this, displacement of the light source causes a transverse shift of the interference pattern on the screen. If all shifts between 0 and 2π occur, the fringe patterns of all point sources are shifted with respect to, and extinguish, each other. In order to observe interference, the maximum phase shift $\Delta_{\max}$ occurring between two point sources of light at a separation of $\Delta a = z_S(\beta - \beta')$ from each other and at a distance z_S from the double slit must not become too large:

$$
\Delta_{\max} = \frac{2\pi d\Delta a}{\lambda z_S} < 1.
$$

This condition is met if the angle $\Omega = \theta - \theta' = \Delta a/z_S$ with which both of the source points are seen is sufficiently small, i.e.

$$
\Omega = \frac{\Delta a}{z_S} < \frac{1}{2\pi}\frac{\lambda}{d}. \tag{5.6}
$$

According to this, for a given wavelength λ and a given distance z_S, the ability to interfere ('interferability') can be achieved through a light source with a sufficiently small point-like area ($\Delta a \le \lambda z_S/2\pi d$) or slit separation ($d \le \lambda z_S/2\pi\Delta a$)

The coherence area of a source is to be determined by changing the slit separation d while the source distance is fixed. The central interference fringe (which is always a maximum) with its adjacent minima should be watched and evaluated according to Eq. (5.5). The distance where the value $V = 1/2$ is obtained is defined as the *transversal coherence length*.

Excursion: Double-slit experiments with matter waves

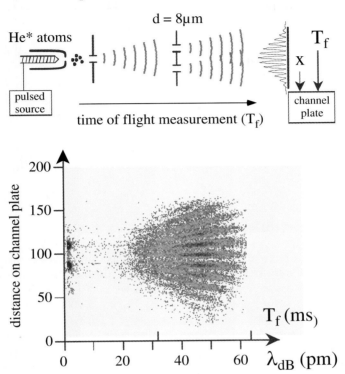

Fig. 5.4: *Diffraction of matter waves at a double slit. By courtesy of J. Mlynek and T. Pfau.*

We have dealt with double-slit interference as a purely wave phenomenon in the previous section, but we have also already referred to the application to other wave phenomena, in particular to matter waves. In this context, there is a very well-known fact that tests our intuition quite hard: an interference pattern is also generated by a single particle, by so-called 'self-interference'. Although we always detect only one particle, its matter wave must have gone through both slits simultaneously! We gather this idea from the way in which quantum mechanics deals with it theoretically. It has been proven by experiments time and again, but stands in bizarre contradiction to our natural, i.e. macroscopic, view of a 'particle'.

The first demonstration of the double-slit experiment with matter waves was given by Moellenstedt [76] using electron beams. For that experiment an electron beam was collimated and sent through an electric field arrangement corresponding to a Fresnel biprism. In recent times atom optics [1] has been established as a new field. With helium atoms a double-slit experiment has been carried out in perfect analogy to Young's experiment [18]. On the one hand, the de Broglie wavelength λ_{deB} of neutral atoms with mass m and velocity v within the atomic beam is very small, $\lambda_{\text{deB}} = h/mv \simeq 20\,\text{pm}$. That is why very tiny slit widths and separations had to be used, $d \leq 1\,\mu\mathrm{m}$, in order to obtain resolvable diffraction. The atomic flux was accordingly very small. On the other hand, helium atoms in the metastable $^3\mathrm{S}$ state can be detected nearly atom by atom by means of channel plates. This high detection sensitivity has made possible the atomic Young's experiment with neutral atoms.

In the lower part of Fig. 5.4 the result of the experiment is shown. The small atomic flux density has even an additional advantage. At a pulsed beam source one can record the velocity of the atom by time-of-flight measurement and watch the change of the interference pattern. This can be directly interpreted as a consequence of the variation of the de Broglie wavelength, which can be immediately calculated from the time-of-flight measurement.

Finally we may turn to the interpretation once more, and consider the light from the point of view of the particles or *photons*. For that we imagine an experiment in which the double slit is illuminated with such a weak intensity that there is only one photon at one time – the condition for self-interference is also met again. Sensitive photon-counting cameras are used to detect the interference pattern. We observe indeed a statistical pattern, which after some time generates a frequency distribution described exactly by the interference of the light waves.

5.3.2   Optical or diffraction gratings

If the number of slits is greatly increased, one obtains an optical grating, an example of multiple beam interference. Optical gratings are used as amplitude, phase or reflective gratings, and are qualitatively introduced in Fig. 5.5. They are specified according to the number of lines per millimetre, typically $1000\,\mathrm{line\,mm^{-1}}$ or more for optical wavelengths. It is remarkable and impressive that even very fine gratings may 'simply' be carved mechanically with diamonds. Optical gratings exhibit typically several *orders of diffraction*. For efficient application, grooves with special shapes are used to concentrate the intensity into a single or a few diffraction orders only, see Fig. 5.5. Such gratings are called *blazed* gratings.

Mechanically manufactured gratings, though, suffer from scattering losses and additional faults with a long period ('grating ghosts'). Better optical quality is offered by components called 'holographic gratings' according to the method of manufacture. They are produced by methods of optical microlithography. A film ('photoresist') on a substrate of optical quality is exposed to a standing light wave. The solubility of the exposed film depends on the dose, and thus a remnant of film is left over at the nodes of the standing wave (see Fig. 5.5). A reflection grating can be manufactured from this structure, e.g. by coating with a reflecting material. One disadvantage, in the case of a holographic grating, is that it is more difficult to control the 'blaze' by properly shaping the grooves.

The condition for interference is identical with that of the double slit. We consider the beams radiating from the N lines of a grating with length L. Two adjacent beams have a path difference Δ that is a function of θ:

$$\Delta(\theta) = (kL/N)\sin\theta. \tag{5.7}$$

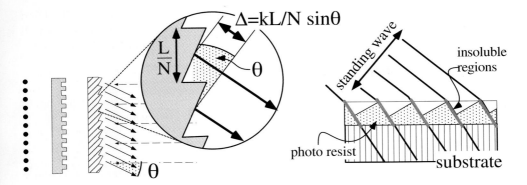

Fig. 5.5: *Left: Amplitude, phase and reflective grating. The blaze of a reflective grating can be chosen in such a way that the diffraction is mostly directed into a certain desired order, for instance by shaping the grooves. Right: Manufacturing of a holographic grating with an asymmetric groove. The photoresist is illuminated by a standing-wave light field. At the nodes it is only weakly affected and thus remains insoluble.*

Under homogeneous illumination, the field amplitude is

$$
\begin{aligned}
E &= E_1 + E_2 + \cdots + E_N \\
&= E_0(1 + e^{-i\Delta} + e^{-2i\Delta} + \cdots + e^{-Ni\Delta})\, e^{-i\omega t} \\
&= E_0 \exp\{-i[\omega t + \tfrac{1}{2}(N-1)\Delta]\} \frac{\sin(N\Delta/2)}{\sin(\Delta/2)}.
\end{aligned}
$$

The diffraction pattern (Fig. 5.6) of the grating has maxima at $\Delta = 2m\pi$, with $m = 0, \pm 1, \pm 2, \ldots$. There the intensity is calculated from $I_0 = c\epsilon_0|E_0|^2$ and $I_{\max} = c\epsilon_0|E(\Delta{=}2m\pi)|^2/2 = N^2 I_0$. Diffraction between the intensity maxima is strongly suppressed, and the grating can be used very advantageously as a dispersive element for spectral analysis.

The first minimum appears at $\Delta = 2\pi/N$, the first secondary maximum at $\Delta = 3\pi/N$. For large N the intensity is limited to

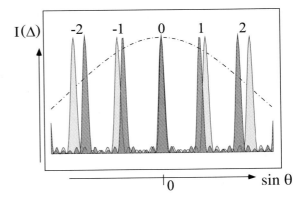

Fig. 5.6: *Diffraction pattern of a grating consisting of six single slits at two different wavelengths. The contribution of the single slit (broken line, width = 0.6 × distance between slits) has been neglected.*

$I(\Delta{=}3\pi/N) \simeq N^2/(3\pi/2)^2 \approx 0.05 I_{\max}$ only. The diffraction grating concentrates the radiation energy into the main maxima.

The spectroscopic resolution is of primary interest. According to the Rayleigh criterion, the main maximum of one wavelength is supposed to fall into the first null of the only just resolvable adjacent wavelength, i.e. according to Eq. (5.7)

$$\Delta(\theta+\delta\theta) - \Delta(\theta) \simeq \frac{2\pi}{\lambda}\frac{L}{N}\cos\theta\,\delta\theta = \frac{2\pi}{N}.$$

The condition for the main maximum varies with the wavelength according to $m\,\delta\lambda = (L/N)\cos\theta\,\delta\theta = \lambda/N$ and so results finally in the resolution

$$\mathcal{R} = \frac{\lambda}{\delta\lambda} = mN.$$

This increases with the number of illuminated slits N and with the order of interference m, as can also be easily seen in Fig. 5.6.

Example: Resolution of an optical grating

We determine the resolution of a grating with a diameter of 100 mm and number of grooves equal to 800 line mm^{-1} at $\lambda = 600$ nm. We get

$$\mathcal{R} = 100\,\text{mm} \times 800\,\text{mm}^{-1} = 0.8 \times 10^5.$$

From that a wavelength can be separated just at a difference of

$$\delta\lambda = \frac{\lambda}{mN} \simeq 7 \times 10^{-3}\,\text{nm}.$$

5.3.3   Monochromators

Grating monochromators are standard equipment in most optical laboratories, and they play an important role by offering one of the simplest instruments of spectroscopy with high resolution. They all have in common the use of reflective gratings, which are technically superior to transmission gratings. They differ only in those structural details dealing with operation or resolution.

As an example we introduce the Czerny–Turner construction (Fig. 5.7). Here, the grating has to be completely illuminated to achieve the highest possible resolution, which is why the input light has to be focused on the entrance slit. The grating simultaneously serves as a mirror that is turned with a linear motion drive. One finds according to eq. (5.7)

$$m\lambda = \frac{L}{N}(\sin\theta - \sin\theta').$$

Because $\theta = \alpha/2 - \theta_G$ and $\theta' = \alpha/2 + \theta_G$ (Fig. 5.7), one gets

$$\lambda = \frac{2L}{mN}\cos(\alpha/2)\sin\theta_G,$$

and hence the wavelength at the exit slit only depends on the rotation angle θ_G.

Spectral resolution depends on angular resolution in this instrument, and it improves with the distance between the slits and the grating. Thus monochromators are

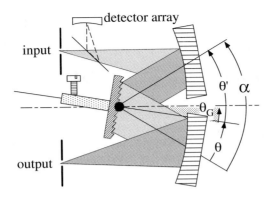

Fig. 5.7: *Principle of the Czerny–Turner monochromator.*

offered with standard lengths of 1/8 m, 1/4 m, 1/2 m, etc., which are a coarse measure of their resolution. Above approximately 1 m they become large, heavy and impracticable, so that resolution exceeding 10^6 is cumbersome to achieve. Through the advent of laser spectroscopy, which we will discuss in Section 11, resolution inconceivable with the conventional methods using grating monochromators has been reached.

5.4   Michelson interferometer: longitudinal coherence

The interferometer arrangement given for the first time by the American physicist M. Michelson (1852–1931) has become very famous. It was developed to identify experimentally the 'ether' postulated in the 19th century to be responsible for the spreading of light. If the ether existed, the speed of light should depend on the relative speed of the light source in that medium.

The results by Michelson and Morley could only be interpreted by assuming that the speed of light was independent of the reference frame – a discovery that led Poincaré, Lorentz and finally Einstein to the development of the theory of special relativity.

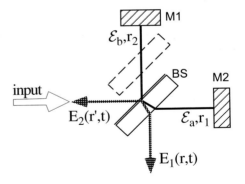

Fig. 5.8: *Michelson interferometer. BS = beam splitter; M1, M2 = mirrors.*

The heart of a Michelson interferometer (Fig. 5.8) is the amplitude beam splitter, mostly consisting of a semitransparent mirror. An incident planar wave $E = \mathcal{E}_{in}\, e^{-i(\omega t - \mathbf{kr})}$ is separated into two partial waves with equal amplitudes $\mathcal{E}_{a,b} = \mathcal{E}_{in}/\sqrt{2}$. Usually the beam splitter consists of a polished glass substrate coated on one side. The reflected and transmitted beams travel along different optical paths. The two paths

are different since the substrate acts on one of the two arms only. For compensation, sometimes an additional glass substrate of the same thickness is inserted into the
other arm in order to make the interferometer arms as symmetric as possible. Using
monochromatic laser light, this does not matter because the difference in the light
pathlengths can simply be geometrically compensated. If the light is polychromatic,
however, dispersion of the glass substrates caused by wavelength-dependent differences
in the light paths is also compensated by the additional substrate.

At the end of the two interferometer branches the two partial waves are reflected
and pass through the beam splitter again. The interferometer generates two separate
output waves E_1 and E_2,

$$
\begin{aligned}
E_1 &= \frac{1}{\sqrt{2}}(\mathcal{E}_a + \mathcal{E}_b) = \frac{1}{2}\mathcal{E}_{in}\, e^{-i(\omega t - kr)}(e^{2ikr_1} + e^{2ikr_2}), \\
E_2 &= \frac{1}{\sqrt{2}}(\mathcal{E}_a - \mathcal{E}_b) = \frac{1}{2}\mathcal{E}_{in}\, e^{-i(\omega t - kr')}(e^{2ikr_1} - e^{2ikr_2}),
\end{aligned}
\tag{5.8}
$$

at its exits. We calculate the intensity there and get from $I = \epsilon_0 c \mathcal{E}\mathcal{E}^*/2$ the results

$$
\begin{aligned}
I_1 &= \tfrac{1}{2}I_0\{1 + \cos[2k(r_1 - r_2)]\}, \\
I_2 &= \tfrac{1}{2}I_0\{1 - \cos[2k(r_1 - r_2)]\}.
\end{aligned}
\tag{5.9}
$$

According to this, the total intensity is distributed on both exits $I_0 = I_1 + I_2$ depending
on the difference in light paths $s = 2(r_1 - r_2)$. Note that, in this arithmetical treatment,
the different signs in the sum of the partial beams ($E_{1,2} = (\mathcal{E}a \pm \mathcal{E}_b)/\sqrt{2}$) are caused
by the reflections at the beam splitter, in one case at the more dense, and in the other
case at the less dense, medium. This $90°$ phase difference is also essential to satisfy
energy conservation.

5.4.1 Longitudinal or temporal coherence

With the Michelson interferometer, the temporal coherence length $\ell_{coh} = c\tau_{coh}$ is
measured by increasing the length of one branch until the interference contrast is
decreased to the half. The coherence length is then twice the difference of the two
branches, $\ell_{coh} = |r_1 - r_2|$ in Fig. 5.8. Usually, the visibility from Eq. (5.5) is again
used as a quantitative measure.

The interference contrast is measured through the field autocorrelation function
$\Gamma_{EE^*}(s/c)$, according to Eqs. (5.2, 5.3) with $\tau = s/c$. This is linked to the spectral
power density

$$
S_E(\omega) = \frac{1}{c}\int_0^\infty \Gamma_{EE^*}(s/c)\, e^{i\omega s/c}\, ds
$$

according to the Wiener–Khintchin theorem (see Appendix A, Eq. (A.9)). So a Fourier
transformation of the interferometer's signal as a function of the path difference delivers
information about the spectral properties of the light source. Analysis of the light
from a sodium vapour lamp with the Michelson interferometer shows this connection
very clearly, as we describe qualitatively in Fig. 5.9. This relation is also the basis
of the Fourier spectrometer, which we mention here for the sake of completeness.

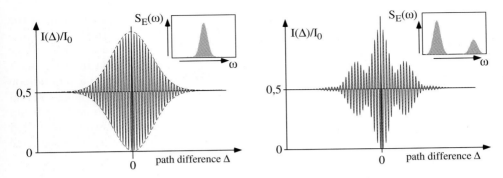

Fig. 5.9: *Interferometer signals of a Michelson interferometer for a single and a double spectral line, like e.g. the yellow D line of the Na vapour lamp. In the upper inset boxes the associated spectra are shown.*

Furthermore, the self-heterodyne method from section 7.3.2 can be considered as a variant on the Michelson interferometer. Here the path difference of the arm lengths even has to be so big that no stable interference can be observed in the time average. This method allows determination of the spectral properties of a narrowband laser light source.

Example: The wavemeter

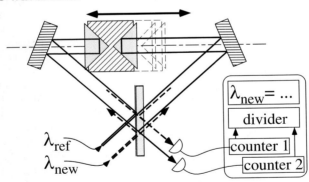

Fig. 5.10: *Wavemeter arrangement for the determination of laser wavelengths. For clarity the laser beam to be measured (broken line) is drawn only at the entrance and at the exit.*

The *wavemeter*, also known as the lambda meter, is a variant of the Michelson interferometer used in many laser laboratories. Monochromatic laser light sources have a very large coherence length ($\ell_{\text{coh}} > 10\,\text{m}$). During continuous variation of the pathlength difference of the interferometer arms, they consequently generate a sinusoidal modulation of the interferometer signal with period proportional to the frequency or inverse wavelength of the laser light according to Eq. (5.9). Comparison of the interferometer signal of an unknown wavelength λ_{new} to a reference laser wavelength λ_{ref} amounts to the determination of the unknown frequency or wavelength by simple division of the number of fringes counted when the reflector trolley slides along its track.

In the wavemeter arrangement, two retro-reflectors are fitted to a mobile carriage (Fig. 5.10), so that the incident and reflected beams of the Michelson interferometer are spatially separated. At one exit the reference beam is directed to a photodiode in order to count the number of interferometer fringes for a certain travelling interval N_{ref}. At the other exit it serves as a tracer beam for a differently coloured laser beam to be measured. Its interferometer fringes are counted on a second photodiode, N_{meas}. Electronic division then yields the unknown wavelength through comparison with the reference laser: $\lambda_{\mathrm{meas}} = \lambda_{\mathrm{ref}} N_{\mathrm{meas}}/N_{\mathrm{ref}}$.

Excursion: Gravity wave interferometer

A particularly unusual variant of the Michelson interferometer with huge dimensions has been constructed at several places around the world. For example, the project at Hannover (Germany) called GEO600 has an arm length of 600 m, while at other places even arm lengths up to 4 km have been realized.

With a Michelson interferometer, as well as with every optical interferometer, minute lengths or variations of length can be measured with an accuracy far below the optical wavelength. Exactly this feature can serve to detect distortions of space caused by gravity waves. Though they were predicted in detail by Einstein's theory of general relativity, they have not been directly observed yet since they only exert an extraordinarily weak force even on big masses.

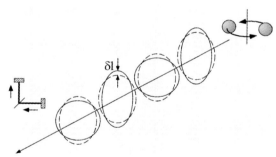

Fig. 5.11: *Gravity waves are predicted to cause quadrupolar distortions of space, e.g. by radiation from a rotating binary star. Astrophysicists are using sensitive Michelson interferometers for the search.*

For a most sensitive proof of a change of length $\delta\ell$ with an interferometer, the instrument itself has to have a length ℓ as large as possible. According to the theory of general relativity, even for strong astronomical 'gravitational wave sources' like e.g. supernova explosions, relative sensitivities of $\delta\ell/\ell \approx 10^{-20}$ are necessary. At a length of 1 km this corresponds to about 100th of a proton radius! Gravity waves spread like electromagnetic waves, they are transverse, but have quadrupolar characteristics (Fig. 5.11).

The sensitivity can be increased by folding the light path in each arm. Narrowband detection of the weaker but continuous and strongly periodic emission of a binary star system (see Fig. 5.11) promises an increase of sensitivity. To achieve sufficient signal-to-noise ratio of the interferometer signal, the use of very powerful laser light sources with superb frequency stability is necessary. At the present time neodymium lasers are preferred for this task.

Not only could the proof of the existence of gravity waves offer the long-sought confirmation of the theory of general relativity, but also, with gravity wave antennas, a new window could be opened for the observation of space. In the face of these expectations, the plans for the *Laser Interferometer Space Antenna* (LISA) [79] do not seem to be completely eccentric.

In this spaceflight project, in about 2008 it is planned to park a Michelson interferometer consisting of three spaceships (two mirrors and a beam splitter with light source) shifted by 20° in Earth orbit around the Sun. This Michelson interferometer will have an arm length of 5×10^6 km!

5.4.2   Mach–Zehnder and Sagnac interferometers

There are numerous variants of the Michelson interferometer that have different methodical advantages and disadvantages. Two important examples are the Mach–Zehnder and the Sagnac interferometers, the latter of which, strictly speaking, forms a class of its own.

Mach–Zehnder interferometer

The Mach–Zehnder interferometer (MZI) is derived from the Michelson interferometer, in which the reflections at the mirrors are no longer carried out at normal incidence and a second beam splitter is used for the recombination of the beams. The MZI is also used for spatially resolved studies of changes in the wavefronts passing objects of interest [42].

The reflection angle at the beam splitters (BS) and mirrors (M1, M2) in Fig. 5.12 (left) is not necessarily limited to 90°. Several times the MZI concept has stimulated ideas for interferometric experiments in particle optics, since there mirrors and beam splitters can often be realized only under grazing incidence, with small deflection angles.

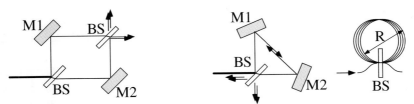

Fig. 5.12: *Mach–Zehnder (left) and Sagnac (right) interferometers. BS = beam splitter. The Sagnac interferometer can be realized with mirrors or with an optical waveguide.*

Sagnac interferometer

The Sagnac interferometer also derives from the Michelson interferometer, in which the light beams are not reflected back to themselves but run back on diametrically opposed paths that are at first identical. But if the interferometer is rotating around an axis perpendicular to its plane, a phase shift between the opposing beams is predicted by the theory of special (and general) relativity. For the sake of simplicity, we consider a circular light path (radius R) in a fibre and with one beam splitter. The round-trip time is $T = L/c = 2\pi n R/c$ with n the refractive index of the fibre. We adopt the result from the theory of special relativity that, in a medium moving at velocity v, the speed of light as measured in the laboratory frame is modified according to [64]

$$c_\pm = c\,\frac{1 \pm nv/c}{n \pm v/c}.$$ (5.10)

In the rotating fibre path (angular velocity $\Omega = v/R$), in one direction the light travels towards the beam splitter, and in the other direction, away. Hence the effective round trip time is increased or decreased corresponding to the path $R\Omega T = vT$ travelled by the beam splitter, yielding the condition $T_\pm = L_\pm/c_\pm = (L \pm vT_\pm)/c_\pm$. From this implicit equation we extract $T_\pm = L/(c_\pm \mp v)$, and with a short calculation using the result Eq. (5.10) we find $1/(c_\pm v) \simeq (n/c)(1 \pm (v/nc))$. Surprisingly, the time difference $T_+ - T_+$ no longer depends on n,

$$T_+ - T_- \simeq 2v/c^2 = 2R\Omega/c^2.$$

For light with frequency ω, we now directly obtain the difference of the light paths or phase difference at the beam splitter from this:

$$\Delta = \omega(T_+ - T_-) = \simeq \omega\frac{4\pi R^2 \Omega}{c^2} = \Omega\frac{4F}{\lambda c}.$$

According to this, the interference signal is proportional not only to the angular velocity Ω but also to the area $F = \pi R^2$ of the Sagnac interferometer. The effective area and with it the sensitivity can be increased by the coil-like winding of a glass fibre (Fig. 5.12).

Example: Phase shift in the Sagnac interferometer

We determine the phase shift generated by the Earth's rotation ($2\pi/24\,\mathrm{h} = 1.8 \times 10^{-6}\,\mathrm{s}^{-1}$) in a Sagnac interferometer. The fibre has a length of $1\,\mathrm{km}$ and is rolled up into an area with a diameter of $2R=10\,\mathrm{cm}$. The interferometer is operated with a diode laser at $\lambda = 780\,\mathrm{nm}$. Thus

$$\Delta = 1.8 \times 10^{-6}\,\frac{\pi \times 4(\cdot 0.1/2)^2(10^3/\pi \times 0.1)}{(0.78 \times 10^{-6}) \times (3 \times 10^8)} = 0.77 \times 10^{-5}\,\mathrm{rad}.$$

This condition requires a high standard of experimental knowledge but can be realized in the laser gyro.

If a laser amplifier is installed in a Sagnac interferometer, one has realized the 'laser gyro'. This is widely used since it allows very sensitive detection of rotary motion and acceleration, but for studies of this we refer the reader to the specialized literature. It should be emphasized, however, that in the laser gyro the waves running around to the left and to the right, respectively, have to have different frequencies.

5.5   Fabry–Perot interferometer

We consider two plane parallel dielectric interfaces illuminated by a light beam at a small angle. Such an optical component can be easily made from a plane parallel glass substrate. In this case it is called a Fabry–Perot etalon (FPE) (from the French *étalon*, meaning 'calibration spacer' or 'gauge'). It is often used for frequency selection in laser resonators or as a simple and very highly resolving diagnostic instrument for laser wavelengths. The light beams are reflected back and forth many times and so exhibit multiple beam interference in the longitudinal direction analogous to the diffraction grating.

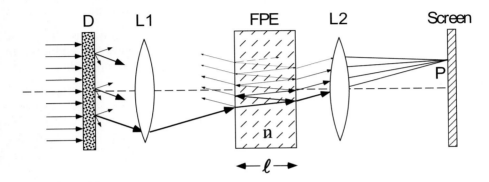

Fig. 5.13: *Multiple beam interference in the Fabry–Perot etalon (FPE). A ground glass substrate acts as a diffuser to generate light beam incident onto the etalon in many directions; The second lens L2 induces parallel light rays to interfere on a screen at the focal plane.*

The surfaces of an FPE are partly silvered and must be very smooth and plane. Furthermore, their relative tilt, or wedge, has to be very small. For precise measurement, also the distance ℓ of the spacing between the reflecting surfaces must be very well known and controlled. The optical length of the FPE depends on the index of refraction n of the substrate,

$$\ell_{\mathrm{opt}} = n\ell,$$

which for a material such as glass changes relatively rapidly with temperature ($dn/dT \simeq 10^{-3}\,\mathrm{K}^{-1}$). Stable, less-sensitive etalons are built with an air gap between glass substrates fixed by spacers with small thermal expansion, e.g. quartz rods. If the distance ℓ of the gap can be varied, e.g. by a piezo-translation, it is called a *Fabry–Perot interferometer*. This type of instrument was used for the first time by C. Fabry and A. Perot in 1899.

The condition for constructive interference can again be determined from the phase difference δ between two adjacent beams. One determines the pathlength A–B–C in Fig. 5.14(a) and finds, with $k = 2\pi/\lambda$,

$$\delta = k\ell_{\mathrm{opt}} = 2nk\ell\cos\theta = 2\pi N, \tag{5.11}$$

where N is the *order* of the interference, usually a large number. This result perhaps contradicts our initial expectation, since, because of the geometry, each individual beam in the interferometer travels along an elongated path $\ell/\cos\theta$ tending to longer wavelengths and smaller frequencies. However, exactly the opposite occurs: tilting of an etalon shifts the interference condition to shorter wavelengths!

Let us now add the individual contributions of each beam, where we now have to account for reflection and transmission. The change of intensity is described by the reflection and transmission coefficients, while the coefficients of the field amplitudes are defined by $r = \sqrt{R}$ and $t = \sqrt{T}$:

$$r, r' = \text{amplitude reflectivity,} \qquad R, R' = \text{reflection coefficient,}$$
$$t, t' = \text{amplitude transmissivity,} \qquad T, T' = \text{transmission coefficient.}$$

Phase jumps during reflection (π phase shift for reflection off the denser material) are included with the total phase shift accumulated after one round trip and given by $e^{i\delta}$. Then the transmitted partial waves contributing to the field amplitude E_{tr} at the interference point P are summed up in a complex geometric series,

$$E_{\text{tr}} = t't E_{\text{in}} + rr' e^{i\delta} tt' E_{\text{in}} + (rr')^2 e^{2i\delta} tt' E_{\text{in}} + \cdots,$$

yielding the result

$$E_{\text{tr}} = \frac{tt' E_{\text{in}}}{1 - rr' e^{i\delta}}. \tag{5.12}$$

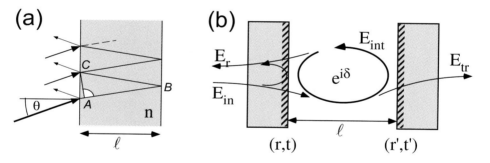

Fig. 5.14: *Phase condition for the Fabry–Perot etalon. (a) Difference in the optical paths of the partial beams. (b) Self-consistent condition for the internal field.*

Alternatively, this result can be derived in a clear and efficient manner by considering only the wave circulating within the etalon right after the first mirror (see Fig. 5.14), because in equilibrium it has to be reconstructed by interference of the internal wave after one full round trip and the incident wave:

$$E_{\text{int}} = e^{i\delta} rr' E_{\text{int}} + t E_{\text{in}}.$$

From this, with $E_{\text{tr}} = t' E_{\text{int}}$, one again and immediately obtains the first result. Already, to satisfy energy conservation, there has to be a reflected wave. From this consideration, the effect of interference becomes still more transparent,

$$E_{\text{r}} = r E_{\text{in}} - r't e^{i\delta} E_{\text{int}} = \frac{r - r' e^{i\delta}}{1 - rr' e^{i\delta}} E_{\text{in}}. \tag{5.13}$$

The minus sign occurs here because in this case there was one reflection – and hence one π phase jump – less compared to the circulating wave.

Let us now explore Eq. (5.12) by considering the transmitted intensity. By taking the modulus, we first get

$$I_{\text{tr}} = I_{\text{in}} \frac{TT'}{|1 - \sqrt{RR'}\, e^{i\delta}|^2}.$$

This can be written more transparently by introducing the *finesse coefficient*, F,

$$F = \frac{4\sqrt{RR'}}{(1 - \sqrt{RR'})^2}, \tag{5.14}$$

from which after a short calculation we get the *Airy function*

$$I_{\text{tr}} = I_{\text{in}} \frac{TT'}{(1 - \sqrt{RR'})^2} \frac{1}{1 + F \sin^2(\delta/2)}. \tag{5.15}$$

According to our calculation the transmitted intensity varies over the range

$$\frac{(1 - R)(1 - R')}{(1 + \sqrt{RR'})^2} \leq \frac{I_{\text{tr}}}{I_{\text{in}}} \leq \frac{(1 - R)(1 - R')}{(1 - \sqrt{RR'})^2}, \tag{5.16}$$

and can even become identical with the incident wave if there are ideal loss-free mirrors with the same reflection coefficients:

$$(R, T) = (R', T') : \qquad I_{\text{tr}} = \frac{I_{\text{in}}}{1 + F \sin^2(\delta/2)},$$

$$\delta = N\pi : \qquad I_{\text{tr}} = I_{\text{in}}.$$

We will learn more about this case in Section 5.6 when we look at optical resonators.

Now let us determine also the accumulated intensity circulating in the etalon along with the reflected intensity,

$$I_{\text{int}} = \frac{1}{T'} I_{\text{tr}},$$

$$I_{\text{r}} = I_{\text{in}} - I_{\text{tr}}.$$

Real resonators are always affected by losses, which should be as low as possible. If we take the losses per revolution simply into account with a coefficient A, we get the generalized finesse coefficient

$$F_A = \frac{4\sqrt{RR'(1 - A)}}{[1 - \sqrt{RR'(1 - A)}\,]^2}, \tag{5.17}$$

by which we can again calculate the transmitted power according to

$$I_{\text{tr}} = \frac{4TT'(1 - A)}{(T + T' + A)^2} \frac{I_{\text{in}}}{1 + F_A \sin^2(\delta/2)}.$$

We can find analogous expressions for the reflected and the coupled power.

Example: Coupling of an optical resonator

Optical resonators, which we shall discuss in more detail in Section 5.6, allow the storage of light energy, albeit only for relatively short times. Thus it is interesting to know the amount of an incident light field that is coupled into the resonator. This can be answered by the recent considerations.

Again, the case of resonance $\delta = 0$ is particularly important. We find these relations for the reflected and the transmitted fractions of the incident intensity:

$$\frac{I_r}{I_{in}} = \left(\frac{T' + A - T}{T' + A + T}\right)^2 \quad \text{and} \quad \frac{I_{tr}}{I_{in}} = \frac{4TT'(1 - A)}{(T' + A + T)^2}.$$

The power circulating in the resonator can also be easily determined according to $I_{res} = I_{tr}/T'$ and is shown in Fig. 5.15 as a function of T/A and for the special but instructive case $T' = 0$.

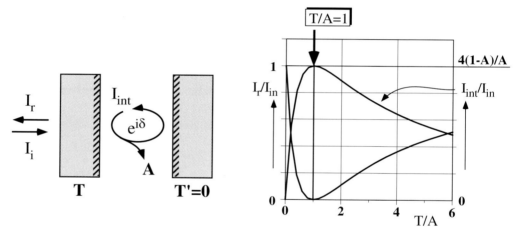

Fig. 5.15: *Influence of losses on the coupling of a Fabry–Perot resonator in the case of resonance $\delta = 0$. The normalized intensities of the reflected (I_r) and stored light field (I_{res}) are displayed for the special case $T' = 0$.*

The maximum of the coupled power is reached at $T/A = 1$. The power circulating there in the resonator is proportional to $1/A$ for low A. In this case the external losses (caused by the coupling mirror) are just equal to the internal losses. This situation is quite well known for resonators: only in the case of perfect 'impedance matching' is the full incident power coupled into the resonator, otherwise it is over- or under-coupled.

5.5.1 Free spectral range, finesse and resolution

According to Eq. (5.11), the Fabry–Perot interferometer delivers a periodic series of transmission lines as a function of the frequency $\omega = ck$ of the incident light field. The distance of adjacent lines corresponds to successive orders N and $N + 1$ and is called the 'free spectral range', Δ_{FSR}:

$$\Delta_{\text{FSR}} = \nu_{N+1} - \nu_N = \frac{c}{2n\ell} = \frac{1}{\tau_{\text{circ}}}. \tag{5.18}$$

The free spectral range also just corresponds to the inverse circulation time τ_{circ} of the light in the interferometer. If the gap between the mirrors is empty ($n = 1$), then we simply have $\Delta_{\text{FSR}} = c/2\ell$. Typically, Fabry–Perot interferometers with centimetre distances are used whose free spectral range is calculated according to

$$\Delta_{\text{FSR}} = \frac{15\,\text{GHz}}{\ell/\text{cm}}$$

and are usually designed for some $100\,\text{MHz}$ up to several GHz.

The Fabry–Perot interferometer can only be used for measurements if the periodicity leading to superpositions of different orders is visible. In that case the resolution between two narrowly adjacent frequencies is determined by the width of the transmission maxima. It can approximately be calculated from Eq. (5.15) taking into account that most interferometers have large F coefficients. Then the sine function can be replaced by the argument,

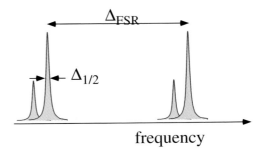

Fig. 5.16: *Free spectral range (FSR) and full width at half-maximum (FWHM) in the Fabry–Perot resonator.*

$$I_{\text{tr}} \simeq \frac{I_{\text{in}}}{1 + F(\delta/2)^2}.$$

If two spectral lines are assumed to be separable if their FWHMs $\Delta\nu_{1/2}$ do not overlap, the lowest resolvable frequency difference is determined from $\delta = 1/F^{1/2} = 2nk\ell$, and for pairs of mirrors with identical reflectivity one obtains

$$\Delta\nu_{1/2} = \Delta_{\text{FSR}} \frac{1 - R}{\pi\sqrt{R}}. \tag{5.19}$$

The ratio $\mathcal{F} = \Delta_{\text{FSR}}/\Delta\nu_{1/2}$ of free spectral range and resolution can be easily read from an oscilloscope screen like in Fig. 5.16. This measure is more common than the finesse coefficient F and is called the *finesse*

$$\mathcal{F} = \frac{\pi}{4}\sqrt{F} = \frac{\pi\sqrt{R}}{1 - R}. \tag{5.20}$$

The interferometric resolution $\nu/\Delta\nu_{1/2}$ is indeed considerably higher with

$$\mathcal{R} = N\mathcal{F}$$

and can easily exceed a value of $\mathcal{R} > 10^8$.

Example: Resolution of Fabry–Perot interferometers

In Tab. 5.2 we have compiled some characteristic specifications for typical Fabry–Perot interferometers, which will play an important role as optical cavities in the next section. In the table it is remarkable that the half-width $\Delta\nu_{1/2}$ always has a similar order of about 1 MHz. The reason for this is the practical applicability to the continuous laser light sources used in the laboratory, which exhibit typical linewidths of 1 MHz.

Tab. 5.2: *Characteristics of Fabry–Perot interferometers.*

ℓ (mm)	$1 - R = T$	Δ_{FSR} (GHz)	$\Delta\nu_{1/2}$ (MHz)	$\mathcal{F}$	Q at 600 nm	τ_{res} (ms)
300	1%	0.5	1.7	300	3×10^8	0.1
10	0.1%	15	5	3 000	10^8	0.03
1	20 ppm	150	1	150 000	5×10^8	0.15
100	20 ppm	1.5	0.01	150 000	5×10^{10}	15

5.6 Optical cavities

Fabry–Perot interferometers are very important as optical cavities, which are necessary for the construction of laser resonators or are widely used as optical spectrum analysers (see details in Section 7.3.2).

5.6.1 Damping of optical cavities

An electromagnetic resonator stores radiant energy. It is characterized, on the one hand, by the spectrum of its resonant frequencies, also known as modes ν_{qmn}, and, on the other, by their decay or damping times τ_{res}, which are related to the stored energy $\mathcal{U} \propto E^2$,

$$\frac{1}{\mathcal{U}}\frac{d\mathcal{U}}{dt} = \frac{2\,dE/dt}{E} = -1/\tau_{\mathrm{res}}.$$

We can work out the loss approximately by evenly spreading the mirror reflectivities $(R = r^2)$ and other losses over one revolution $\tau_{\mathrm{circ}} = \Delta_{\mathrm{FSR}}^{-1}$,

$$\frac{\Delta E}{E\tau_{\mathrm{circ}}} \simeq \frac{1}{2}\ln[(1 - A)RR'] = \ln\sqrt{(1 - A)RR'}.$$

From this, the relation

$$\tau_{\mathrm{res}} = -\frac{\tau_{\mathrm{circ}}}{\ln\sqrt{(1 - A)RR'}} \simeq \frac{\tau_{\mathrm{circ}}}{1 - \sqrt{(1 - A)RR'}}$$

is obtained, which is again related to the Q value or *quality* factor and the half-width $\Delta\nu_{1/2}$ by

$$\Delta_{1/2} = \frac{\nu}{Q} = \frac{1}{2\pi\tau_{\mathrm{res}}}.$$

For $A \rightarrow 0$ and $R = R'$, this result reproduces Eq. (5.19). The resonator's damping time τ_{res} determines the transient as well as the decay behaviour of optical cavities. In Tab. 5.2 we have given some Q values and oscillation damping times τ_{res}. It is assumed that the absorptive losses can be neglected compared to the decoupling.

5.6.2   Modes and mode matching

For stability reasons, resonators no longer use plane mirrors in their construction, but curved ones.[2] With our knowledge of Gaussian beams from section 2.3, we can understand immediately how an appropriate resonator mode has to be constructed according to the following principle: The surfaces of the mirrors must fit exactly the curvature of the wavefronts (see Fig. 5.17).

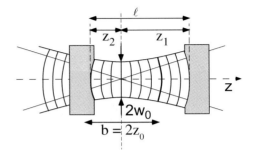

Fig. 5.17: *Gaussian wave and resonator mirrors.*

Whether resonators work stably or unstably can again be investigated by means of the ABCD law of ray or Gaussian beam optics. A pair of mirrors is completely equivalent to the periodic lens system from Section 1.9.5, if we replace the focal lengths by the radii $R_1/2$ and $R_2/2$. Thus from Eq. (1.27) we obtain the stability diagram (Fig. 1.21 on p. 23) for optical cavities according to

$$0 \leq \left(1 - \frac{\ell}{R_1}\right)\left(1 - \frac{\ell}{R_2}\right) \leq 1. \tag{5.21}$$

The characteristic parameters of an optical cavity consisting of two mirrors are their radii R_1 and R_2 and their separation ℓ. Between the mirrors, a Gaussian standing wave with confocal parameter $b = 2z_0$ and beam waist w_0 is excited. The surfaces of the mirrors are at a separation ℓ corresponding exactly to the length of the cavity,

$$\ell = z_1 + z_2.$$

The full solution of the Gaussian modes is described according to Eqs. (2.27) and (2.36),

$$
\begin{aligned}
E_{mn}(x, y, z) \;=\; & \mathcal{E}_0 \frac{w_0}{w(z)} \mathcal{H}_m(\sqrt{2}\,x/w(z)) \mathcal{H}_n(\sqrt{2}\,y/w(z)) \\
& \times \exp\{-[(x^2 + y^2)/w(z)]^2\} \, \exp[ik(x^2 + y^2)/2R(z)] \quad (5.22) \\
& \times \exp\{-i[kz - (m + n + 1)\eta(z)]\}.
\end{aligned}
$$

In the middle line the geometric form of the Gaussian general solution is given, which is characterized by $(R(z), w(z))$ and (z_0, w_0), respectively. Higher modes cause a transverse modulation $\mathcal{H}_{m,n}$ of this basic form (upper line). Along the z axis the phase is solely determined by the Gouy phase, the last line in Eq. (5.22). That is why we can at first concentrate on the geometric adjustment of the wavefronts which are described by $R(z)$ according to eq. (2.22).

---

[2]Unstable resonators are also used, e.g. for the construction of high-power lasers [95].

At $z_{1,2}$ in Fig. 5.17 the radii of the wavefronts have to match the radii of curvature of the mirrors exactly, so

$$R_{1,2} = \frac{1}{z_{1,2}}(z_{1,2}^2 + z_0^2) = z_{1,2} + \frac{z_0^2}{z_{1,2}}.$$

By means of

$$z_{1,2} = \tfrac{1}{2}R_{1,2} \pm \sqrt{R_{1,2}^2 - 4z_0^2},$$

we can then express the parameters of the Gaussian wave (z_0, w_0) by the cavity parameters (R_1, R_2, ℓ),

$$\begin{aligned}
z_0^2 &= \frac{-\ell(R_1 + \ell)(R_2 - \ell)(R_2 - R_1 - \ell)}{(R_2 - R_1 - 2\ell)^2}, \\
w_0^2 &= \frac{\lambda z_0}{n\pi}.
\end{aligned} \tag{5.23}$$

Exploration of this formula has to take into account that, according to the conventions for ABCD matrices (p. 19), mirror surfaces with their centre to the left and right of the surface, respectively, have different signs.

For the excitation of a cavity mode, the Gaussian beam parameters (z_0, w_0) have to be precisely tuned to the incident wave. If this, the *mode matching* condition, is not met, only that share of the field is coupled in which corresponds to the overlap with the resonator mode.

5.6.3   Resonance frequencies of optical cavities

A resonator is characterized by the spectrum of its resonance frequencies. From the Fabry–Perot resonator, we expect an equidistant pattern of transmission lines at the distance of the free spectral range Δ_{FSR}. For a more exact analysis, we have to take into account the phase factor (the Gouy phase, last line of Eqs. (5.22) and (2.23)), respectively. The phase difference must again be an integer multiple of π,

$$\Phi_{mn}(z_1) - \Phi_{mn}(z_2) = q\pi k(z_1 - z_2) - (m + n + 1)[\eta(z_1) - \eta(z_2)]. \tag{5.24}$$

With $\ell = z_1 - z_2$ and $\eta(z) = \tan^{-1}(z/z_0)$ we at first find

$$k_{qmn}\ell = q\pi + (m + n + 1)\left[\tan^{-1}\left(\frac{z_1}{z_0}\right) - \tan^{-1}\left(\frac{z_2}{z_0}\right)\right].$$

The resonance frequencies ν_{qmn} are determined from $k_{qmn}\ell = 2\pi n\nu_{qmn}\ell/c = \pi\nu_{qmn}/\Delta_{\mathrm{FSR}}$. We introduce the resonator Gouy frequency shift

$$\Delta_{\mathrm{Gouy}} = \left[\tan^{-1}\left(\frac{z_1}{z_0}\right) - \tan^{-1}\left(\frac{z_2}{z_0}\right)\right]\frac{\Delta_{\mathrm{FSR}}}{\pi}, \tag{5.25}$$

which varies between 0 and Δ_{FSR}. We obtain the transparent result

$$\nu_{qmn} = q\Delta_{\mathrm{FSR}} + (m + n + 1)\Delta_{\mathrm{Gouy}}.$$

It shows a mode spectrum with a rough division into the free spectral range Δ_{FSR}. The fine structure is determined by resonance lines at the distance Δ_{Gouy}.

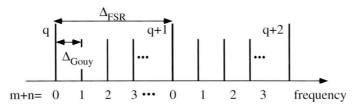

Fig. 5.18: *Mode or frequency spectrum of a Fabry–Perot resonator.*

5.6.4   Symmetric optical cavities

We are now going to investigate the special case of a symmetric optical cavity consisting of two identical mirrors, $R_2 = R = -R_1$. In this important special case the form of eq. (5.23) is strongly simplified and can be interpreted as

$$z_0^2 = \frac{(2R - \ell)\ell}{4} \qquad \text{and} \qquad w_0^2 = \frac{\lambda}{2\pi n}\sqrt{(2R - \ell)\ell}. \tag{5.26}$$

The length of the symmetric cavity can be varied from $\ell = 0$ to $2R$ before the region of stability is left.

The parameters of the Gaussian wave in a symmetric optical cavity, (z_0, w_0), are shown in Fig. 5.19, normalized to the maximum values $z_{0\text{max}} = R/2$ and $w_{0\text{max}} = (\lambda R/4\pi n)^{1/2}$. The instability of the plane–plane and the concentric cavity is here also expressed by the sensitive dependence of the mode parameters on the ℓ/R ratio.

In the symmetric cavity the Gouy phase (5.25) depends on the length and the radius of curvature according to

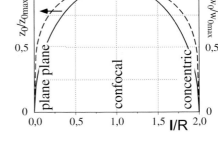

Fig. 5.19: *Rayleigh length and beam waist for a symmetric optical cavity.*

$$\Delta_{\text{Gouy}} = \Delta_{\text{FSR}} \frac{2}{\pi} \tan^{-1}\left(\frac{\ell}{2R - \ell}\right)^{1/2}.$$

5.6.5   Optical cavities: important special cases

The three special cases $\ell/R = 0, 1, 2$ deserve particular attention because they exactly correspond to the plane parallel, confocal and concentric cavities.

Plane parallel cavity: $\ell/R = 0$

The Fabry–Perot interferometer or etalon described in the previous sections exactly corresponds to this extreme case. As we know from Fig. 1.21, it is an extreme case in terms of stability. In practical use it is also important that polished flat surfaces always have a slight convex curvature for technical reasons, so that an FPE consisting of two plane air-spaced mirrors always tends to instability.

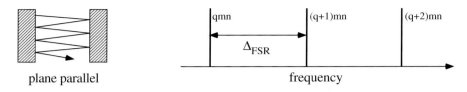

plane parallel   frequency

Fig. 5.20: *Path of rays and resonance frequencies of the plane parallel cavity.*

Confocal cavity: $\ell/R = 1$

If the focal length of the cavity mirrors coincide($f_1 + f_2 = R_1/2 + R_2/2 = \ell$), the configuration of the confocal cavity is obtained. In the symmetric case we have $R_1 = R_2 = \ell$.

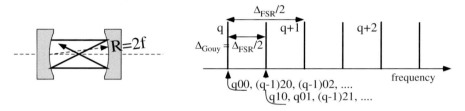

Fig. 5.21: *Path of rays and resonance frequencies of the confocal cavity.*

In this case, the modes are arranged at two highly degenerate frequency positions at a separation of

$$\Delta_{\text{FSR}}^{\text{confoc}} = c/4n\ell. \tag{5.27}$$

The high degeneracy has its ray optical analogue in the fact that paraxial trajectories are closed after two revolutions.

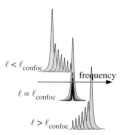

Fig. 5.22: *When confocal cavities are slightly displaced from their perfect mirror separation at $\ell/R = \ell_{\text{confoc}}/R = 1$, the degeneracy of transverse modes is lifted.*

If the confocal cavity is irradiated by a laser beam without mode matching, many transverse modes are excited and the frequency separation $c/4n\ell$ (Eq. (5.27)) can be observed as an effective free spectral range and not as $c/2nl$. It is instructive to observe the emergence of transverse modes to the left or right of the fundamental modes if the length of the confocal cavity is slightly displaced from the perfect position $\ell/R = 1$, as indicated in Fig. 5.22.

The high degeneracy makes the confocal cavity particularly insensitive in terms of handling and convenient for practical spectral analysis (see Section 7.3.2). In general, a larger linewidth will be observed than is to be expected according to the simple relation of Eq. (5.19).

This broadening is caused by the higher modes which suffer from stronger damping and show exact degeneracy only within the paraxial approximation.

Concentric cavity: $\ell/R = 2$

Obviously, this cavity is very sensitively dependent on the exact positions of the mirrors, but it leads to a very sharp focusing, which reaches the diffraction limit. In laser resonators, nearly concentric components are used to concentrate the pump laser as well as the laser beam into a small volume where large amplification density is realized.

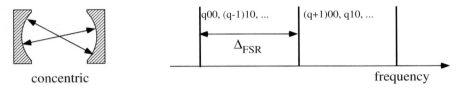

concentric   frequency

Fig. 5.23: *Path of ray and resonance frequencies of the concentric cavity.*

Excursion: Microcavities
In recent times there has been great interest in miniaturized devices of optical cavities with dimensions of few μm. Since the radiation field is stored in a very small volume, a strong coupling of radiation field and matter can be obtained there.

The external coupling is not simple in such cavities since the direction of the emission is not simply controllable. In this context, there have recently been investigations on oval[3] cavities [80], which help to solve this problem by their shape.

In Fig. 5.24 the calculated intensity distributions for a cylindrically symmetric, an elliptical and an oval cavity are shown. The connection with concepts borrowed from ray optics can be seen in particular for the oval cavity.

  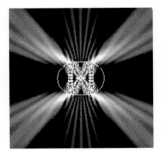

Fig. 5.24: *Distribution of light in circular, elliptical and oval microcavities. By courtesy of Dr J. Noeckel [80].*

---

[3]These are not elliptical cavities, which can be treated analytically and show a discrete spectrum.

5.7   Thin optical films

Thin optical films play an important role for applications, since dielectric coatings to reduce or enhance optical reflections have found their way into everyday life. We shall limit ourselves to the interference phenomena associated with thin optical films, and we shall ignore almost completely the important aspects of material science for their manufacture.

Metallic mirrors cause losses of 2–10% when reflecting visible wavelengths. That is more than many laser systems can tolerate just to overcome the threshold. With a wealth of transparent materials, dielectric film systems with a structured refractive index can be manufactured making predictable reflectivities between 0 and 100% possible. For the highest reflectivity, both transmission and absorption are specified, which in the best case are only a few ppm!

5.7.1   Single-layer films

We consider the single film from Fig. 5.25 and determine the reflected wave that is the result of the superposition of reflections from the first and second interfaces, i.e. $E_r = r_1 E_i + t_1 r_2 E_i$. It is straightforward to check using the known formulae for the reflection coefficient from Section 3.1.1 that, for perpendicular incidence, the amplitude of the reflected wave obeys

$$
\begin{aligned}
E_r &= \left(\frac{1-n_1}{1+n_1} + \frac{4n_1}{(1+n_1)^2} \frac{n_1-n_2}{n_1+n_2} e^{i2kn_1 d} e^{i\Phi} \right) E_i \\
&\simeq \left(\frac{1-n_1}{1+n_1} + \frac{n_1-n_2}{n_1+n_2} e^{i2kn_1 d} e^{i\Phi} \right) E_i.
\end{aligned}
\tag{5.28}
$$

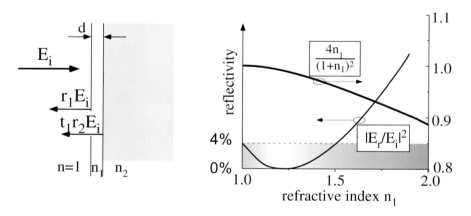

Fig. 5.25: *Reflection at a single thin film. Left: film system. Right: factor $4n_1/(1+n_1)^2$ and the effect of a single film on glass with $n = 1.5$ and optimal film thickness $d = \lambda/4n_1$.*

For reflection from the denser medium at the second interface, with $n_1 < n_2$, we have $\Phi = \pi$, and for $n_1 > n_2$ (less dense medium), $\Phi = 0$. The simplification in the second line of Eq. 5.28 is made possible by the negligible deviation of the transmission factor $4n_1/(1 + n_1)^2$ from unity in the technically important range between $n = 1.3$ and $n = 2$ (see Fig. 5.25). Even with a single-layer thin dielectric film, good results in terms of the coating of optical glasses can be obtained. For technically advanced applications, though, systems consisting of many layers are necessary.

Minimal reflection: AR coating, AR layer, $\lambda/4$ film

The thin film is designed as a single-layer $\lambda/4$ film with $d = \lambda/4$. In addition, we choose $n_1 < n_2$, so that we have $\exp(\Phi=0) = 1$ because of the reflection at the denser medium, and $\exp(2ikd) = -1$ causing destructive interference of the partial waves. In comparison with the substrate, the film shows low refraction and hence is called an *L-film*. For perfect suppression of optical reflection, the condition

$$\frac{1 - n_1}{1 + n_1} = \frac{n_1 - n_2}{n_1 + n_2}$$

has to be met, which is equivalent to

$$n_1 = \sqrt{n_2}. \tag{5.29}$$

The simple anti-reflection (AR) films used for 'coating' of spectacles and windows reduce the reflection of the glass from 4% to typically 0.1–0.5%. A commonly used material is MgF_2, which quite closely fulfils condition (5.29) when used on glass ($n = 1.45$).

Reflection: highly reflective films

In this case we first choose a highly refractive film or *H-film* on a substrate with a lower refractive index, i.e. $n_1 > n_2$. The 180° or π phase jump during the reflection at the less dense medium now causes constructive interference of the two partial waves, and the total reflectivity is enhanced. A single TiO_2 $\lambda/4$ film on glass, for example (see refractive indices in Tab. 5.3), increases the reflectivity from 4% to more than 30% (see Fig. 5.27).

Tab. 5.3: *Refractive index of materials for thin dielectric films.*

MgF_2	SiO	TaO_2	TiO_2
1.38	1.47	2.05	2.30

5.7.2   Multiple-layer films

As a model example of a multiple-layer film, we are going to study a periodic film stack consisting of N identical elements [59]. We have to consider the splitting of the waves at each interface (Fig. 5.26):

$$E_j^+ = t_{ij}E_i^+ + r_{ji}E_j^-,$$
$$E_i^- = t_{ji}E_j^- + r_{ij}E_i^+.$$

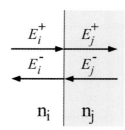

Fig. 5.26: *Interface within a multiple-layer optical film.*

To be specific, we take a wave incident from the left. However, we elaborate the transformation in the opposite direction, since there is no wave travelling to the left behind the last interface in the direction of propagation $(E_N^- = 0)$.

The set of equations can be solved and conveniently represented in a matrix if we also use $r_{ij} = -r_{ji}$ and $|t_{ij}t_{ji}| + |r_{ij}r_{ji}| = 1$. Thus

$$\mathbf{E}_i = \mathbf{G}_{ji}\mathbf{E}_j = \frac{1}{t_{ij}} \begin{pmatrix} 1 & r_{ij} \\ r_{ij} & 1 \end{pmatrix} \begin{pmatrix} E_j^+ \\ E_j^- \end{pmatrix}.$$

Before getting to the next interface, the wave undergoes a phase shift $\varphi = \pm n_j k d$ for the wave running to the right and to the left, respectively. In this case the total transformation from one interface to the other is

$$\mathbf{E}_j = \mathbf{\Phi}_{ji}\mathbf{G}_{ji}\mathbf{E}_i = \mathbf{S}_{ji}\mathbf{E}_i \qquad \text{with} \qquad \mathbf{\Phi}_{ji} = \begin{pmatrix} e^{-i\varphi} & 0 \\ 0 & e^{i\varphi} \end{pmatrix},$$

and in particular for N interfaces

$$\mathbf{E}_1 = \mathbf{S}_{1,2}\mathbf{S}_{2,3}\cdots\mathbf{S}_{N-2,N-1}\mathbf{S}_{N-1,N} \begin{pmatrix} E_N^+ \\ 0 \end{pmatrix} = \begin{pmatrix} R_{11} & R_{12} \\ R_{21} & R_{22} \end{pmatrix} \begin{pmatrix} E_N^+ \\ 0 \end{pmatrix}.$$

Thus the relation between incident, reflected and transmitted waves is uniquely determined. In particular, the reflectivity can be calculated from $|R_{11}|^2/|R_{21}|^2$ if $\mathbf{R}$ is known. While an analytical solution remains laborious, numerical solution by computer is straightforward. In Fig. 5.27 the evolution of reflectivity from a single-layer pair film to a highly reflective multiple-layer film is shown.

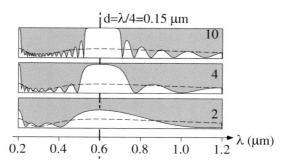

Fig. 5.27: *Wavelength-dependent reflectivity of multiple films with two, four and 10 layer pairs of films. In this example a stack of TiO_2 and glass layers each with a thickness of $0.15\,\mu m$ is assumed. The dashed line marks the reflectivity of a single film. The 10-stack has a reflectivity $R > 99\%$ between 0.55 and $0.65\,\mu m$.*

5.8 Holography

One of the most remarkable and attractive capabilities of optics is image formation, to which we have already dedicated an entire chapter (Chapter 4). Among the various methods, usually holography (from the Greek *holo*, meaning 'complete' or 'intact') arouses the greatest astonishment. The attraction is mostly caused by the completely three-dimensional reconstruction of a recorded object! Here, we shall restrict ourselves to the interferometric principles of holography, and refer the reader to the specialist literature for more intensive studies [41].

5.8.1 Holographic recording

For a conventional record of a picture, whether by using an old-fashioned film or a modern charge-coupled device (CCD) camera, always the spatial distribution of the light intensity is saved on the film or in digital memory. For a hologram, both the amplitude and the phase of the light field are recorded instead by superimposing the light field scattered off the object, the *signal wave* of amplitude distribution

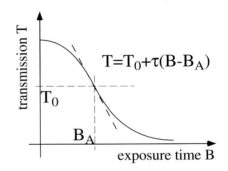

$$E_S(x,y) = \tfrac{1}{2}[\mathcal{E}_S(x,y)\,e^{-i\omega t} + \text{c.c.}],$$

with a coherent *reference wave*

Fig. 5.28: *Holography uses the linear part of the blackening of the film.*

$$E_R(x,y) = \tfrac{1}{2}[\mathcal{E}_R(x,y)\,e^{-i\omega t} + \text{c.c.}].$$

One thus produces an interferometric record of an object – information about the image is truly contained in the interference pattern! The intensity distribution recording this information is generated by the superposition of signal and reference waves:

$$2I(x,y)/c^2\epsilon_0 = |E_S + E_R|^2 = |\mathcal{E}_S|^2 + |\mathcal{E}_R|^2 + \mathcal{E}_S\mathcal{E}_R^* + \mathcal{E}_S^*\mathcal{E}_R. \tag{5.30}$$

For this we have already assumed that the signal and reference waves have a sufficiently well defined phase relation, since they originate from the same coherent light source. Otherwise the mixed terms would suffer from prohibitive temporal fluctuations.

The illumination intensity on the film material – which usually has non-linear properties, see Fig, 5.28 – is adjusted such that a linear relation between the transmission and the intensity distribution is obtained, i.e.

$$T(x,y) = T_0 + \tau I(x,y). \tag{5.31}$$

Today, since lasers with a large coherence length are readily available, the holographic record is typically taken according to the *off-axis* method of Leith–Upatnieks shown in Fig. 5.29.

Historical experiments in the 1940s by D. Gábor (1900–1979, Nobel prize-winner in 1971), though, were obtained as *in-line* holograms, since there the requirements for

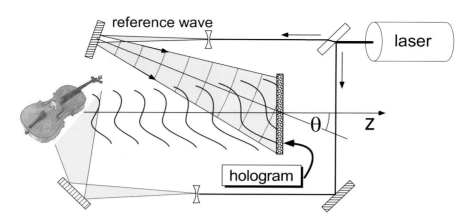

Fig. 5.29: *Record of a hologram according to the Leith–Upatnieks method.*

the coherence of the light source are not so stringent. The monochromatic signal wave propagates in the z direction in the arrangement of Fig. 5.29, and the transverse phase distribution is caused by the illuminated object,

$$E_{\mathrm{S}} = \mathcal{E}_{\mathrm{S}}\, e^{-i\omega t}\, e^{ikz}\, e^{i\phi(x,y)}.$$

The (almost) plane reference wave has identical frequency ω and travels at an angle θ towards the z axis. The wavevector k has components $k_z = k\cos\theta$ and $k_y = k\sin\theta$, and thus

$$E_{\mathrm{R}} = \mathcal{E}_{\mathrm{R}}\, e^{-i\omega t}\, e^{ik_z z}\, e^{ik_y y}.$$

Following Eq. (5.30) at plane P with $\phi_0 = k_z z_0$, we obtain the intensity distribution

$$I_{\mathrm{P}}(x,y) = I_{\mathrm{S}} + I_{\mathrm{R}} + \mathcal{E}_{\mathrm{S}}\mathcal{E}_{\mathrm{R}}^*\, e^{i\phi_0}\, e^{-i[k_y y + \phi(x,y)]} + \text{c.c.} \tag{5.32}$$

All contributions cause a blackening of the film material. The reference wave usually corresponds in good approximation to a plane wave, and so generates homogeneous blackening. The blackening caused by the signal wave, which for simplification has been assumed to have a constant amplitude, usually generates an inhomogeneous intensity distribution since there are no plane wavefronts emanating from an irregular object. In a different situation, this phenomenon is also known as laser speckle and is discussed in more detail in section 5.9.

5.8.2   Holographic reconstruction

The major fascination of holography is manifest in the actual image reconstruction process, since the holographic film itself – the hologram – does not contain any information for the human eye. For reconstruction, the object is removed and the hologram is illuminated once again with the reference wave. By diffraction, the secondary waves shown in Fig. 5.30 are generated.

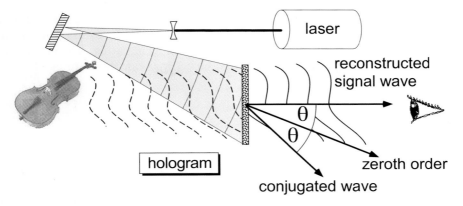

Fig. 5.30: *Image reconstruction from a hologram with secondary waves (see also Fig. 5.29).*

Formally we can derive the secondary wave by considering the field distribution immediately after passing through the hologram. We can recognize four different diffracted waves, U_0, U_0^{H}, U_{+1} and U_{-1},

$$
\begin{aligned}
E_{\mathrm{recon}} &= T(x,y)E_{\mathrm{R}} \\
&= T_0 E_{\mathrm{R}} + \tau E_{\mathrm{R}} I_{\mathrm{R}} + \tau E_{\mathrm{R}} I_{\mathrm{S}} + \tau |E_{\mathrm{R}}|^2 E_{\mathrm{S}} + \tau E_{\mathrm{R}}^2 E_{\mathrm{S}}^* \\
&= U_0(x,y) + U_0^{\mathrm{H}}(x,y) + U_{+1}(x,y) + U_{-1}(x,y),
\end{aligned}
$$

which we are going to consider in detail. Actually, it is quite complicated to determine the diffraction field of a complex hologram. Fortunately, we can identify every term with a known waveform naturally continued from the local field distribution.

Zeroth order

$$
U_0(x,y) = (T_0 + \tau I_{\mathrm{R}})\mathcal{E}_{\mathrm{R}}\, e^{-i\omega t}\, e^{i(k_y y + k_z z)}
$$

This term propagates in zeroth order because its wavevector is identical to that of the incident reference wave, which is continued and plainly multiplied by a constant factor $(T_0 + \tau I_{\mathrm{R}}) < 1$ due to attenuation.

Halo

$$
U_0^{\mathrm{H}}(x,y) = \tau I_{\mathrm{S}}\mathcal{E}_{\mathrm{R}}(x,y)\, e^{-i\omega t}\, e^{i(k_y y + k_z z)}
$$

As mentioned above, the signal wave usually causes inhomogeneous blackening. The secondary wave also propagates in zeroth order, but the diffraction of the speckle pattern leads to broadening compared to the transmitted reference wave and is sometimes called a 'halo'.

Reconstructed signal wave

$$
U_{+1}(x,y) = \tau \mathcal{E}_{\mathrm{S}}\, e^{i\phi(x,y)}\mathcal{E}_{\mathrm{R}}\mathcal{E}_{\mathrm{R}}^*\, e^{-i\omega t}\, e^{ikz}
$$

Obviously, with this contribution, the signal wave is exactly reconstructed except for a constant factor! The reconstructed signal wave propagates in the z direction, which we are going to call the first order in analogy to diffraction by a grating. The virtual image contains all information of the reconstructed object and can therefore be observed – within the light cone – from all sides.

Conjugated wave

$$U_{-1}(x,y) = \tau \mathcal{E}_R^2 \mathcal{E}_S \, e^{-i\omega t} \, e^{-i\phi(x,y)} \, e^{i[2k_y y + (2k_z - k)z]}$$

In a vector diagram we can determine the propagation direction of the so-called conjugated wave. For small angles $\theta = k_y/k_z$ we have $2k_z - k \simeq k_z$ and $k_{\text{conj}}^2 = 4k_y^2 + (2k_z - k)^2 \simeq k^2$. That is why the axis of the conjugated ray runs at angle 2θ to the z axis and disappears at $\theta = \pi/4$ at the latest. Writing it as

$$U_{-1}(x,y) = \tau \mathcal{E}_R^2 (\mathcal{E}_S \, e^{i\phi(x,y)})^* \, e^{-i\{\omega t - k[\sin(2\theta)y + \cos(2\theta)z]\}},$$

the 'phase-conjugated' form of this ray in comparison to the object wave becomes transparent. From a physical point of view, the curvature of the wavefronts is inverted, so the wave seems to run backwards in time. Again, following the analogy to diffraction by a grating, this wave is also called the minus-one order of diffraction.

Compared to an in-line hologram, the three secondary waves of interest can be easily separated geometrically and observed in off-axis holography (Fig. 5.30).

5.8.3   Properties

Holograms have many fascinating properties out of which we have selected only a few here.

Three-dimensional reconstruction

Since the signal wave coming from the object is reconstructed, the virtual image looked at by the observer through the holographic plate appears three-dimensional as well. It is even possible to look behind edges and corners if there exists a line of sight connection with the illuminated areas.

Partial reconstruction

The complete object can be reconstructed from each fragment of a hologram. This seems to be inconsistent at first, but becomes clear in direct analogy to diffraction by a grating. There, the diffraction pattern observed from more and more reduced fragments always stay the same as well. However, the width of each diffraction increases, i.e. the resolution of the grating is reduced due to the decreasing number of illuminated slits. In a similar way the resolution declines in reconstructing from a holographic fragment. The finer structures of the image disappear, while the gross shape of the signal wave and hence the object is preserved.

Magnification

If, in reconstructing an object, light of a different wavelength is used, the scale of the image is correspondingly changed.

5.9   Laser speckle (laser granulation)

When a dim wall or a rough object is illuminated with laser light, the observer distinguishes a granular, speckled structure, which does not appear in illumination with a conventional light source and is obviously caused by the coherence properties of the laser light. In fact, coherent phenomena, i.e. diffraction and interference, can also be observed using incandescent light sources, but the invention of the laser has really granted us a completely new sensory experience. Newton had already recognized that the 'twinkling' of the stars, having been poetically raised by our ancestors, is a coherence phenomenon caused by the inhomogeneities of the atmosphere, and thus directly related with speckle patterns.

The granular irregular structure is called 'laser granulation' or *speckle pattern*. Reflected off the rough, randomly shaped surface of a large object, a coherent wave acquires a complex wavefront like after passing through a ground glass screen. For simplification, we can imagine that the light beams from a large number of accidentally arranged slits or holes interfere with each other. In each plane there is thus a different statistical interference pattern. Indeed, every observer sees a different but spatially stable pattern as well.

Formal treatment of the speckle pattern requires some expense using the mathematical methods of statistics. We briefly discuss this phenomenon, since it is nearly ubiquitous wherever laser light is used. Although laser granulation at first appears an undesirable consequence of interference, it contains substantial information about the scattering surfaces, and it is even suitable for interferometric application in the measurement techniques for the determination of tiny surface changes [46].

5.9.1   Real and virtual speckle patterns

Speckle patterns can be observed, for example, when we expand a laser beam and project it from a diffuse reflector onto a screen. On the wall there is a fixed granular pattern that only changes with a different reflector. This pattern is only determined by the microstructure of the reflector and is called a *real or objective speckle pattern* [62]. It can be recorded by direct exposure of a film.

When it is imaged, however, it is transformed by the imaging process itself. A *subjective or virtual speckle pattern* is generated, with properties determined by the aperture of the imaging optics, e.g. the size of the pupil of our eye. This property can be easily understood just with a laser pointer illuminating a white wall. If we form a small hole, some kind of artificial pupil, with our hand, the granulation speckles grow rougher the smaller the diameter of the hole.

Detailed consideration of the coherent wave field is not usually of interest. Here we shall limit ourselves to sensible physical questions, concerning the intensity distribution in the statistical wave fields leading to laser granulation and the characteristic dimensions of the speckle grains.

5.9.2   Speckle grain sizes

The sizes of speckle grains can be estimated by a simple consideration [62]. The lens of an imaging objective is illuminated by a wave field of a granulation pattern with a time-invariable but spatially random phase distribution. The characteristic scale d of the interference pattern is determined by the resolution of the image and reaches the Rayleigh criterion as in Eq. (2.48). If wavefronts from a large distance are incident on the lens, the beams coming from a certain direction are superimposed in the focal plane at a distance f. For a circular lens with aperture D, the diameter of the focal spots cannot become smaller than

$$d = 1.22\lambda f/D.$$

With a decrease in the aperture size, a roughening of the speckle pattern is to be expected. This phenomenon is shown in Fig. 5.31, where an effective aperture is formed by focusing of the laser beam onto a ground glass substrate.

Fig. 5.31: *Speckle pattern of a focused helium–neon laser beam after passing through a ground glass substrate, showing the statistical pattern. From left to right, the focus was shifted more and more into the substrate. Stronger focusing leads to coarser interference structures.*

6 Light and matter

An electromagnetic wave accelerates electrically charged particles in gases, liquids and solids, and in so doing generates polarizations and currents. The accelerated charges for their part again generate a radiation field superimposed onto the incident field. To understand macroscopic optical properties, it is necessary to describe the polarization properties of matter microscopically, which can only be done by means of quantum theory. Despite that, classical theoretical physics has been able to explain numerous optical phenomena by phenomenological approaches.

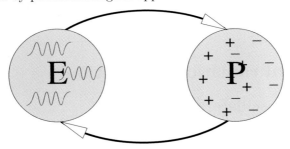

Fig. 6.1: *An electromagnetic field E generates a polarization P in matter consisting of positive and negative charges. The accelerated charges generate a radiation field and so react upon the fields.*

The quantum theoretical description of matter has led to the development of 'quantum electronics' (see Tab. 6.1), in which the electromagnetic radiation fields are still taken into account classically, i.e. with a well-defined phase and amplitude. This kind of treatment of the radiative interaction is also called 'semiclassical'.

Tab. 6.1: *Treatment of light and matter by theoretical physics*.*

	Matter	Light	Atomic motion
Classical optics	C	C	C
Quantum electronics	Q	C	C
Quantum optics	Q	Q	C
Matter waves	Q	Q	Q

*C = classical physics; Q = quantum theory.

Ultimately also electromagnetic fields have to be dealt with in a quantum theoretical way when phenomena such as the famous 'Lamb shift' are to be understood. Today

Optics, Light and Laser. Dieter Meschede
Copyright © 2004 Wiley-VCH Verlag GmbH & Co. KGaA
ISBN: 3-527-40364-7

'quantum electrodynamics' (QED) is considered a model case of a modern physical field theory. In 'quantum optics' in a narrower sense,[1] in particular the quantum properties of optical radiation fields are dealt with [66, 84], e.g. the spectrum of resonance fluorescence or so-called photon correlations. However, these topics go beyond the scope chosen here.

Since the beginning of the 1980s it has been possible to influence the motion of atoms by radiation pressure of light, or laser cooling. The kinetic energy in a gas cooled in such a way can be decreased so much that atomic motion can no longer be comprehended like that of classical, or point-like, particles. Instead, their centre-of-mass motion has to be dealt with according to quantum theory and can be interpreted in terms of matter waves. In the excursion on p. 135 we have already used this explanation for the diffraction of atomic beams. The hierarchy of theoretical concepts for light–matter interaction is summarized in Tab. 6.1.

When the effect of a light field on dielectric samples is to be described, generally the electric dipole interaction is sufficient since it is stronger than all other couplings, such as magnetic effects and higher-order terms, which can be neglected. The concepts of optics can also be extended without any problems if such phenomena are to be treated theoretically.

6.1   Classical radiation interaction

6.1.1   Lorentz oscillators

A simple yet very successful model for the interaction of electromagnetic radiation with polarizable matter goes back to H. Lorentz (1853–1928). In this model, electrons are considered that are harmonically bound to an ionic core like little planets with a spring and oscillating at optical frequencies ω_0. The classical dynamics of such a system is well known. The influence of a light field shows up as driving electrical or magnetic forces adding to the binding force $\mathbf{F}_B = -m\omega_0^2\mathbf{x}$.

Additionally, we assume that damping of the oscillator is caused by release of radiation energy. Although this concept cannot be fully explained by classical electrodynamics without some contradictions, in an approximation it leads to the *Abraham–Lorentz equation*, in which, besides the binding force, a damping force $\mathbf{F}_R = -m\gamma(d\mathbf{x}/dt)$ occurs causing weak damping ($\gamma \ll \omega_0$). At this stage the limits of classical electrodynamics become evident [83], because a consistent and correct calculation of γ can only be obtained by means of quantum electrodynamics [108]. For our purposes, however, it is sufficient to consider γ as the phenomenological damping rate.

For simplification we use complex quantities to write the orbit radius, $\mathbf{x} \rightarrow r = x + iy$. We consider the equation of motion of the driven oscillator,

$$\ddot{r} + \gamma\dot{r} + \omega_0^2 r = \frac{q}{m}\mathcal{E}\,e^{-i\omega t}, \qquad (6.1)$$

---

[1]The term 'quantum optics' is in general not very precisely defined.

under the influence of a driving light field $\mathcal{E}\,e^{-i\omega t}$, which is circularly polarized. With the trial function $r(t) = \rho(t)\,e^{-i\omega t}$, the equilibrium solution $\rho(t) = \rho_0 = \text{const.}$ and

$$\rho_0 = \frac{q\mathcal{E}/m}{(\omega_0^2 - \omega^2) - i\omega\gamma}$$

can be found from the secular equation $\rho(-\omega^2 - i\omega\gamma + \omega_0^2) = q\mathcal{E}/m$.

For the near-resonant approximation, we can replace $(\omega_0^2 - \omega^2) \simeq 2\omega_0(\omega_0 - \omega) = -2\omega_0\delta$ with detuning δ, introduce the maximum radius $\rho_{\max} = -q\mathcal{E}/m\omega_0\gamma$ and obtain

$$\rho_0 = \rho_{\max}\frac{\gamma/2}{\delta + i\gamma/2}.$$

For the x and y coordinates of the driven oscillator, we have

$$r(t) = x + iy = \rho_{\max}\frac{\gamma}{2}\frac{\delta - i\gamma/2}{\delta^2 + (\gamma/2)^2}\,e^{-i\omega t}. \tag{6.2}$$

We will see that, in terms of the propagation of light in polarizable matter, x and y give exactly the 'dispersive' (x) and the 'absorptive' (y) components of the radiation interaction. The shape of the *dispersion curve* and the *Lorentz profile* of absorption are presented in Fig. 6.2. Here the term 'normal dispersion' refers to the dominant positive slopes of the dispersion curve. This situation is typically found for transparent optical materials, which have electronic resonance frequencies beyond the visible domain in the UV. Negative slopes of dispersion are called 'anomalous dispersion'.

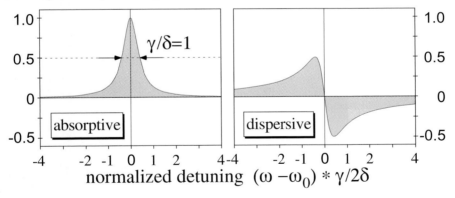

Fig. 6.2: *'Quadrature' components of the Lorentz oscillator, which are oscillating in phase (x, absorptive) and 90° out of phase (y, dispersive) with the driving field, respectively. The swing is normalized to the maximum value in the case of resonance at $\delta = 0$.*

It is known that an accelerated charge radiates, and so a charged harmonic oscillator has to lose energy. The damping thus caused is called 'radiation reaction' and has already been accounted for phenomenologically in Eq. (6.1). From a shortened version of the derivation leading to the well-known Abraham–Lorentz equation [83], we can infer a damping rate depending on elementary atomic quantities only, which provides further insight into radiative properties. We have to keep in mind, however, that a suitable theory of damping must invoke the full quantum theory of light.

By multiplication with $\dot{r}$ we can introduce the rate of energy change into eq. (6.1),

$$\frac{d}{dt}\left(\frac{m\dot{r}^2}{2} + \frac{m\omega_0^2 r^2}{2}\right) + m\gamma\dot{r}^2 = 0 = P_{\text{rad}} + m\gamma\dot{r}^2.$$

If damping is weak ($\omega_0 \gg \gamma$), we can assume that, during one revolution period $2\pi/\omega_0$, the amplitude change (r) is negligible, and so we can replace $\ddot{r} = \omega_0^2 r$ and $\dot{r} = \omega_0 r$. Then we may identify the radiation power with the power dissipated through friction ($m\gamma\dot{r}^2$), which was introduced phenomenologically before. We obtain

$$\gamma = \frac{q^2\omega_0^2}{6\pi\epsilon_0 c^3 m} \qquad \text{and} \qquad \rho_{\text{max}} = \frac{3\epsilon_0\lambda^3}{4\pi^2 q}\mathcal{E}. \tag{6.3}$$

This result is frequently used to introduce the so-called *classical electron radius* [31, 83],

$$r_{\text{el}} = \frac{e^2/4\pi\epsilon_0}{2mc^2} = 1.41 \times 10^{-15}\,\text{m} \qquad \text{with} \qquad \gamma = \frac{4}{3}\frac{r_{\text{el}} c}{\lambda^2}.$$

As far as we know from scattering experiments in high-energy physics, the electron is point-like down to 10^{-18} m, and thus this quantity does not have physical significance.

In this way we can obtain the complex dipole moment of a single particle from $d = q\rho_0$ according to Eqs. (6.2) and (6.3),

$$d(t) = q\rho_0 = -\frac{3\lambda^3}{4\pi^2}\frac{i - 2\delta/\gamma}{1 + (2\delta/\gamma)^2}\epsilon_0\mathcal{E}\,e^{-i\omega t}. \tag{6.4}$$

Often the polarizability α is used as well. It is defined by

$$d(t) = \alpha\mathcal{E}\,e^{-i\omega t},$$

and the coefficient α is easily extracted from comparison with Eq. (6.4).

In the x as well as in the y component, there is a phase delay ϕ between the electric field and the dipole moment, which is only dependent on the damping rate γ and the detuning $\delta = \omega - \omega_0$ (Fig. 6.3),

$$\phi = \arctan(\gamma/2\delta). \tag{6.5}$$

The so-called phase lag shows the known behaviour of a driven harmonic oscillator, i.e. in-phase excitation at low ('red') frequencies, out of phase or 90° following in the case of resonance, and opposite phase at high ('blue') frequencies.

From Eq. (6.1) we can furthermore infer the time-dependent equation for $\rho(t)$. We assume that the oscillation amplitude $\rho(t)$ changes only slowly in comparison with the oscillation itself, i.e. $\ddot{\rho} \ll \omega\dot{\rho}$, etc. We then approximately obtain

$$\dot{\rho} + \left(i\delta + \frac{\gamma}{2}\right)\rho = -i\frac{q\mathcal{E}}{2m\omega}, \tag{6.6}$$

by furthermore applying $i\omega + \gamma/2 \simeq i\omega$ as well. This complex equation provides an interesting analogy with the result of quantum mechanics discussed on p. 181. There we will find Bloch vector components exhibiting strong formal similarity with dipole quadrature components (u, v), which we introduce here by letting $\rho = u + iv$.

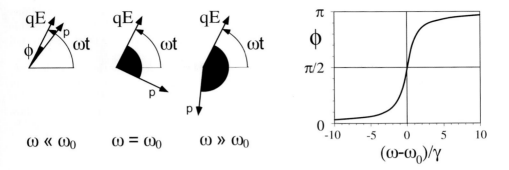

Fig. 6.3: *Phase lag of the Lorentz oscillator in steady state. At low frequencies, the driving field and the dipole oscillate in phase; in the case of resonance, the dipole follows the field out of phase at 90°; and at high frequencies, it oscillates with opposite phase.*

Decomposition of the complex equation of motion (6.6) into a system of real equations yields

$$
\begin{aligned}
\dot{u} &= \delta v - \frac{\gamma}{2} u, \\
\dot{v} &= -\delta u - \frac{\gamma}{2} v - \frac{q\mathcal{E}}{2m\omega}.
\end{aligned}
\tag{6.7}
$$

We can furthermore complement this equation by

$$
\frac{d}{dt}(u^2 + v^2) = -\gamma(u^2 + v^2) - \frac{q\mathcal{E}}{m\omega} v,
\tag{6.8}
$$

and thus obtain a relation describing the excitation energy of the system. This is analogous to the third optical Bloch equation for the w component of the difference of the occupation numbers (see Eq. (6.28)).

Excursion: Lorentz oscillator in a magnetic field
If a magnetic field influences the motion of a charge, a Lorentz force is added to the equation of motion (6.1), which is $\mathbf{F}_{\mathrm{Lor}} = q\dot{\mathbf{x}} \times \mathbf{B}$ and results in an extra term $iq\dot{\rho}B$ when the replacement $\mathbf{x} \to r = x + iy \to \rho e^{-i\omega t}$ is carried out. If its influence on the dynamics is low, $|q\mathbf{B}|/m \ll \omega_0$, then the components of the magnetic field in the xy plane cause a rotation of the orbital plane, while the z component modifies the eigenfrequency of the oscillator. The complete equation of motion is now

$$
\ddot{\rho} + \gamma\dot{\rho} + \omega_0^2\rho = \frac{q}{m}(\mathcal{E} + i\dot{\rho}B_z)\, e^{-i\omega t}.
\tag{6.9}
$$

We seek solutions using the same procedures as before, and, with the *Larmor frequency*

$$
\omega_{\mathrm{L}} = qB_z/2m,
$$

we obtain the equilibrium solution

$$
\rho_0 = \frac{\rho_{\mathrm{max}}\gamma/2}{(\omega_0 - \omega_{\mathrm{L}} - \omega) - i\gamma}.
\tag{6.10}
$$

In Eqs. (6.2)–(6.8) and (6.13), the eigenfrequency ω_0 has only to be replaced by the modified value $\omega_0 - \omega_L$. Otherwise, the results can be taken over. With this theory, H. Lorentz was able to interpret the Zeeman effect, the shift and splitting of atomic resonance lines by external magnetic fields.

We finish by studying the effect of a transverse magnetic field on the motion of an electron. For this purpose we take the vector product of Eq. (6.9) by $\mathbf{x}\times$ (replacing $\rho \to \mathbf{x}$) and obtain a new equation for the electronic angular momentum $\mathbf{L} = m\mathbf{x}\times\dot{\mathbf{x}}$. Strictly speaking, this should be $m\mathbf{x}\times(\dot{\mathbf{x}}\times\mathbf{B}) = \mathbf{L}\times\mathbf{B} + m\dot{\mathbf{x}}\times(\mathbf{x}\times\mathbf{B})$, but in static fields the second term vanishes, and in alternating fields it is equivalent to a relativistic correction of first order, $(v/c)\mathbf{d}\times\mathbf{E}$, and can be neglected. So:

$$\frac{d}{dt}\mathbf{L} + \gamma\mathbf{L} = \mathbf{d}\times\mathbf{E} + \frac{q}{m}\mathbf{L}\times\mathbf{B}.$$

It can be recognized from this equation that a circularly polarized electric light field as well as a transverse static field ($\mathbf{B} \perp \mathbf{L}$) can cause rotation of the electronic angular momentum. The former case is usually called 'optical pumping' in spectroscopy [40], and the latter case occurs in the Hanle effect [21].

6.1.2   Macroscopic polarization

The macroscopic polarization $\mathbf{P}(\mathbf{r}, t)$ has already been introduced in section 2.1.2 in order to describe the propagation of electromagnetic waves in a dielectric medium. From the microscopic point of view, a sample consists of the microscopic dipole moments of atoms, molecules or lattice elements. The 'near field' of the microscopic particle does not play a role in the propagation of the radiation field, which is always a 'far field'. If there are N_{at} atomic or other microscopic dipoles in a volume, the macroscopic polarization is obtained by averaging, $\mathbf{P} = N_{at}\mathbf{p}/V$. Here the volume V is chosen much larger than molecular length scales, e.g. $d_{mol} < 5$ Å, and the average volume of a single particle as well. If the microscopic polarization density $\mathbf{p}(\mathbf{r})$ is known, there is the more exact form:

$$\mathbf{P}(\mathbf{r}, t) = \frac{N_{at}}{V}\int_V \mathbf{p}(\mathbf{r} - \mathbf{r}', t)\, d^3r'. \tag{6.11}$$

In our classic model the Fourier amplitudes of the polarization $\mathcal{P} = \mathcal{F}\{\mathbf{P}\}$ and of the driving field $\mathcal{E}$ are linearly connected,

$$\mathcal{P}(\omega) = \epsilon_0 \chi(\omega)\mathcal{E}(\omega), \tag{6.12}$$

and the susceptibility $\chi(\omega) = \chi'(\omega) + i\chi''(\omega)$ can be given using the results of Eq. (6.4),

$$\begin{aligned}
\chi'(\delta) &= \frac{N_{at}}{V}\frac{3\lambda^3}{4\pi^2}\frac{2\delta/\gamma}{1 + (2\delta/\gamma)^2}, \\
\chi''(\delta) &= \frac{N_{at}}{V}\frac{3\lambda^3}{4\pi^2}\frac{1}{1 + (2\delta/\gamma)^2}.
\end{aligned} \tag{6.13}$$

Since the temporal behaviour of the polarization is also characterized by transient processes, it usually depends on the field intensity also at earlier times. This becomes more apparent in the time-domain expression

$$\mathbf{P}(\mathbf{r}, t) = \epsilon_0 \int_{-\infty}^{\infty} \chi(t - t')\mathbf{E}(\mathbf{r}, t')\, dt', \tag{6.14}$$

which requires $\chi(t - t') = 0$ for $(t - t') < 0$ in order not to violate causality. Here as well the literal meaning of 'susceptibility' or 'after-effect' shows up. But for our purposes we assume that we are allowed to neglect relaxation processes occurring in solid materials within picoseconds or less, and therefore we can restrict our treatment to an instantaneous interaction.[2] According to the convolution theorem of Fourier transformation, the relation is, however, much simpler in the frequency domain following Eq. (6.12).

To be more exact, the 'dielectric function' (Eq. (2.4)) $\epsilon_0 \kappa(\omega) = \epsilon_0 [1 + \chi(\omega)]$ and the susceptibility are second-rank tensors, e.g. $\chi_{ij} = \partial P_i / \partial \mathcal{E}_j$, and reflect the anisotropy of real materials. The magnetic polarization can mostly be neglected for optical phenomena ($\mu_r \sim 1$), since the magnetic field $\mathbf{B}$ and the $\mathbf{H}$ field are identical except for a factor, $\mathbf{H} = \mathbf{B}/\mu_0$.

Only in an isotropic ($\nabla \cdot \mathbf{P} = 0$) and, according to Eq. (2.4), linear medium does the wave equation take on a simple form. This is, however, an important and often realized special case where the polarization obviously drives the electric field:

$$\nabla^2 \mathbf{E} - \frac{1}{c^2} \frac{\partial^2}{\partial t^2} \mathbf{E} = \frac{1}{\epsilon_0 c^2} \frac{\partial^2}{\partial t^2} \mathbf{P}. \tag{6.15}$$

Linear polarization and macroscopic refractive index

If the polarization depends linearly on the field intensity according to Eq. (6.12), then the modification of the wave velocity within the dielectric, $c^2 \to c^2 / \kappa(\omega)$, can be taken into account using the macroscopic refractive index $n(\omega)$ (see eq. (2.12)):

$$\nabla^2 \mathbf{E} - \frac{n^2(\omega)}{c^2} \frac{\partial^2}{\partial t^2} \mathbf{E} = 0. \tag{6.16}$$

According to Eq. (6.12) we have $\mathcal{E} + \mathcal{P}/\epsilon_0 = [1 + \chi(\omega)]\mathcal{E} = n^2(\omega)\mathcal{E}$ with

$$n^2(\omega) = \kappa(\omega) = 1 + \chi(\omega).$$

Here the relation between the complex index of refraction $n = n' + in''$ and the susceptibility χ becomes simpler in a significant way, if, for example in optically thin (dilute) matter like a gas, the polarization is very low, $|\chi(\omega)| \ll 1$:

$$n' \simeq 1 + \chi'/2 \qquad \text{and} \qquad n'' \simeq \chi''/2,$$

or

$$n \simeq 1 + \frac{N}{V} \frac{\lambda^3}{8\pi^2} \frac{i + 2\delta/\gamma}{1 + (2\delta/\gamma)^2}. \tag{6.17}$$

Thus, by measuring the macroscopic refractive index, the microscopic properties of the dielectric requiring theoretical treatment by quantum mechanics can be determined. Using $(N/V)\lambda^3/(8\pi^2) \geq 0.1$, we can also estimate the density of particles

---

[2] The methods of femtosecond spectroscopy developed in the 1990s now also allow us to study such fast relaxation phenomena with excellent time resolution.

where we ultimately leave the limiting case of optically thin media. For optical wavelengths ($\lambda \simeq 0.5\,\mu\text{m}$), this transition occurs already at the relatively low density of $N/V = 2 \times 10^{14}\,\text{cm}^{-3}$, which at room temperature for an ideal gas corresponds to a vacuum pressure of only $10^{-2}\,\text{mbar}$.

The solution for a planar wave according to Eq. (6.16) is then

$$\mathbf{E}(\mathbf{r}, t) = \mathbf{E}_0 \, e^{-i(\omega t - n'\mathbf{k}\cdot\mathbf{r})} \, e^{-n''\mathbf{k}\cdot\mathbf{r}}.$$

Propagation not only takes place with a modified phase velocity $v_{\text{ph}} = c/n'$ but also is exponentially damped according to Beer's law in the z direction with absorption coefficient $\alpha = 2n''k_z$,

$$I(z) = I(0) \, e^{-2n''k_z z} = I(0) \, e^{-\alpha z}. \tag{6.18}$$

We have chosen $n'', \chi'' > 0$ for normal dielectrics according to Eq. (6.17); as we will see, in a 'laser medium' one can create $n'', \chi'' < 0$ as well, realizing amplification of an optical wave.

Let us briefly study the question of whether a single microscopic dipole can generate a refractive index, i.e. whether it could cause noticeable absorption or dispersion of an optical wave. For this consideration we again rewrite the absorption coefficient as

$$\alpha = 2n''k = \frac{N}{V} \frac{3\lambda^2}{2\pi} \frac{1}{1 + (2\delta/\gamma)^2} = \frac{N}{V} \frac{\sigma_Q}{1 + (2\delta/\gamma)^2}. \tag{6.19}$$

Therefore, the effect of a single atom is determined by a resonant cross-section of

$$\sigma_Q = 3\lambda^2/2\pi \qquad \text{at} \qquad \delta = 0, \tag{6.20}$$

which is much larger than the atom itself. If we succeed in limiting a single atom to a volume with this wavelength as diameter ($V \simeq \lambda^3$), then a laser beam focused on this volume will experience strong absorption. Such an experiment has in fact been carried out with a stored ion [110]. Dispersion is observed for nonzero detuning only, but for small values $\delta = \pm\gamma/2$ a single atom is predicted to cause a measurable phase shift $\delta\Phi = \pm 1/(8\pi)$ as well.

Absorption and dispersion in optically thin media

Sometimes it is useful to consider directly the effect of polarization on the amplitude of an electromagnetic wave propagating in a dielectric medium. For this, we take the one-dimensional form of the wave equation (6.15),

$$\left(\frac{\partial^2}{\partial z^2} - \frac{1}{c^2} \frac{\partial^2}{\partial t^2} \right) \mathcal{E}(z) \, e^{-i(\omega t - kz)} = \frac{1}{\epsilon_0 c^2} \frac{\partial^2}{\partial t^2} \, \mathcal{P}(z) \, e^{-i(\omega t - kz)},$$

we fix the frequency $\omega = ck$ for $|\mathbf{E}(z, t)| = \mathcal{E}(z) \, e^{-i(\omega t - kz)}$ and additionally we assume that the amplitude changes only slowly (on the scale of a wavelength) during propagation. Thus:

$$\left| \frac{\partial^2 \mathcal{E}(z)}{\partial z^2} \right| \ll k \left| \frac{\partial \mathcal{E}(z)}{\partial z} \right|.$$

Then with $\partial^2/\partial z^2\,[\mathcal{E}(z)\,e^{ikz}] \simeq e^{ikz}[2ik\,\partial/\partial z - k^2]\mathcal{E}(z)$, the wave equation is approximately

$$\left[2ik\frac{\partial}{\partial z} - k^2 + \frac{\omega^2}{c^2}\right]\mathcal{E}(z) = -\frac{\omega^2}{\epsilon_0 c^2}\mathcal{P}(z),$$

which with $k = \omega/c$ further simplifies to

$$\frac{\partial}{\partial z}\mathcal{E}(z) = \frac{ik}{2\epsilon_0}\mathcal{P}(z). \tag{6.21}$$

Now we consider the electromagnetic wave with a real amplitude and phase, $\mathcal{E}(z) = \mathcal{A}(z)\,e^{i\Phi(z)}$, and calculate

$$\mathcal{E}(z)\frac{d\mathcal{E}^*(z)}{dz} = \mathcal{A}\frac{d\mathcal{A}}{dz} + i\mathcal{A}^2\frac{d\Phi}{dz} = \frac{-ik}{2\epsilon_0}\mathcal{P}^*(z)\mathcal{E}(z),$$

$$\mathcal{E}^*(z)\frac{d\mathcal{E}(z)}{dz} = \mathcal{A}\frac{d\mathcal{A}}{dz} - i\mathcal{A}^2\frac{d\Phi}{dz} = \frac{ik}{2\epsilon_0}\mathcal{P}(z)\mathcal{E}^*(z).$$

From this we can determine the change of the intensity $I(z) = \frac{1}{2}c\epsilon_0 \mathcal{A}^2$ of an electromagnetic wave while propagating within a polarized medium according to

$$\frac{d}{dz}I(z) = \frac{\omega}{2}\,\mathfrak{Im}\{\mathcal{E}(z)\mathcal{P}^*(z)\}$$

and the phase shift according to

$$\frac{d}{dz}\Phi(z) = \frac{\omega}{2I(z)}\,\mathfrak{Re}\{\mathcal{E}(z)\mathcal{P}^*(z)\}.$$

The absorption coefficient α and the real part of the refractive index n' can be calculated in an obvious way from

$$\begin{aligned}
\alpha &= \frac{1}{I(z)}\frac{dI(z)}{dz} = \frac{\omega}{2I(z)}\,\mathfrak{Im}\{\mathcal{E}(z)\mathcal{P}^*(z)\}, \\
n' - 1 &= \frac{1}{k}\frac{d\Phi(z)}{dz} = \frac{c}{2I(z)}\,\mathfrak{Re}\{\mathcal{E}(z)\mathcal{P}^*(z)\}.
\end{aligned} \tag{6.22}$$

We naturally reproduce the results from the section on the linear refractive index, if we assume the linear relation according to Eq. (6.12). The form developed here also allows us to investigate nonlinear relations, and will be useful in the chapter on nonlinear optics (Section 12).

Dense dielectric media and near fields

Certainly, in a dilute, optically thin medium, we do not make a big mistake by neglecting the field additionally generated in the sample by polarization. But this is no longer the case in the liquid or solid states. In order to determine the 'local field' of the sample, we cut out a fictitious sphere with a diameter $d_{\mathrm{atom}} \ll d_{\mathrm{sph}} \ll \lambda$ with 'frozen' polarization from the material (Fig. 6.4).

To determine the microscopic local field $\mathbf{E}_{\mathrm{loc}}$ at the position of a particle, we decompose it into various contributions, $\mathbf{E}_{\mathrm{loc}} = \mathbf{E}_{\mathrm{ext}} + \mathbf{E}_{\mathrm{surf}} + \mathbf{E}_{\mathrm{Lor}} + \mathbf{E}_{\mathrm{near}}$, which depend on the different geometries and structures of the sample and are in total called the 'depolarizing field' since they usually weaken the external field $\mathbf{E}_{\text{ext}}$:

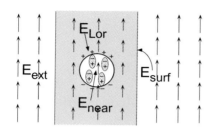

Fig. 6.4: *Contributions to the local electric field in an optically dense medium. For a transverse wave the contribution of the surface vanishes in the case of normal incidence.*

$\mathbf{E}_{\text{surf}}$:
The field of the surface charges generated by the surface charge density $\rho_{\text{surf}} = \mathbf{n} \cdot \mathbf{P}(\mathbf{r}_{\text{surf}})$. It vanishes for a wave at normal incidence.

$\mathbf{E}_{\text{Lor}}$:
The field of the surface of a fictitious hollow sphere cut out from the volume (also known as the 'Lorentz field'). For homogeneous polarization, one finds $\mathbf{E}_{\text{Lor}} = \mathbf{P}/3\epsilon_0$.

$\mathbf{E}_{\text{near}}$:
The field of the electric charges within the sphere. In the case of isotropic media, this contribution vanishes, $\mathbf{E}_{\text{near}} = 0$.

From $\mathbf{P} = \epsilon_0 \chi \mathbf{E}_{\text{loc}} = \epsilon_0 \chi (\mathbf{E} + \mathbf{P}/\epsilon_0)$, we then obtain by insertion of $\mathbf{E}_{\text{loc}} = \mathbf{E}_{\text{Lor}} = \mathbf{P}/3$ the macroscopic volume susceptibility χ^{V} of an isotropic and linear but dense material,

$$\chi^{\text{V}}_{ij}(\omega) = \frac{1}{\epsilon_0} \frac{P_i}{E_j} = \frac{\chi}{1 - \chi/3}.$$

From this by rearrangement can be obtained the *Clausius–Mossotti equation*, which describes the influence of the depolarizing field on the refractive index (density $\mathcal{N} = N/V$),

$$3 \frac{n^2 - 1}{n^2 + 2} = \chi = \frac{\mathcal{N}q^2}{\epsilon_0 m} \frac{1}{(\omega_0^2 - \omega^2) - i\omega\gamma}. \tag{6.23}$$

For small polarizations, $\chi/3 \ll 1$, Eq. (6.23) again turns into eq. (6.17).

Realistic polarizable substances, though, do not have just one degree of freedom like the Lorentz oscillator described here but lots of them. We can extend the Lorentz model for a not too strong field by linearly superimposing many oscillators with different resonance frequencies ω_k and damping rates γ_k and weighting them with their relative contribution, their 'oscillator strength' f_k:

$$3 \frac{n^2 - 1}{n^2 + 2} = \frac{\mathcal{N}q^2}{\epsilon_0 m} \sum_k \frac{f_k}{(\omega_k^2 - \omega^2) - i\omega\gamma_k}.$$

Even if the field intensity becomes quite large, we can still use the concepts described here if we introduce a nonlinear susceptibility. This case is dealt with in Section 12 on nonlinear optics.

The dimensionless oscillator strength allows a simple transition to the quantum mechanically correct description of the microscopic polarization [97]. For this, only the matrix element of the dipole transition between the ground state $|\phi_g\rangle$ and excited states $|\phi_k\rangle$ of the system, $q\mathbf{r}_{kg} = q\langle\phi_k|\mathbf{r}|\phi_g\rangle$ has to be used:

$$f_{kg} = \frac{2m\omega_{kg}|\mathbf{r}_{kg}|^2}{\hbar}.$$

We do not need to require anything specific about the nature of these states. They can be atomic or molecular excitations but also, for example, optical phonons or polaritons within solid states. Strictly speaking, the success of the classical Lorentz model for single atoms is justified by this relation. In atoms the oscillator strengths follow the Thomas–Reiche–Kuhn sum rule $\sum_k f_{kg} = 1$; already for low atomic resonance lines such as, for example, the well-known doublets of the alkali spectra, we have $f \sim 1$; therefore the other resonance lines have to be significantly weaker.

6.2 Two-level atoms

6.2.1 Are there any atoms with only two levels?

In quantum mechanics we describe atoms by their states. In the simplest case a light field couples a ground state $|g\rangle$ to an excited state $|e\rangle$. This model system can be theoretically dealt with well, and is particularly useful for understanding the interaction of light and matter. However, even simple atoms such as the alkali and alkali earth atoms, which are technically easy to master and widely used for experimental investigations, present a complex structure with a large number of states even in the ground level.[3]

Occasionally, though, it is possible to prepare atoms in such a way that no more than two states are effectively coupled to the light field. The calcium atom, for instance, has a non-degenerate singlet ground state ($^1S_0$, $\ell = 0$, $m = 0$). By using a light field with a wavelength of 423 nm and proper choice of the polarization ($\sigma^\pm$, π), three different two-level systems can be prepared by coupling to the ($^1P_1$, $\ell = 1$, $m = 0, \pm 1$) states.

The famous yellow doublet of the sodium atom ($\lambda = 589$ nm) is another example that has played a central role in experimental investigations, though it has large total angular momenta $F = 1, 2$ even in the $^2S_{1/2}$ ground state doublet due to its nuclear spin of $I = 3/2$ and presents a wealth of magnetic substructure. By so-called 'optical pumping' [40] with σ^+-polarized light, all the atoms in a gas can be prepared in, for example, the state with quantum numbers $F = 2$, $m_F = 2$. This state is then coupled only to the $F' = 3$, $m_{F'} = 3$ sub-state of the excited $^2P_{3/2}$ state by the light field.[4]

These effective 'two-level atoms', the list of which can easily be extended, play an enormously important role in physical experiments since they provide the simplest

---

[3]The wealth of structure is generated by the coupling of the magnetic orbital and spin momenta of electrons and core. For low states the splittings are about 100–1000 MHz. Details can be found in textbooks about quantum mechanics [20] or atomic physics [111].

[4]In reality, the circular polarization is never perfect. Small admixtures of σ^- light to the σ^+ light cause, for example, occasional excitations with $\Delta m_F = -1$ and therefore limit the 'quality' of the two-level atom.

Calcium

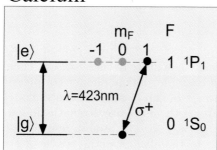

Sodium

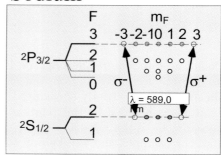

Fig. 6.5: *Abstract and realistic two-level atoms. Left: Calcium atom. A σ^+ polarized light field couples only states with angular momentum quantum numbers $|g\rangle = |F, m_F\rangle = |0,0\rangle$ and $|e\rangle = |1,1\rangle$. Right: Sodium atom. A circularly polarized light field ($\sigma^{\pm}$) is 'pumping' the sodium atoms to the outer $|F, m_F\rangle = |2, \pm 2\rangle$ states, which with $\sigma^{\pm}$ light are coupling only to the $|3, \pm 3\rangle$ states.*

models of a polarizable physical system and radiative interaction is reduced to its most fundamental case.

6.2.2 Dipole interaction

The 'free' two-level atom with total mass M is now reduced to a Hamiltonian H_{at} having only a ground state $|g\rangle$ and an excited state $|e\rangle$.[5] To complete the picture, we allow for an arbitrary centre-of-mass energy $E_0 = P^2/2M$. Thus

$$H_{\mathrm{at}} = \frac{P^2}{2M} + \frac{\hbar \omega_0}{2}(|e\rangle\langle e| - |g\rangle\langle g|). \tag{6.24}$$

The energy of the atom is $E_{\mathrm{e}} = \langle e|H_{\mathrm{at}}|e\rangle = E_0 + \hbar \omega_0/2$ in the excited state and $E_{\mathrm{g}} = E_0 - \hbar \omega_0/2$ in the ground state. The resonance frequency presents the energy separation of the two states, $\omega_0 = (E_{\mathrm{e}} - E_{\mathrm{g}})/\hbar$.

The dipole operator $\hat{V}_{\mathrm{dip}}$ is obtained by an analogy with classical electrodynamics, i.e. by converting the classical energy of a dipole subject to an electric field into an operator. For the electron position operator $\hat{\mathbf{r}}$, we obtain[6]

$$\hat{V}_{\mathrm{dip}} = -q\hat{\mathbf{r}}\mathbf{E}.$$

In a realistic experiment, we always have to take the exact geometric orientation of atom and electric field into account. For the consideration of the two-level atom,

---

[5]We assume that the reader is familiar with the basic principles of quantum mechanics. Quantum states are given in Dirac notation, where state vectors $|i\rangle$ are associated with complex wavefunctions $\psi_i(\mathbf{r})$. The expectation values of an operator $\hat{\mathcal{O}}$ are thus calculated from $\langle \hat{\mathcal{O}} \rangle = \langle f|\hat{\mathcal{O}}|i\rangle = \int_V d\tau\, \psi_f^*(\mathbf{r})\hat{\mathcal{O}}\psi_i(\mathbf{r})$.

[6]Rigorous analysis according to quantum mechanics results in the product of electron momentum and electromagnetic vector potential $\hat{\mathbf{p}}\mathbf{A}$, but it can be shown that in the vicinity of resonance frequencies $\hat{\mathbf{r}}\mathbf{E}$ leads to the same result [90].

however, we neglect this geometric influence and restrict the problem to one dipole coordinate $\hat{d} = q\hat{r}$ only,

$$\hat{V}_{\text{dip}} = -\hat{d}\mathcal{E}_0 \cos \omega t.$$

Using the completeness theorem of quantum mechanics, we can project the position operator onto the states involved ($\langle i|\hat{d}|i\rangle = 0$):

$$\hat{d} = |e\rangle\langle e|\hat{d}|g\rangle\langle g| + |g\rangle\langle g|\hat{d}|e\rangle\langle e|.$$

We use the matrix element $d_{eg} = \langle e|\hat{d}|g\rangle$ of the dipole operator. Using the definition of atomic raising and lowering operators, $\sigma^\dagger = |e\rangle\langle g|$ and $\sigma = |g\rangle\langle e|$, we write

$$\hat{d} = d_{eg}\sigma^\dagger + d_{eg}^*\sigma.$$

With those operators we can already express the atomic Hamiltonian and the dipole operator very compactly:

$$H_{\text{at}} \quad = \quad \frac{p^2}{2m} + \hbar\omega_0(\sigma^\dagger\sigma - \tfrac{1}{2}),$$

$$\hat{V}_{\text{dip}} \quad = \quad -(d_{eg}\sigma^\dagger + d_{eg}^*\sigma)\mathcal{E}_0 \cos \omega t.$$

From linear combinations of the atomic field operators, Pauli operators can be generated, which are known to describe a spin-1/2 system with only two states:

$$\sigma_x \quad = \quad \sigma^\dagger + \sigma,$$
$$\sigma_y \quad = \quad -i(\sigma^\dagger - \sigma),$$
$$\sigma_z \quad = \quad \sigma^\dagger\sigma - \sigma\sigma^\dagger \quad = \quad [\sigma^\dagger, \sigma].$$

Therefore, the two-level atom can be described like a pseudo-spin system. We will see that we can interpret the expectation values of σ_x and σ_y as components of the atomic polarization and σ_z as the difference of occupation numbers or 'inversion'. The concept of this theoretical description is completely analogous to the dynamics of the corresponding spin-1/2 system and is borrowed from there.

The Pauli operators have the form

$$\sigma_x = \begin{pmatrix} 0 & 1 \\ 1 & 0 \end{pmatrix}, \qquad \sigma_y = \begin{pmatrix} 0 & -i \\ i & 0 \end{pmatrix}, \qquad \sigma_z = \begin{pmatrix} 1 & 0 \\ 0 & -1 \end{pmatrix},$$

in the matrix representation and follow the generally useful relation $[\sigma_i, \sigma_j] = 2i\sigma_k$ on cyclic permutation of the coordinates xyz. In addition we have

$$\sigma^\dagger \quad = \quad \tfrac{1}{2}(\sigma_x + i\sigma_y),$$
$$\sigma \quad = \quad \tfrac{1}{2}(\sigma_x - i\sigma_y).$$

The operators' equation of motion is obtained from the Heisenberg equation

$$\dot{\sigma}_i = \frac{\partial}{\partial t}\sigma_i + \frac{i}{\hbar}[H, \sigma_i].$$

For this, the Hamiltonian is usefully written in the form

$$H = \tfrac{1}{2}\hbar\omega_0\sigma_z - \tfrac{1}{2}(d_{eg} + d_{eg}^*)\mathcal{E}_0 \cos(\omega t)\sigma_x - \tfrac{1}{2}i(d_{eg} - d_{eg}^*)\mathcal{E}_0 \cos(\omega t)\sigma_y.$$

Often real values can be chosen for d_{eg}. Then the third (σ_y) term is omitted and it can simply be written as

$$H = \tfrac{1}{2}\hbar\omega_0\sigma_z - d_{eg}\mathcal{E}_0\cos(\omega t)\sigma_x.$$

If the operators are not explicitly time-dependent, the result is an equation system known as *Mathieu's differential equations*,

$$\dot{\sigma}_x = -\omega_0\sigma_y,$$

$$\dot{\sigma}_y = \omega_0\sigma_x - \frac{2d_{eg}\mathcal{E}_0}{\hbar}\cos(\omega t)\sigma_z, \qquad\qquad (6.25)$$

$$\dot{\sigma}_z = \frac{2d_{eg}\mathcal{E}_0}{\hbar}\cos(\omega t)\sigma_y.$$

It can easily be shown that only the orientation, but not the magnitude of the angular momentum, is changed under the effect of the light field; we have $\sigma_x^2 + \sigma_y^2 + \sigma_z^2 = 1$ as for the Pauli matrices.

6.2.3   Optical Bloch equations

Until now we have considered the development of atomic operators under the influence of a light field. For the semiclassical consideration we can replace them by expectation values[7] $S_i = \langle\sigma_i\rangle$ and again obtain the equation system (6.25) only now for classical variables [2]. To produce transparent solutions it is advantageous to consider the evolution of variables in a new coordinate system rotating with the light frequency ω around the z axis, i.e. with the polarization,

$$S_x = u\cos\omega t - v\sin\omega t,$$
$$S_y = u\sin\omega t + v\cos\omega t,$$
$$S_z = w.$$

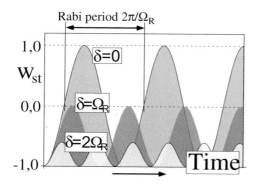

Fig. 6.6: *Rabi frequency and occupation number oscillation for different detunings.*

This often used approximation is called the 'rotating wave approximation' (RWA). The variables (u, v) describe the sine and cosine components of the induced electric dipole moment, and w is the difference in occupation numbers. The close relation of these variables with the classical Lorentz model in Section 6.1.1 and their physical interpretation will be explained in more detail in Section 6.2.5.

With detuning $\delta = \omega - \omega_0$ we obtain after some algebra

---

[7]There are no operator products that could cause typically quantum mechanical signatures due to non-commutativity.

$$\dot{u} = \delta v - \frac{d_{eg}\mathcal{E}_0}{\hbar} \sin(2\omega t)w,$$

$$\dot{v} = -\delta u - \frac{d_{eg}\mathcal{E}_0}{\hbar}[1 + \cos(2\omega t)]w,$$

$$\dot{w} = \frac{d_{eg}\mathcal{E}_0}{\hbar} \sin(2\omega t)u + \frac{d_{eg}\mathcal{E}_0}{\hbar}[1 + \cos(2\omega t)]v.$$

For typical optical processes the contributions oscillating very rapidly with $2\omega t$ play only a small role (they cause the so-called *Bloch–Siegert shift*) and are therefore neglected. We introduce the *Rabi frequency*

$$\Omega_R = |d_{eg}\mathcal{E}_0/\hbar| \tag{6.26}$$

and get the undamped optical Bloch equations,

$$\begin{aligned} \dot{u} &= \delta v, \\ \dot{v} &= -\delta u + \Omega_R w, \\ \dot{w} &= -\Omega_R v, \end{aligned} \tag{6.27}$$

originally found for magnetic resonance by F. Bloch (1905–1983, nobel prize 1952) in order to describe there the interaction between of a magnetic moment with spin 1/2 in a strong homogeneous magnetic field exposed to a high-frequency field.

Since optical two-level systems obey an identical set of equations, almost all the concepts of coherent optics are borrowed from electron and nuclear spin resonance. The system of equations (6.26) can also be written in a shorter way by using the vectors $\mathbf{u} = (u, v, w)$ and $\boldsymbol{\Omega} = (-\Omega_R, 0, -\delta)$: $\dot{\mathbf{u}} = \boldsymbol{\Omega} \times \mathbf{u}$.

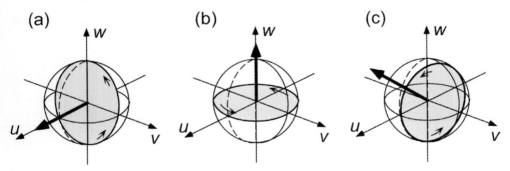

Fig. 6.7: *Dynamics of the Bloch vector: (a) $\boldsymbol{\Omega} = (-\Omega_R, 0, 0)$; (b) $\boldsymbol{\Omega} = (0, 0, -\delta)$; and (c) $\boldsymbol{\Omega} = (-\Omega_R, 0, -\delta)$.*

6.2.4   Precession and Rabi nutation

Let us now consider special solutions and extensions of the optical Bloch equations (6.26) to analyse the physical dynamics of the two-level system.

Undamped dynamics

At thermal ambient energies, an atom with an optical excitation frequency usually resides in its ground state, and therefore we normally have $w(t=0) = -1$. The resonance case is particularly easy to determine: here the detuning vanishes, $\delta = 0$. The occupation number w and the v component of the polarization perform an oscillation with the *Rabi frequency* according to eq. (6.26),

$$
\begin{aligned}
v(t) &= -\sin(\Omega_R t), \\
w(t) &= -\cos(\Omega_R t).
\end{aligned}
$$

The system of equations (6.26) describes the behaviour of a magnetic dipole transition, e.g. between the hyperfine states of an atom, in an excellent approximation. Even if the detuning δ does not vanish, there is a generalized Rabi oscillation with frequency

$$
\Omega = \sqrt{\delta^2 + \Omega_R^2},
$$

though the amplitude of the occupation number oscillations decreases with increasing detuning, as shown in Fig. 6.6.

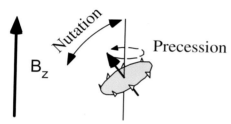

Fig. 6.8: *Precession and nutation of a magnetic dipole moment associated with a gyro. The pseudo-spin of the two-level atom performs an analogous motion in which the longitudinal z direction is associated with the difference in occupation numbers w, while the transverse directions are identified with the polarization components u and v.*

The dynamical Bloch vector evolution for resonant excitation is shown in Fig. 6.7(a). For the so-called π-pulse $\Omega_R t = \pi$, the atomic system rotates into the completely 'inverted' state with $w = 1$! This pseudo-spin rotation should properly be called 'Rabi nutation' in analogy with conventional gyromagnets (Fig. 6.8).

We have learned that the motion is undamped according to (6.26) while an optical atomic excitation is damped by multiple processes. Among them there is the radiative decay but also collisions and other phenomena. We thus introduce the 'longitudinal relaxation rate' $\gamma = 1/T_1$ phenomenologically. It describes the energy loss of the two-level system characterized by the difference in occupation numbers, and the z coordinate of the Bloch vector, respectively. In equilibrium without a driving light field, the stationary thermal value of the inversion $w_0 = -1$ must be reproduced.

Damping

The *transverse relaxation rate* $\gamma' = 1/T_2$ describes the damping of polarization, i.e. of the x and y components of the Bloch vector. In an ensemble the macroscopic polarization can also get lost because each particle precesses with a different speed and so the particles lose their original phase relation (in the precession angle). For pure radiation damping we have $T_2 = 2T_1$. The polarization vanishes as well, when the light field is switched off.

The complete *optical Bloch equations* are

$$
\begin{aligned}
\dot{u} &= \delta v - \gamma' u, \\
\dot{v} &= -\delta u - \gamma' v + \Omega_R w, \\
\dot{w} &= -\Omega_R v - \gamma(w - w_0).
\end{aligned}
\tag{6.28}
$$

Their similarity to the classical equations (6.7) and (6.8), a result of the Lorentz model, cannot be overlooked any more. Apparently the ratio of the Rabi frequency Ω_R to the damping rates γ, γ' determines the dynamics of the system: we expect oscillating properties as in the undamped system only then, when they are sufficiently large, i.e. the system is driven sufficiently strongly.

Often the optical Bloch equations are written in the more compact complex notation using the language of the density matrix theory from quantum mechanics (see Appendix B.2). With $\rho_{eg} = u + iv$ and $w = \rho_{ee} - \rho_{gg}$ we find

$$
\begin{aligned}
\dot{\rho}_{eg} &= -(\gamma' + i\delta)\rho_{eg} + i\Omega_R w, \\
\dot{w} &= -\Im\{\rho_{eg}\}\Omega_R - \gamma(w - w_0).
\end{aligned}
\tag{6.29}
$$

6.2.5   Inversion and polarization

We consider the situation when transients have settled, i.e. a time $t \gg T_1, T_2$ has passed since switching on the light field. Then (6.28) has the following stationary solutions, which are related to the inversion w_0 without a driving light field:

$$
w_{\text{st}} = \frac{w_0}{1 + \dfrac{\Omega_R^2}{\gamma\gamma'} \dfrac{1}{1 + (\delta/\gamma')^2}} = \frac{w_0}{1 + s}.
\tag{6.30}
$$

In a light field with intensity I, the 'saturation parameter'

$$
s = \frac{I/I_0}{1 + (\delta/\gamma')^2}
\tag{6.31}
$$

determines the significance of coherent processes with dynamics determined by the Rabi frequency and

$$
s(\delta = 0) = s_0 = \frac{I}{I_0} = \frac{\Omega_R^2}{\gamma\gamma'},
\tag{6.32}
$$

in comparison with incoherent damping processes determined by the relaxation rates γ, γ'. Owing to $\Omega_R^2 = |-d_{eg}E_0/\hbar|^2 = |-d_{eg}/\hbar|^2(2I/c\epsilon_0)$, the saturation intensity I_0 can be calculated:

$$
I_0 = \frac{c\epsilon_0}{2} \frac{\hbar^2 \gamma\gamma'}{d_{eg}^2}.
\tag{6.33}
$$

From Fig. 6.9 it is clear that the saturation intensity sets the typical scale for the onset of coherent processes.

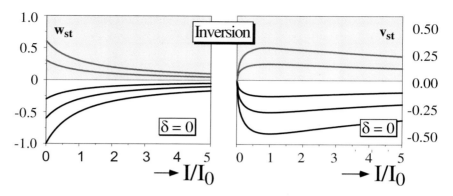

Fig. 6.9: *Effect of a light field on the equilibrium values of the difference in occupation numbers ('inversion') w_{st} and the polarization components u_{st} and v_{st} as functions of the saturation parameter $s(\delta=0) = I/I_0$ according to Eqs. (6.30) and (6.36). With the light field vanishing, the values $w_0 = -1$, -0.6, -0.3, $+0.3$ and $+0.6$ have been used.*

Using the known result for spontaneous emission $\gamma = d_{eg}^2 \omega^3 / 3\pi\hbar\epsilon_0 c^3$ (6.41), the saturation intensity can be determined just by knowing the resonance wavelength λ and the transverse relaxation rate γ'. We get a useful correlation with the resonant cross-section of the absorption σ_Q from Eq. (6.20),

$$I_0 = \frac{2\pi hc\gamma'}{3\lambda^3} = \frac{\hbar\omega\gamma'}{\sigma_Q}, \tag{6.34}$$

which can be interpreted in the following way. Apparently, at the saturation intensity, the energy of just one photon flows through the resonant absorption cross-section σ_Q during the transverse coherence time $T' = 1/\gamma'$.

If only radiative decay is possible, as e.g. in dilute gases or atomic beams, then the saturation intensity with $\gamma' = \gamma/2$ depends only on the properties of the free atom and is given by

$$I_0 = \frac{\pi hc\gamma}{\lambda^3}.$$

As an example, we present the saturation intensity for several important atoms. They can be realized technically using continuous wave laser light sources without special effort except for the case of the hydrogen atom. The 'strength' of the transition is characterized by the decay rate $\gamma/2\pi = \Delta_{1/2}$, given in the table in units of the natural linewidth. It is perhaps strange at first glance that the saturation intensity becomes smaller with decreasing linewidth and therefore weaker lines, but it has to be taken into account that coherent coupling needs more and more time to reach excitation.

According to Eq. (6.30) the inversion can also be expressed by its dependence on the saturation intensity

$$w_{st} = w_0 \frac{\delta^2 + \gamma'^2}{\delta^2 + \gamma'^2(1 + I/I_0)}.$$

Tab. 6.2: *Saturation intensity of some important atomic resonance lines.*

Atom	H	Na	Rb	Cs	Ag	Ca	Yb
Transition	$1S \rightarrow 2P$	$3S \rightarrow 3P$	$5S \rightarrow 5P$	$6S \rightarrow 6P$	$5S \rightarrow 5P$	$4S \rightarrow 4P$	$6S \rightarrow 6P$
$\gamma/2\pi \ (10^6 \, \text{s}^{-1})$	99.5	9.9	5.9	5.0	20.7	35.7	0.18
λ (nm)	121.6	589.0	780.2	852.3	328.0	422.6	555.8
I_0 (mW cm^{-2})	7242	6.34	1.63	1.06	76.8	61.9	0.14

Here it is worth introducing another new parameter,

$$\gamma_{\text{sat}} = \gamma' \sqrt{1 + I/I_0}, \tag{6.35}$$

to describe the v polarization component of a single particle in equilibrium,

$$v_{\text{st}} = \frac{w_0}{1+s} \frac{\gamma' \Omega_R}{\delta^2 + \gamma'^2} = w_0 \frac{\gamma' \Omega_R}{\delta^2 + \gamma'^2_{\text{sat}}}. \tag{6.36}$$

Sometimes it is technically more convenient to express the Rabi frequency according to (6.32) again by the saturation intensity. For u_{st} and v_{st} we then get

$$v_{\text{st}} = \sqrt{\frac{\gamma}{\gamma'}} \frac{w_0 \sqrt{I/I_0}}{1 + I/I_0 + (\delta/\gamma')^2}, \tag{6.37}$$

$$u_{\text{st}} = \frac{\delta}{\gamma'} v_{\text{st}}.$$

The intensity dependence of the polarization is presented in Fig. 6.9 as a function of the normalized intensity I/I_0 in the special case of perfect resonance at $\delta = 0$. It increases very rapidly and decreases at high intensities as $1/\sqrt{I/I_0}$, i.e. with the amplitude of the driving field. For low intensities we find again the limiting classical case. Then $(u_{\text{st}}, v_{\text{st}})$ correspond with the (u, v) coordinates of the Lorentz oscillator from Eq. (6.7) and of course present as well the frequency characteristic from Fig. (6.2).

To determine the macroscopic polarization density P of a sample, we have now to introduce the particle density N/V and obtain for an ensemble of identical particles, exactly as in the classical case,

$$P = \frac{N}{V} d_{eg}(u + iv). \tag{6.38}$$

In contrast to the classical case (see Section 6.1.2) now (u, v) depend nonlinearly on the electric amplitude of the light field. In a physical system in thermal equilibrium the average occupation number of states decreases with increasing energy according to the Maxwell–Boltzmann formula $n_{\text{th}} \sim e^{-E/kT}$. Therefore the equilibrium value for the difference in occupation numbers of the upper (N_e) and lower (N_g) states is always negative, $w_0 = (N_e - N_g)/N < 0$. Since optical transitions with energies of several eV have much more energy than thermal excitations with only $1/40$ eV, we can usually assume $w_0 = -1$. The treatment according to quantum mechanics yields the same result because according to (6.37) $v_{\text{st}} < 0$ for arbitrary field intensities. Thus the polarization follows the field and always causes absorption.

6.3   Stimulated and spontaneous radiation processes

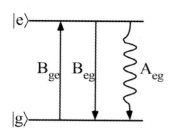

Fig. 6.10: *Two-level quantum system with Einstein coefficients, stimulated and spontaneous radiation processes.*

In the previous section we investigated the coupling of an atom to a monochromatic light wave. Three different radiation processes were identified:

1.   By coupling to the driving field, an atom can be promoted from the ground state to the excited state. This process is called 'stimulated absorption' and can only take place if there is an applied external field.

2. An analogous process takes place as well from the excited to the ground state, and is called 'stimulated emission'. The stimulated processes describe the coherent evolution of the atom–field system, i.e. phase relations play an important role.

3. If an atom is in the excited state, it can decay to the ground state by 'spontaneous emission'. This process is incoherent, always takes place (apart from the exceptions in the excursion on p. 187) and has been taken into account phenomenologically in Eq. (6.28) by introducing the damping constants.

Excursion: The spectrum of black bodies

Just before the end of the 19th century, the spectrum of black bodies was very carefully studied at the Physikalisch-Technische Reichsanstalt, Berlin (the historic German National Laboratory for Standards and Technology). At that time, light bulbs for public lighting had been only very recently introduced, and the intention was to control their output and increase their efficiency. This blackbody spectrum has since played an outstanding role for modern physics in general and our understanding of light sources in particular. During these investigations, it turned out that the formula given by Wien for low frequencies, $S_E(\omega) \propto \omega^3 \exp(-\hbar\omega/kT)$, no longer matched the experimental results. At the same time, in England, Lord Rayleigh gave a different, more appropriate, radiation formula for low frequencies, $S'_E(\omega) \propto \omega^2 T$.

Max Planck arrived at his famous radiation formula by a clever interpolation, here in the modern notation

$$S_E(\omega) = \frac{8\pi}{c^3} \frac{\hbar\omega^3}{exp(\hbar\omega/kT) - 1}. \tag{6.39}$$

Today we know that this formula is derived from the product of the density of states of the radiation field at frequency ω and the occupation probability according to Bose–Einstein statistics. This formula, published for the first time by Planck in Berlin on 14 December 1900, was the beginning of a sequence of ideas leading to modern physics. Thermal light sources, the concepts of optics and a problem of truly applied research – the efficiency of light bulbs – all played an important role in the birth of quantum physics!

The unbroken fascination of radiation physics has recently also been confirmed by radioastronomical measurements. It must be noted that the most exact measurement of a black body is now obtained from the spectrum of the cosmic background radiation. The difference

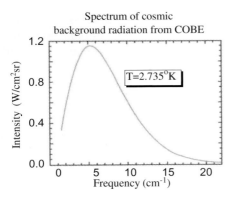

Spectrum of cosmic
background radiation from COBE

$T=2.735°K$

Fig. 6.11: *Left: Spectrum of the 2.7 K cosmic background radiation. Right: Celestial maps of the intensity fluctuations. Top: Dipole asymmetry. Bottom: Residual fluctuations with maximum $\Delta T/T \simeq 10^{-5}$. After [78].*

between the measured values and the theoretical curve in Fig. 6.11 is in fact not visible! The average temperature of this radiation, often interpreted as the 'afterglow' of the now very much cooled down Big Bang, can be determined as $T = 2.726 \pm 0.005$ K.

The measurements by the *Cosmic Background Explorer* (COBE) satellite [78, 12] are so exact that the temperature fluctuations of radiation related to the average detected from a certain direction can be mapped on a celestial map. The spectacular result shows a dipole-like asymmetry of the order $\Delta T/T \simeq 10^{-3}$, which can be explained by the proper motion of our Galaxy relative to a homogeneous radiation background. Beyond that, the microwave radiation is isotropic except for smaller spatial fluctuations of about $\Delta T/T \simeq 10^{-3}$. It is assumed [12] that those small fluctuations reflect the density fluctuations of the early Universe and have acted as seeds for the observable matter, which is not homogeneously distributed across the Universe.

The terms 'stimulated emission' and 'spontaneous emission' were developed by Einstein in relation to thermal broadband light sources, since both types were necessary for thermodynamic reasons. Coherent coupling of light fields and atoms was neither conceptually nor experimentally conceivable at that time.

6.3.1 Stimulated emission and absorption

Let us now investigate how we can obtain the limiting case of a broadband incoherent light field from the Bloch equations. For this purpose we use the complex form in Eq. (6.29) and assume $|\rho_{eg}| \ll w \simeq -1$. With the equilibrium value ρ_{eg} we obtain without difficulty

$$\dot{w} = -\frac{\gamma' \Omega_R^2}{\gamma'^2 + \delta^2} w - \gamma(w - w_0).$$

We are interested in the first term, containing the stimulated processes (emission and absorption) because of $\Omega_R^2 = d_{eg}^2 \mathcal{E}_0^2/\hbar^2 \propto I$, and we take the broadband spectrum into account by integrating over all detunings δ and defining $\mathcal{E}_0$ to be the mean quadratic field amplitude,

$$\dot{w} = \pi(d_{eg}^2/3)\mathcal{E}_0^2/\hbar^2 - \gamma(w+1).$$

The coupling of unpolarized field and atomic dipole generates a factor $1/3$ by averaging over the space directions, and with $\rho_{ee} = (w+1)/2$ we find the form

$$\dot{\rho}_{ee} = \frac{\pi d_{eg}^2}{3\epsilon_0 \hbar^2}u(\nu_0)w - \gamma(w+1)/2 = B_{eg}u(\nu_0)(\rho_{gg}-\rho_{ee}) - \gamma\rho_{ee}. \qquad (6.40)$$

Here we call $u(\nu_0) = \epsilon_0 \mathcal{E}_0^2/2$ the energy density at the resonance frequency ν_0. The coefficient B_{eg} is called the *Einstein B coefficient* and determines the rate of stimulated emission and absorption, respectively.

6.3.2   Spontaneous emission

Rigorous calculation of the spontaneous emission rate requires a treatment according to the rules of quantum electrodynamics, i.e. with the help of a quantized electromagnetic field. In fact the calculation of the spontaneous emission rate by V. Weisskopf and E. Wigner [108] in 1930 was the first major success of this, then very new, theory.

We here choose a much shorter way by using the result of the Larmor formula known from classical electrodynamics. It says that the radiation power of an accelerated electric charge is proportional to its squared acceleration,

$$P = \gamma h\nu_0 = \frac{2}{3c^3}\frac{e^2}{4\pi\epsilon_0}\ddot{x}.$$

We assume that, during the characteristic decay time γ^{-1}, just the excitation energy $h\nu_0$ is emitted. From quantum mechanics we adopt the result $\ddot{x} = x\omega_0^2$, and by ad hoc multiplication with the factor 2 delivered only by quantum electrodynamics, we obtain the result

$$\gamma = A_{eg} = 2 \times \frac{P}{h\nu_0} = \frac{d_{eg}^2\omega_0^3}{3\pi\hbar\epsilon_0 c^3} \qquad (6.41)$$

for the *Einstein A coefficient*.

By comparison with Eq. (6.40), we confirm the result that Einstein obtained from purely thermodynamic reasoning,

$$\frac{A}{B} = \frac{\hbar\omega_0^3}{\pi^2 c^3} = \hbar\omega\frac{\omega^2}{\pi^2 c^3}. \qquad (6.42)$$

With $\rho(\omega) = \omega^2/\pi^2 c^3$, the latter form contains just the state density of the radiation field for the frequency ω (see Appendix B.3). If the driving field contains photons in a certain mode $\bar{n}_{\mathrm{ph}}$, then the ratio of spontaneous and stimulated emission rate in this mode has to be

$$A : B = 1 : \bar{n}_{\mathrm{ph}}.$$

Excursion: Suppression of 'natural' decay

The natural decay of an excited atomic or molecular state appears to be inevitable and fundamental, though, in an environment with conductive surfaces, the decay rate can be modified or even switched off. As an example we consider an atom, which we imagine in a simplified way as a microscopic dipole antenna between two metallic reflecting walls distance d apart. The atomic radiation is reflected by the walls and re-affects the atom. Depending on its position, which determines the phase lag due to the round trip, the reflected radiation is either reabsorbed and hinders decay or causes an even faster decay of the excited atom by amplifying the dipole oscillation, or equivalently by constructive interference.

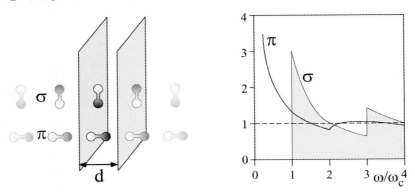

Fig. 6.12: *Suppression of spontaneous emission between plane mirrors. Left: Image charge model. For small distances, the interference field of the image dipoles leads to extinction in σ position, and to intensification in π position. Right: Density of states for σ- and π-polarized radiation fields.*

The reflected radiation field can be analysed following the intuitive method of image charges (Fig. 6.12). Then the modification of the atomic free space decay rate can be calculated equivalently to the radiation field of a chain of atomic image dipoles [71]. Alternatively, we may consider waves allowed to propagate between the two mirrors. They form a primitive waveguide with cut-off frequency at

$$\omega_c = \pi c/d,$$

at least for electric fields polarized perpendicular to the surface normal (σ polarization). Thus, if the separation of the mirrors falls below half of the resonance wavelength, $d < \pi c/\omega_{at} = \lambda_{at}/2$, then the atomic radiation field is no longer able to propagate at this wavelength or frequency, and the spontaneous decay of the excited state is completely suppressed! This fact is theoretically reflected in the density of states of the radiation field as shown in Fig. 6.12. Though atomic resonance wavelengths are very small (they only have µm dimensions), exactly this type of experiment has been carried out to demonstrate the dependence of the spontaneous decay on the environment of the microscopic radiator [71]. This example impressively shows that the radiative properties of a microscopic particle are influenced by its environment. In highly reflective cavities, the modification is particularly radical. This topic has been extensively investigated under the name of 'cavity QED' for many years [9].

6.4   Inversion and optical gain

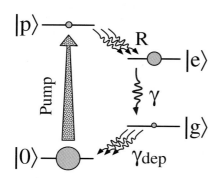

Fig. 6.13: *Four-level system with inversion between the upper ($|e\rangle$) and lower ($|g\rangle$) levels. The circles indicate the population of the levels in dynamic equilibrium.*

It is impossible to obtain inversion by optically exciting a two-level system, as outlined in Section 6.2.5. In a system with four states, however, we can build a dynamic equilibrium that generates a stationary inversion between two of the four levels by supplying energy, and so fulfil the requirement for running a laser. An inversion (and thus the requirement for laser operation) can be obtained with three levels as well. However, the four-level system causes a strict separation of the states directly contributing to the pumping process and the laser process. That is why it is preferred for a transparent treatment.

The idealized system is presented in Fig. 6.13. The pumping process promoting particles from the ground state $|0\rangle$ to the pump level $|p\rangle$ at a total rate $R = V\mathcal{R}$ can be driven by electron impact in a discharge, by absorption from the light of an incandescent lamp or a laser, or by other mechanisms. We will get to know some of them in the chapter about lasers. Our focus is on the two levels $|e\rangle$ and $|g\rangle$, which from now on are to be referred to as 'laser levels'.

6.4.1   Inversion

We consider the rate equations for the occupation numbers n_0, n_p, n_e and n_g. We focus on weak pumping processes, where most of the atoms remain in the ground state, and we may keep $n_0 \simeq 1$ to a good approximation. By a short consideration or calculation, it can be found that for these conditions the rate equation system can effectively be limited to the laser states $|e\rangle$ and $|g\rangle$. The dynamics is determined by the population rate R of the upper state, by its decay rate γ, by the partial transition rate $\gamma_{eg} \leq \gamma$ falling to the lower laser level, and finally by the depopulation rate of the lower laser level γ_{dep}:

$$\begin{aligned}
\dot{n}_e &= R - \gamma n_e, \\
\dot{n}_g &= \gamma_{eg} n_e - \gamma_{\mathrm{dep}} n_g.
\end{aligned} \tag{6.43}$$

The stationary solutions $n_e^{\mathrm{st}} = R/\gamma$ and $n_g^{\mathrm{st}} = \gamma_{eg} R / \gamma \gamma_{\mathrm{dep}}$ are found, and the difference in the occupation numbers can be calculated in equilibrium but in the absence of any light field that could cause stimulated emission:

$$n_0 = n_e^{\mathrm{st}} - n_g^{\mathrm{st}} = \frac{R}{\gamma} \left(1 - \frac{\gamma_{eg}}{\gamma_{\mathrm{dep}}} \right). \tag{6.44}$$

If the depopulation rate γ_{dep} of the lower state is larger than the decay rate of the upper state, then apparently an *inversion*, $n_0 > 0$, is maintained in this system because

$\gamma_{eg}/\gamma_{\text{dep}} < 1$. The inversion is a non-equilibrium situation from the thermodynamic point of view and requires an energy flow through the system.

Since the imaginary part of the polarization is now also *positive* (eq. (6.36)), we expect the polarization not to be absorbed by the field causing it but in contrast to be intensified! A field growing stronger, though, reduces this inversion according to Eq. (6.37), but maintains the amplifying character (Fig. 6.9). With this system, the requirements for an optical amplifier are met. It is known that an amplifier excites itself by feedback and works as an oscillator. We call these devices 'lasers'.

6.4.2 Optical gain

If inversion occurs ($w_0 > 0$), then, because of the positive $v_{\text{st}} > 0$, a negative absorption coefficient is caused (Eq. (6.18)), also known as the optical gain coefficient. Its unit is cm^{-1} as well. According to Fig. (6.9), it is clear that the inversion – and so the gain – is reduced under the influence of a light field. For laser operation, we will call this *saturated gain*. In the case of very low intensities $I/I_0 \ll 1$, the gain is constant and it is called *small signal gain*. This value is usually given for a laser material.

7 The laser

The laser has become an important instrument, not only in physical research but for almost all fields of everyday life. In this chapter we are going to introduce some particularly important laser systems with technical details. A theoretical description of their most important dynamic physical properties is given in Chapter 8.1.

The word *laser* has become a well-recognized word in everyday language, and is derived from its predecessor, the *maser*, the acronym 'maser' meaning 'microwave amplification by stimulated emission of radiation'. The laser is based on identical physical principles and is an optical maser, the abbreviation 'laser' meaning 'light amplification by stimulated emission of radiation'. We basically regard a laser as a source of an intensive coherent light field. Laser light appears absolutely artificial to us, and our ancestors certainly never happened to experience the effect of a coherent light beam,[1] though in the cosmos there are several natural sources of maser radiation. The example with the shortest wavelength is probably the hydrogen gas surrounding a star named MWC349 in the Cygnus constellation, which is excited to luminescence by the ultraviolet radiation of this hot star. The hydrogen gas arranged in a disc amplifies the far-infrared radiation of the star at the wavelength of $169\,\mu$m several million times, such that it can be detected on Earth.

The laser has historical roots in high-frequency and gas discharge physics. It was known from the maser that it was possible to construct an amplifier and oscillator for electromagnetic radiation with an inverted molecular or atomic system. In a famous publication A. Schawlow (1921-1999, nobel prize 1981) and C. Townes [91] (1915-, nobel prize 1964) had theoretically predicted the properties of an 'optical maser', later called a laser.

The optical properties of atomic gases had already been studied in discharges for a long time. The question was raised whether an inversion and thus amplification of light could be achieved by a suitable arrangement. So it becomes understandable that the first continuous-wave laser realized by the American physicist Ali Javan (1928-) in 1960 [50] with an infrared wavelength of $1.152\,\mu$m was a surprisingly complex system consisting of a gaseous mixture of helium and neon atoms.

The laser bears a close analogy to an electronic amplifier that is excited to oscillations by a positive feedback. Its oscillation frequency is determined by the frequency characteristic of gain and feedback (Fig. 7.1). It is known that an amplifier oscillates

---

[1] Though interference and coherence phenomena can be observed even in our everyday environment – for example, take a piece of thin, fine fabric and watch some distant, preferably coloured lights through it, e.g. the rear lights of a car.

Optics, Light and Laser. Dieter Meschede
Copyright © 2004 Wiley-VCH Verlag GmbH & Co. KGaA
ISBN: 3-527-40364-7

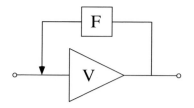 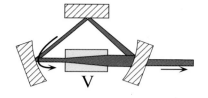

electronic oscillator   laser oscillator

Fig. 7.1: *Analogy between a laser and an electronic amplifier, which becomes an oscillator by feedback coupling. The oscillator frequency can be selected, e.g. by a filter (F) in the feedback path. For the laser, the feedback is achieved using resonator mirrors. For clarification, a ring resonator with three mirrors has been chosen. The spectral properties of the amplifying medium as well as the wavelength-dependent reflectivity of the resonator mirrors determine the frequency of the laser.*

with positive feedback if the gain becomes greater than the losses,

> oscillation condition: gain ≥ losses.

The amplitude grows more and more until the losses by outcoupling or within the oscillator circuit just compensate the gain. The effective gain then decreases. This is called a 'saturated gain' (see Section 8.1.2).

As we already know from the chapter about light and matter, an inversion of the laser medium is necessary to achieve an intensification of a light wave. Using the simplest picture there always have to be many more atoms in the upper excited state than in the lower one. If this condition is not fulfilled, the laser oscillations die or do not even start. An ideal laser is supposed to deliver a gain as large as possible and independent of the frequency. Since such a system has not yet been found, a multitude of laser systems is used. The most important variants roughly divided into classes (Tab. 7.1) will now be introduced with their technical concepts, strengths and weaknesses.

Tab. 7.1: *Laser types.*

	Gaseous	Liquid	Solid state
Fixed frequency	neutral atoms		rare-earth ions
	ions		3d ions
Multiple frequency	molecules		
Tunable		dyes	3d ions
			colour centres

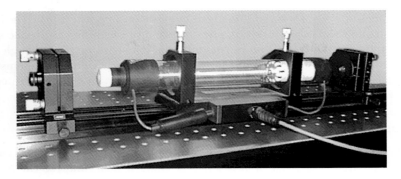

Fig. 7.2: *Helium–neon laser in an open experimental set-up. The current is supplied to the discharge tube by the two cables. The resonator mirrors and the laser tube are mounted on finely tunable bearings.*

7.1   The classic system: the He–Ne laser

The helium–neon laser (He–Ne laser) has played an unsurpassed role in scientific research on the physical properties of laser light sources, e.g. experimental investigations of coherence properties. Just for this alone it is the 'classic' of all laser systems. We shall introduce several important laser features using this system as an example.

1.1.1 Construction

The amplifier

The helium–neon laser obtains its gain from an inversion in the metastable atomic excitations of the Ne atom (the luminescence of Ne atoms is also known due to the proverbial neon tubes).

In Fig. 7.3 the relevant atomic levels with some important features and some selected laser wavelengths ('lines') are presented. Since the gas mixture is quite dilute, we can easily understand the He–Ne laser using the picture of independent atoms. The Ne atoms are excited not directly by the discharge but by energy transfer from He atoms, which are excited to the metastable $^1S_0$ and $^3S_1$ levels by electron impact. The Ne atom has nearly resonant energy levels

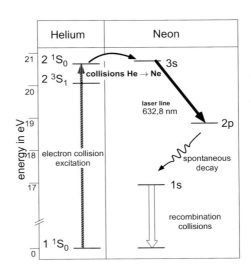

Fig. 7.3: *Energy levels of He and Ne atoms with the most prominent optical transition at 632.8 nm. For nomenclature, the spectroscopic terms are used. For the energy levels, the lifetimes are given as well.*

so that an efficient energy transfer is enabled by resonant impacts. In the He–Ne laser, the excitation and the laser transition are split up into two different atomic systems, which is helpful for the realization of the desirable four-level system, though there is a problem at the lower laser level of the Ne atoms (Fig. 7.3), which is metastable as well and cannot be emptied by radiative decay. In a narrow discharge tube, collisions with the wall lead to efficient depopulation of the lower laser level.

Operating conditions

The inversion can only be maintained in a rather dilute gas mixture compared to the atmospheric environment. The He pressure p is some 10 mbar, and the He : Ne mixing ratio is about 10 : 1. The He discharge is operated at a current of several milliamps and a voltage of 1–2 kV, and it burns in a capillary tube with a diameter of $d \le 1$ mm (Fig. 7.2). At its walls the metastable Ne atoms (Fig. 7.3) fall back to the ground state again due to collisional relaxation and are available for another excitation cycle. The discharge is ignited by a voltage pulse of 7–8 kV.

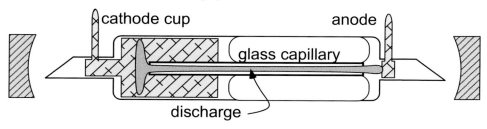

Fig. 7.4: *Schematic representation of a helium–neon laser. The big cathode cup prevents fast erosion caused by the discharge. The Brewster windows at the ends of the laser tube reduce reflection losses at the windows and uniquely determine the laser polarization.*

All He–Ne lasers have this construction principle in common, only the construction length and the gas filling pressure being slightly different depending on the application. Optimum conditions for the product of pressure p and diameter d are empirically found at

$$p \cdot d \simeq 5 \,\mathrm{mbar\,mm}.$$

The output power of commercial He–Ne lasers varies between 0.5 mW, which will just not damage the eyes, and 50 mW. The power depends on the discharge current and the length of the tube. Both can only be extended to a certain amount. The gain is proportional to the density of inverted Ne atoms, but this already reaches a maximum at a few tens of milliamps since increasing electron collisions de-excite the atoms. The length of the tube cannot be significantly expanded over $\ell = 1$ m, for the following reasons. On the one hand, the diameter of the Gaussian modes grows with increasing mirror distance and does not fit into the capillary tube any more. Furthermore, with a larger construction length, the 3.34 μm line starts oscillating as a superradiator even without mirrors, and thus withdraws energy from laser lines competing for the same reservoir of excited atoms.

The laser resonator

The resonator mirrors can be integrated into the discharge tube and may be once and forever adjusted during manufacturing. Especially for experimental purposes, an external resonator with manually adjustable mirrors is used. The tube has windows at its ends. In the simplest case the resonator only consists of two (dielectric) mirrors and the discharge tube. To avoid losses, the windows are either anti-reflection coated or inserted at the Brewster angle.

Example: Radiation field in the He–Ne laser resonator

The laser mirrors determine the geometry of the laser radiation field according to the rules of Gaussian optics (see Section 2.3). They have to be chosen such that the inverted Ne gas in the capillary tube is used as optimally as possible. For a symmetric laser resonator with mirror radii $R = 100\,\text{cm}$ (reflectivity 95% and 100%), and separated by $\ell = 30\,\text{cm}$, one obtains for the red $633\,\text{nm}$ line a TEM_{00} mode with the parameters:

$$
\begin{aligned}
\text{confocal parameter} \quad & b = 2z_0 & = & \quad 71\,\text{cm}, \\
\text{beam waist} \quad & 2w_0 & = & \quad 0.55\,\text{mm}, \\
\text{divergence} \quad & \Theta_{\text{div}} & = & \quad 0.8\,\text{mrad}, \\
\text{power inside/outside} \quad & P_{\text{i}}/P_{\text{o}} & = & \quad 20\,\text{mW}/1\,\text{mW}.
\end{aligned}
$$

There are no problems of fitting the laser beam over the complete length to the typical cross-section of the plasma tube of about $1\,\text{mm}$. Even in a distance of $10\,\text{m}$ it has just a cross-section of about $4\,\text{mm}$.

7.2   Mode selection in the He–Ne laser

We devote the next two sections to the physical properties of the He–Ne laser (mode selection and spectral properties) because it is a good model to present the most important general laser properties and since for the He–Ne laser the physical properties have been investigated particularly thoroughly.

The aim of mode selection in every continuous wave ('cw') laser is the preparation of a light field oscillating both in a single spatial, or transverse mode, and in a single longitudinal mode, i. e. at just a single optical frequency ω. The methods used with the He–Ne laser can be applied to all other laser types with slight modifications. The desirable transverse mode is mostly a TEM_{00} or closely related mode. It has fewer losses due to its comparatively small cross-section and thus is often intrinsically preferred anyway. In case of doubt the relevant spatial mode can be selected by using a suitable adjustment of the resonator or insertion of an additional aperture.

7.2.1   Laser line selection

If several laser lines of the neon atom have a common upper laser level (e.g. 2s, Fig. 7.3), only that one with the highest gain can be observed. If the laser line couples completely different levels (e.g. 2s–2p and 3s–3p), then the lines can be activated simultaneously.

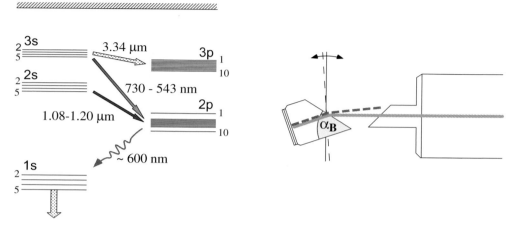

Fig. 7.5: *Wavelength selection in the He–Ne laser. Left: Part of the energy level scheme of the neon atom showing the important laser transitions. The common notation does not follow the customary singlet/triplet convention according to the LS coupling scheme. The notation used here goes back to Paschen, who simply numbered the levels consecutively. An s level splits into 4, and a p level into 10 angular momentum states. Right: Littrow prism as a dispersive end mirror for wavelength selection.*

Owing to the helium discharge in the neon gas, the 2s state as well as the 3s are populated with the occupation of the uppermost $3s_2$ substate dominating in the 3s group. The largest gain factors are obtained at the wavelengths of 0.633, 1.152 and 3.392 µm. Transitions with a low gain can be excited if the feedback coupling by the resonator selectively favours or suppresses certain frequencies by means of some suitable optical components. In general, all dispersive optical components – such as optical gratings, prisms and Fabry–Perot etalons – are appropriate. One of the simplest methods is the installation of a Littrow prism as shown in Fig. 7.5. The Littrow prism is a Brewster prism divided in half so that the losses for p-polarized light beams are minimized. The backside of the Littrow prism is coated with dielectric layers to make a highly reflective mirror. Since the refraction angle depends on the wavelength, the laser line can be selected by tilting the Littrow prism.

Another specialty is the extremely high gain coefficient of the infrared 3.34 µm transition (typically $10^3 \, \text{cm}^{-1}$), causing the line to start oscillating almost every time. It can be suppressed by using infrared-absorbing glass and limiting the length of the plasma tube. The latter fact is unfortunate since it imposes a technical limit to the output power, which otherwise increases with length.

7.2.2 Gain profile and laser frequency

For the example of the He–Ne laser, we are now going to ask how the oscillation frequency of a laser line depends on the combined properties of the gas and the resonator. The spectral width of the fluorescence spectrum (the width of the optical resonance line) is determined by the Doppler effect caused by the neon atoms moving at thermal speed of several $100\,\mathrm{m\,s^{-1}}$ in a gas at room temperature. This broadening is called *inhomogeneous* (see Section 11.3.2), since atoms with different velocities have different spectra. With regard to the laser process this especially means that the coupling of the neon atoms to the laser light field depends very strongly on their velocity. For the red laser line at 633 nm, the Doppler linewidth at room temperature is about $\Delta\nu_{\mathrm{Dopp}} = 1.5\,\mathrm{GHz}$ according to Eq. (11.8) and can just be resolved with a high-resolution spectrometer (e.g. Fabry–Perot).

In Fig. 7.6 the gain profile and its significance for the laser frequency is presented. It makes the laser begin oscillating if the gain is higher than the losses. Within the gain profile the laser frequency is determined by the resonance frequencies of the laser resonator (here indicated by the transmission curve showing maxima at frequencies separated by the free spectral range Δ_{FSR}). At these 'eigenfrequencies' the laser may start oscillating, as we shall investigate more deeply in Section 8.1. True lasers are slightly shifted off the resonances of the empty resonator, an effect that is called *mode pulling*.

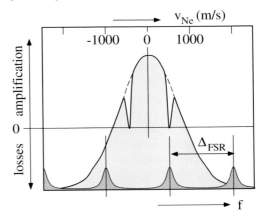

Fig. 7.6: *Gain profile of a He–Ne laser in operation. Without the laser field, the small signal gain corresponds to the neon Doppler profile (dashed line); the laser operation modifies the gain profile by so-called spectral or Bennett holes (see text). Two holes symmetric to $v = 0\,\mathrm{m\,s^{-1}}$ occur because in a standing wave the Doppler shift acts in two directions.*

7.2.3 Spectral holes

At the eigenfrequencies of the He–Ne laser one can observe spectral holes (the so-called *Bennett holes*). The atoms make up the difference between their rest-frame frequency ν_0 and the laser frequency ν_{L} by their velocity v_z in the direction of the resonator axis, and the atoms of a gas laser contribute to the gain only within their homogeneous, i.e. their natural, linewidth. This situation is called *saturated gain*. In equilibrium the effective gain is reduced at these frequencies to that value just corresponding to the losses (including the decoupling at the resonator mirrors).

The small signal gain from Fig. 7.6 can be measured by sending a very weak tunable probe beam through the He–Ne laser and measuring the gain directly. Since in a resonator with standing waves atoms can couple to the light field in both directions, two spectral holes can be observed at

$$\nu_L = \nu_0 \pm k v_z.$$

This observation also indicates that two different velocity groups of atom contribute to the gain of the backward and forward running intra cavity wave, respectively. Thus a very interesting case occurs when both holes are made to coincide by e.g. changing the resonator frequency by length variations with the help of a piezo-mirror. At $v_z = 0$ a lower gain than outside the overlap region of the holes is available and the output power of the laser decreases. This collapse is called *Lamb dip* after Willis E. Lamb (born 1913),[2], and it initiated the development of Doppler-free saturation spectroscopy.

7.2.4   The single-frequency laser

In Fig. 7.6 only one resonator frequency lies within the gain profile such that the laser threshold is exceeded. Since the free spectral range $\Delta_{FSR} = c/2\ell$ of the He–Ne laser exceeds the width of the Doppler profile at $\lambda = 633\,\mathrm{nm}$ below $10\,\mathrm{cm}$, for typical, i.e. larger, construction lengths, generally 2–4 frequencies start to oscillate because in the inhomogeneous gain profile there is no competition between the modes about the available inversion. But we can still insert additional (and low-loss) dispersive elements into the resonator which modulate the spectral properties of the gain profile in a suitable way to filter the desired laser frequency from the available ones. To discriminate between adjacent resonator modes, highly dispersive elements such as Fabry–Perot etalons are required.

Example: Gain modulation with intra-cavity etalons
Etalons cause a modulation of the effective laser gain which is periodic with Δ_{FPE}, the free spectral range (Eq. 5.18) separating adjacent transmission maxima. The periodicity can be chosen by the etalon thickness or length ℓ. Tilting the etalon causes the transmission maxima to shift, and from geometrical considerations following section 5.5 one obtains

$$\nu_{max} = N\Delta_{FPE}(\alpha) = N\frac{c}{2\ell\sqrt{n^2 - \sin^2\alpha}} \simeq N\frac{c}{2\ell n}\left[1 - \frac{1}{2}\left(\frac{\alpha}{n}\right)^2\right].$$

Small tilts change the free spectral range only slightly but are enough due to the high order N to tune the centre frequency efficiently. The monolithic thin etalon presented in Fig. 7.7 is simple and intrinsically stable, but when tilted it leads to a *walk-off* and thus increased losses. The air-spaced etalon (the variable air gap is cut at the Brewster angle to keep the resonator losses as small as possible) has a more complex mechanical construction but avoids the walk-off losses of the tilted device.

---

[2]W. E. Lamb has become immortal in physics mainly by the discovery of the *Lamb shift* also named after him, for which he received the Nobel prize in 1955. He also contributed significantly to the pioneering days of laser spectroscopy.

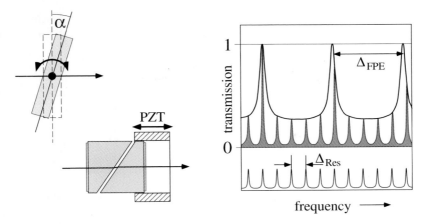

Fig. 7.7: *Frequency selection by intra-cavity etalons. Left: 'Thin' etalon and 'thick' etalon. For rough tuning the thin etalon is tilted where the walk-off caused by the tilt is tolerable. The thick etalon is constructed with an air gap at the Brewster angle, which can be varied in length by a piezo-translator. Right: Combined effect of an etalon (Δ_{FPE}) and the laser resonator (Δ_{Res}) on the transmission (and thus on the gain). For this example the length of the etalon is about one-fifth of that of the resonator.*

The intra-cavity etalons do often not need any additional coating. Even using only the glass–air reflectivity of 4% a modulation depth of the total gain of about 15% is obtained according to Eq. (5.16). With regard to the small gain, in many laser types this is completely sufficient.

7.2.5 Laser power

We shall now investigate how to optimize the laser output power, i.e. for practical reasons generally to maximize it. The properties of the amplifier medium are physically determined and thus can be influenced only by a suitable choice, length, density, etc. The losses can be kept as small as possible by design and choice of the components of the resonator. Finally there is only the choice of the mirror reflectivity as the remaining free parameter, which also acts as a loss channel. We consider a Fabry–Perot interferometer with gain as a model. For this we make use of the considerations about dissipative resonators from Section 5.5. The laser is always operated in the resonance

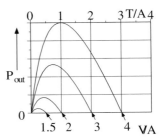

Fig. 7.8: *Output power of a laser as a function of the transmission T of the outcoupling mirror, and the gain V normalized to the resonator losses A.*

case. We here anticipate the correlation between gain V, losses A and transmission T of the only decoupling mirror more accurately dealt with in Section 8.1, eq. (8.17),

$$I_{\text{out}} = I_0 \frac{T(V - A - T)}{A + T},$$

and investigate it graphically as a function of the variable transmission T in Fig. 7.8.

7.3   Spectral properties of the He–Ne laser

7.3.1   Laser linewidth

Until now we have taken monochromatic optical light fields for granted, i.e. we have assumed that the optical wave can be described by a single exactly defined frequency ω. In the chapter about the laser theory we will see that laser light comes closer to this deeply classical idea of a perfect harmonic oscillation than almost any other physical phenomenon. The physical limit for the spectral width of a laser line measured according to the so-called *Schawlow–Townes limit* (Section 8.4.4) amounts to several hertz or even less! This physical limit is imposed due to the quantum nature of the light field. It has been mentioned already by the authors in that paper proposing the laser in 1958. According to this the linewidth of the laser is (see Eq. (8.32))

$$\Delta\nu_{\text{L}} = \frac{N_2}{N_2 - N_1} \frac{2\pi h \nu_{\text{L}} \gamma_{\text{c}}^2}{P_{\text{L}}},$$

with ν_{L} is the laser frequency, $\gamma_{\text{c}} = \Delta\nu_{\text{c}}$ is the damping rate or linewidth of the laser resonator, P_{L} is the laser power and $N_{1,2}$ are the occupation numbers of the upper and lower laser level, respectively.

Example: Schawlow–Townes linewidth of the He–Ne laser

We consider the He–Ne laser from the previous example. The laser frequency is $\nu_{\text{L}} = 477\,\text{THz}$, the linewidth of the resonator is $\Delta\nu_{\text{c}} = 8\,\text{MHz}$ (according to the data from the example on p. 195), while all internal resonator losses are neglected according to Eq. (5.19). The He–Ne laser is a four-level laser so that we have $N_1 \simeq 0$. For an output power of $1\,\text{mW}$ we calculate a laser linewidth of just

$$\Delta\nu_{\text{L}} \simeq \frac{2\pi h \times 477\,\text{THz}\,(8\,\text{MHz})^2}{1\,\text{mW}} = 0.13\,\text{Hz}.$$

The extremely small Schawlow–Townes linewidth of the red He–Ne line corresponds to a Q value $\nu/\Delta\nu \simeq 10^{15}$! Even today laser physicists think of this limit as a thrilling challenge because it promises to make the laser the ultimate precision instrument wherever a physical quantity can be measured by means of optical spectroscopy.

From the beginning the He–Ne laser has played an extraordinary role for precision experiments, and it is indeed a challenge even just to measure this linewidth! It is thus useful to illustrate the methods used to measure the linewidth of a laser.

7.3.2   Optical spectral analysis

The spectrum of a laser oscillator, like that of any other oscillator, can be investigated using several methods. A Fabry–Perot interferometer is used as a narrow-band filter. Its mid-frequency is tuned over the area of interest. A photodiode measures the total power transmitted within the filter passband.

Alternatively the laser beam can be superimposed with a second coherent light field (*local oscillator*) onto a photodiode. The photodiode generates the difference frequency or 'heterodyne' beat signal. The superposition signal in turn can be analysed by radio-frequency methods or Fourier analysis.[3]

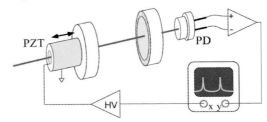

Fig. 7.9: *Scanning Fabry–Perot interferometer for the spectral analysis of laser radiation.*

The Fabry–Perot spectrum analyser

In the simplest, and therefore very often used, method a tunable optical filter is used, also called in brief a 'scanning Fabry–Perot', usually a confocal optical resonator. One of its mirrors can be displaced ('scanned') by several $\lambda/4$ corresponding to several free spectral ranges with the help of a piezo-translator. The resolution of the optical filter usually reaches some megahertz and therefore be used only for rough analysis or as a laser with a large linewidth (like e.g. diode lasers, see Section 9).

If the linewidth of the laser is smaller than the width of the transmission curve of the Fabry–Perot interferometer, some information about the frequency fluctuations can still be obtained by setting its frequency to the wing of the filter curve and using this as a frequency discriminator (Fig. 7.10). So any frequency variations are converted to amplitude fluctuations, which in turn can be analysed by means of radio-frequency techniques or by Fourier transformations.

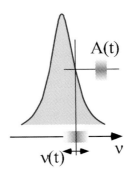

Fig. 7.10: *Transmission curve used as a frequency-to-amplitude converter.*

The heterodyne method

When applying the heterodyne method it is important to closely match the wavefronts of both light fields and enter the photodiode with excellent flatness so that the detector is exposed to the same phase everywhere. Otherwise the beating signal is strongly reduced.

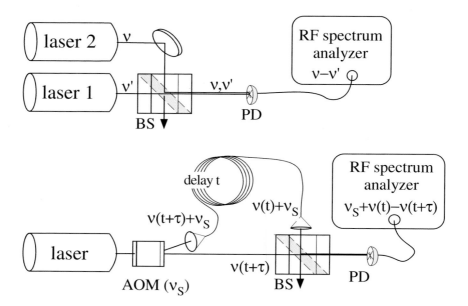

Fig. 7.11: *Heterodyne method for the determination of the linewidth. Top: Superposition with a laser used as a local oscillator. Bottom: Superposition according to the autocorrelation method. PD = photodiode; BS = beam splitter; AOM = acousto-optical modulator.*

In Fig. 7.11 the scheme is presented according to which a heterodyne signal can be achieved at radio frequencies. A second laser can be used as a local oscillator. Its frequency should be much more 'stable' than the laser to be tested. Furthermore, it must not deviate too far from the test frequency, since above 1–2 GHz high-speed photodiodes become more and more unwieldy (the active area shrinks more and more to avoid parasitic capacitances and to gain bandwidth) and expensive.

The autocorrelation method is an alternative in which the laser in a way 'pulls itself out of the mud'. One part of the laser light is split off by an AOM (acousto-optical modulator) and at the same is time shifted by the frequency ν_S, which is typically some 10 MHz. One of the two light beams is now delayed over a long optical fibre so that there is no longer any phase correlation ('coherence') between the light waves. Both light waves are superposed onto a photodiode as before and the mixed signal is investigated using a radio-frequency spectrum analyser.

The method is completely analogous to a Michelson interferometer. It is operated at a path difference larger than the coherence length. There is no visibility (i.e. the average of the interference signal vanishes) but a fluctuating beat signal that delivers a good measure of the spectral properties of the laser.

---

[3]Superposition of two electromagnetic waves with different frequencies is usually called 'hetero-dyning', while superposition of identical frequencies is called 'homodyning'. Heterodyning is generally preferred since the noise properties of both detectors and receivers are favourable at higher frequencies.

7.4   Applications of the He–Ne laser

For the production of He–Ne laser tubes, the manufacturing technology of radio tubes was very suitable. Radio tubes were being replaced by transistors in the 1960s, so a high production capacity was available when the He–Ne laser was developed. Historically, its rapid distribution was significantly supported by this fact.

The best-known wavelength of the He–Ne laser is the red laser line at 632 nm, which is used in countless adjusting, interferometric and reading devices. Though the use of the red He–Ne laser is declining rapidly since mass produced and hence cheap red diode lasers (Chapter 9) have become available, that can be operated with normal batteries, are very compact and meanwhile offer very acceptable TEM_{00} beam profiles, too. The He–Ne laser still plays an important role in metrology (the science of precision measurements). The red line is used, for example, to realize length standards. The infrared line at 3.34 μm constitutes a secondary frequency standard if it is stabilized on a certain resonance of the methane molecule.

7.5   Other gas lasers

Stimulated by the success of the helium–neon laser, many other gas systems have been investigated for their suitability as a laser medium. Gas lasers have a small gain bandwidth and are *fixed-frequency lasers* when their small tunability within the Doppler bandwidth is neglected. Like the He–Ne laser, they play a role as instrumentation lasers provided they have reasonable physical and technical properties, such as e.g. good beam quality, high frequency stability and a low energy consumption. Some gas lasers are in demand because they deliver large output power, not in pulsed, but in continuous-wave (cw) operation. In Tab. 7.2 are listed those gas lasers which nowadays have practical significance. It is technically desirable to have a substance already gaseous at room temperature. That is why the rare gases are particularly attractive. Among them argon- and, even more so, krypton-ion lasers have achieved technical significance.

7.5.1   The argon laser

The argon-ion laser plays an important role since it is among the most powerful sources of laser radiation and is commercially available with output powers of several 10 watts. However, the technical conversion efficiency, i.e. the ratio of electric power consumption and effective optical output power, is typically 10 kW : 10 W. For many applications this is absolutely unacceptable. In addition, it is also burdened with the necessity to annihilate most of the energy spent by a costly water cooling system. Therefore it can be observed that the frequency-doubled solid-state lasers (e.g. Nd:YAG, see Section 7.8.2) have become more and more favourable replacements of the argon laser. In the ultraviolet range there is no competitor to the Ar-ion laser in sight yet.

Tab. 7.2: *Overview: gas lasers.*

Laser	Short form	CW/P*	Laser lines	Power
Neutral-atom gas lasers				
Helium–neon	He–Ne	CW	633 nm	50 mW
			1.152 nm	50 mW
			3.391 nm	50 mW
Helium–cadmium	He–Cd	CW	442 nm	200 mW
			325 nm	50 mW
Copper vapour	Cu	P	511 nm	60 W
			578 nm	60 W
Gold vapour	Au	P	628 nm	9 W
Noble-gas ion lasers				
Argon-ion laser	Ar^+	CW	514 nm	10 W
			488 nm	5 W
			334–364 nm	7 W
Krypton-ion laser	Kr^+	CW	647 nm	5 W
			407 nm	2 W
Molecular gas lasers				
Nitrogen	N_2	P	337 nm	100 W
Carbon monoxide	CO	CW	4–6 µm	100 W
Carbon dioxide	CO_2	CW	9.2–10.9 µm	10 kW
Excimers	F_2, ArF, KrF, XeCl, XeF	P	0.16–0.35 µm	250 W

*CW = continuous wave; P = pulsed.

The amplifier

Excitation of the high-lying Ar^+ states is obtained by stepwise electron impact. That is why there is a very much higher current density necessary than in a He–Ne laser. The upper laser level can be populated from the Ar^+ ground state as well as from other levels above or below it. The krypton laser follows a quite similar concept, but it has achieved less technical importance.

Operating conditions

In the 0.5–1.5 m long tubes a discharge is operated maintaining an argon plasma. The cross-section through the plasma tube in Fig. 7.12 indicates the elaborate technology that is necessary due to the high plasma temperatures. The inner bores of the plasma tube are protected by robust tungsten discs, which are inserted into copper discs in order to rapidly remove the heat. A magnetic field additionally focuses the plasma current onto the axis to protect the walls against erosion. Since, as a result of diffusion, the argon ions move to the cathode, the copper discs have holes for the compensating

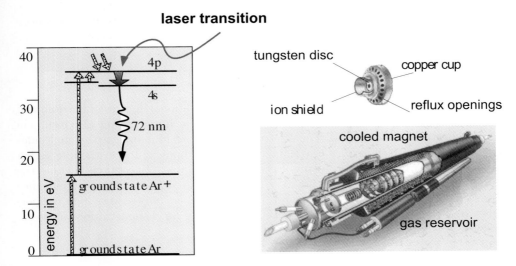

Fig. 7.12: *Lasing process of the argon-ion laser and a cross-section of the plasma tube. A magnetic field supports concentration of the plasma current near the axis. The copper discs have additional bores to allow reflux of the argon ions.*

current. An argon laser consumes gas since the ions are implanted into the walls. Therefore, commercial ion lasers are equipped with an automatic reservoir. The gas pressure is 0.01–0.1 mbar.

Features and applications

Ar ions have several optical transitions up to the ultraviolet spectral range which can be operated with a high output power. For this reason they have dominated the market for fixed-frequency lasers with high pump power until recently, but are now being severely challenged by frequency-doubled solid state lasers. Most laser laboratories cannot be imagined without high power visible laser sources because they are used to excite or, more loosely, to 'pump' other, tunable lasers like e.g. dye lasers and Ti–sapphire lasers.

7.5.2   Metal-vapour lasers

The copper- and gold-vapour lasers are commercially successful because they offer attractive specifications for lots of purposes: they are pulsed lasers but with very high repetition rates of about 10 kHz – to our eyes they are perfectly continuous. The pulse length is some 10 ns and the average output power can be 100 W. The most important wavelengths are the yellow 578 nm line and the green 510 nm line ($^2P_{1/2,3/2} \rightarrow {}^2D_{3/2,5/2}$) of the copper atom.

   The physical reason for this success, which cannot necessarily be expected given the high operating temperature of the metal vapour of about 1500°C, are the large excitation probability by electron impact (the discharge is supported e.g. by a neon buffer gas) and the high coupling strength of the dipole-allowed transitions.

7.6   Molecular gas lasers

In contrast to atoms, molecules have vibrational and rotational degrees of freedom and thus a much richer spectrum of transition frequencies, which in principle also results in a varied spectrum of laser lines. The electronic excitations of many gaseous molecules are at very short wavelength, where the technology is complex, and in the interesting visible spectral range not very many systems are available. Exceptions include the sodium-dimer laser (Na_2), which however has not achieved any practical significance since sodium vapour contains a reasonable density of dimers only at very high temperatures, and the nitrogen laser, which is today used only for demonstration purposes. However, it is quite simple to construct!

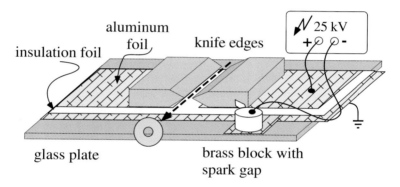

Fig. 7.13: *Simple home-made 'air laser'. Good parallel alignment of the knife edges is essential for successful operation.*

Excursion: Can a laser be operated just with air?
The short answer is: Yes! The 78% nitrogen fraction of the air can be used as a laser amplifier. And even better, a primitive 'air laser' is so simple to construct that with some skill (and caution because of the high voltage!) it can be copied in school or in scientific practical training. The original idea of a simple nitrogen laser was already presented with an instruction in *Scientific American* in 1974 [100]. It is still costly in so far as a vacuum apparatus is necessary for the control of the nitrogen flow. In a practical project carried out by high-school students [113], the laser – with slightly reduced output power – was operated directly with the nitrogen from the air.

In the simplest version a spark discharge along the knife edges in Fig. 7.13 is employed. The spark is generated according to the circuit diagram in Fig. 7.14. First, the knife edges are charged to the same high voltage potential. The breakdown of the air then takes place at the sharp tip of the spark gap so that between the two knife edges the full voltage is abruptly switched on. The discharge runs along the knife edges and also turns off rapidly again due to the high-voltage sources used with large source impedance. Suitable high-voltage sources are available in many institutions, but a small voltage multiplier can also be built by oneself without any great expense. The central experimental challenge is a reproducible and stable discharge as experience shows.

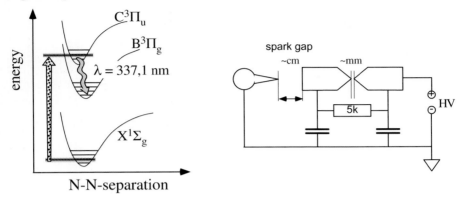

Fig. 7.14: *Left: Molecular potentials in the nitrogen molecule (schematic). Right: Circuit diagram of the air laser.*

Between the knife edges a line-shaped occupation inversion of nitrogen molecules is generated, which leads to laser emission on the 337.1 nm UV line. In Fig. 7.14 relevant molecular levels are shown along with their designations. The lower laser level is emptied only very slowly since the two involved states belong to the triplet system of the molecule (parallel electronic spins of paired electrons), which has no dipole transitions to the singlet ground state. Therefore, the inversion of energy levels cannot be maintained by a continuous discharge either, and laser operation breaks off after few nanoseconds.

Strictly speaking, the 'mirror-less air laser' is not a laser but a so-called 'superradiator'. In a superradiator spontaneous emission is amplified along the line-shaped inversion distribution (*amplified spontaneous emission*, ASE) and emitted as a coherent and well-directed flash of light.

7.6.1   The CO_2 laser

The most important examples of the molecular gas laser are the carbon monoxide and dioxide (CO and CO_2) lasers, which are based on infrared transitions between vibration–rotation energy levels. The CO_2 laser is one of the most powerful lasers in general and thus plays an important role for material processing with lasers [47].

Gain

The molecular states involved in the laser process of the CO_2 laser can be found in Fig. 7.15: a symmetric (v_1) and an antisymmetric (v_3) stretching vibration as well as a bending vibration (v_2). Vibrational quantum states of the CO_2 molecule are identified by quantum numbers (v_1, v_2, v_3).

The (001) level decays by dipole transitions, which are very slow due to the ω^3 factor of the Einstein A coefficient, however, allowing convenient build-up of inversion

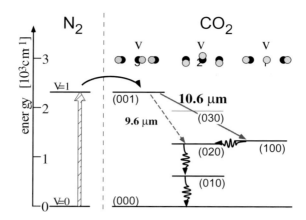

Fig. 7.15: *Transitions relevant for the CO₂ laser.*

in this level. The most important laser transition takes place between the (001) and the (100) level.

The CO_2 lasers are excited by a discharge. The occupation of the upper laser level is possible directly, but is much more favourable with the addition of nitrogen. Metastable N_2 levels not only can be excited very efficiently in a discharge but also can transfer the energy to the CO_2 molecules very profitably as well.

The (100) level is emptied very rapidly by collisional processes. In addition, it is energetically adjacent to the (020) level, which itself ensures quick relaxation to thermal equilibrium also with the (000) and (010) levels. For this, the so-called vv-relaxation plays an important role, which is based on processes of the $(020) + (000) \rightarrow (010) + (010)$ type. The heating of the CO_2 gas associated with these processes is not desirable because it increases the occupation of the lower laser level. It can be significantly reduced by adding He as a medium for thermal conductivity.

In Fig. 7.15 we have completely neglected the rotational levels of the molecule, though they cause a fine structure of the vibrational transitions leading to many closely adjacent laser wavelengths (Fig. 7.16). A typical CO_2 laser makes available about 40 transitions from the P and S branches of the rotation–vibration spectrum. The gain bandwidth of each line (50–100 MHz) is very small since the Doppler effect no longer plays a significant role at low infrared frequencies. The laser lines of a CO_2 laser can be selected by a grating used as one of the laser mirrors.

Operating conditions

The CO_2 laser is among the most powerful and robust of all laser types. It makes available a high and focusable energy density that is highly favourable for contactless material processing and laser machining. Owing to the strong application potential, multiple technically different CO_2 laser types have been developed. The operation of the CO_2 laser is disturbed by induced chemical reactions. Thus the laser gas needs to

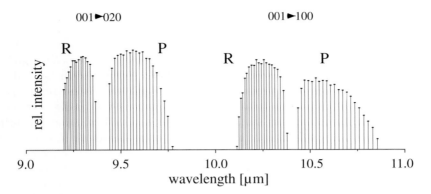

Fig. 7.16: *Emission lines of the CO_2 laser on the 9.6 µm and the 10.6 µm lines. If one resonator mirror is replaced by a grating (with mirror-like reflection in −1st order), tuning can be quite easily achieved by rotating the grating. The terms 'R' and 'P' branch are taken from molecular spectroscopy. At the R branches of the spectrum the rotational quantum number J of the molecule is decreased by 1, and it is increased by 1 at the P branches, $J \to J\pm1$.*

be regenerated, either by maintaining a continuous flow through the laser tube or by adding some suitable catalysts to the gas, e.g. a small amount of water, which oxidizes the undesired CO molecules back to CO_2. Output powers of some 10 kW are routinely achieved in larger laser systems.

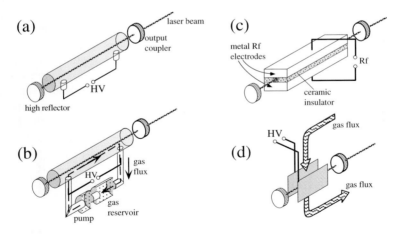

Fig. 7.17: *Important configurations of CO_2 lasers. The conventional laser (a) is operated with a sealed tube and longitudinal discharge. To increase the output power, a longitudinal gas flow (b) or a radio-frequency waveguide laser (c) can be used. The highest power can be achieved if the gas flow as well as the discharge are operated transversely to the laser beam (transversely excited, TE-laser) (d).*

7.6.2   The excimer laser

Excimer lasers play an important role for applications since they offer very high energy and furthermore the shortest UV laser wavelengths, although in the pulsed mode only. The term *excimer* is a short form of 'excited dimer', and means unusual diatomic molecules (dimers) that exist in an excited state only. Today the term has been transferred to all molecules that exist only excitedly, e.g. ArF or XeCl to mention just two examples important for laser physics. The level scheme and the principle of the excimer laser are presented in Fig. 7.18. Since the lower state is intrinsically unstable, the inversion condition is always fulfilled once the excimer molecules have come into existence. In order to generate them, the gas is pre-ionized with UV light to increase the conductivity and thus to increase the efficiency of excitation in the following discharge. The lifetime of the excimer molecules is typically about 10 ns, which also determines the pulse period of this laser type.

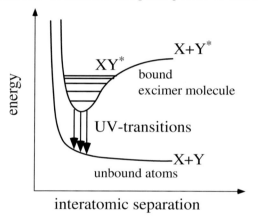

Fig. 7.18: *Laser process within the excimer laser.*

Generation and handling of a gas of excimer molecules are quite costly. The gas is corrosive and the laser medium ages after some thousands or millions of pulses (with typical repetition rates of 10–1000 pulses per second). That is why selected materials and sophisticated gas exchange systems are used. The strong demand for excimer lasers for medical applications and their increasing use as a light source for optical lithography in the semiconductor industry (see 'excursion' in Section 4.3.2) have already made the KrF laser at 248 nm a mature product. In the near future, the ArF (193 nm) laser as well as even the laser with the shortest commercially available wavelength, the F_2 laser, will probably follow.

7.7   The workhorses: solid-state lasers

The world's first laser, constructed by T. Maiman (1927-) [69], was a pulsed ruby laser. Its red light ($\lambda = 694.3$ nm) was emitted by the chromium dopant ions of the $Cr:Al_2O_3$ crystal, and therefore it was a solid-state laser. Though today the ruby laser plays a role for historical reasons only, solid-state lasers have received increasing attention since many types can be excited by the more and more competitive diode lasers. Furthermore, electrical power inserted into the system can be converted into light power with an efficiency up to 20%. Solid-state lasers are thus among the preferred laser light sources because of their robust construction and economical operation.

7.7.1   Optical properties of laser crystals

Optically active ions can be dissolved in numerous host lattices, and such systems can be considered as a frozen gas if the concentration of the former is not higher than a few per cent at most. Nevertheless the density of these impurity ions within the crystal is much higher than the particle density within a gas laser and thus allows a higher gain density if there are suitable optical transitions. Of course, the host lattices have to have a high optical quality since losses by absorption and scattering impair laser oscillation. Impurity ions can be inserted into a host crystal particularly easily if they can replace a chemically similar element. Therefore many materials contain yttrium, which can be replaced very easily by rare-earth metals.

Tab. 7.3: *Selected host materials [7].*

Host		Formula	Thermal conductivity (W cm^{-1} K^{-1})	$\partial n/\partial T$ * (10^{-6} K^{-1})	Ions
Garnet	YAG	Y$_3$Al$_5$O$_{12}$	0.13	7.3	Nd, Er, Cr
Vanadate	YVO	YVO$_4$			Nd, Er, Cr
Fluoride	YLF	LiYF$_4$	0.06	−0.67(o) −2.30(e)	Nd
Sapphire	Sa	Al$_2$O$_3$	0.42	13.6(o) 14.7(e)	Ti, Cr
Glass		SiO$_2$	typ. 0.01	3–6	Nd

* (o),(e): Ordinary (extraordinary) index of refraction in birefringent materials.

Another important property of the host crystals is their thermal conductivity, since a large amount of the excitation energy is always converted into heat within the crystal. Inhomogeneous temperature distributions within the laser crystal cause e.g. lensing effects due to the temperature sensitivity of the index of refraction, which may alter the properties of the Gaussian resonator mode significantly. Since few laser media fulfil all requirements simultaneously, the growth of new and improved laser crystals is still an important field of research in laser physics.

In the simplest case the properties of the free ions are modified only slightly when the dopant ions are dissolved in the host material. The energy levels of the erbium ion in different systems may be taken as an example (Fig. 7.19). This laser can be described very well using the concept of a 'frozen' glass laser.

A most important group of dopant elements are the rare-earth ions. Their unusual electron configuration makes them very suitable for laser operation. Another group is formed by the ions of simple transition metals, which allow one to build laser systems tunable over large wavelength ranges. These are the so-called vibronic lasers including colour centre and Ti–sapphire lasers.

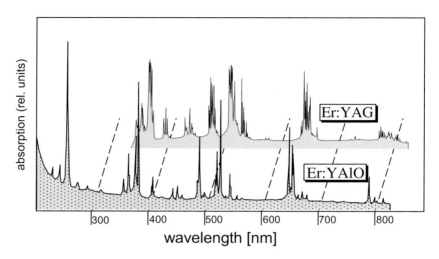

Fig. 7.19: *Absorption spectrum of the erbium ion Er^{3+} in the host materials YAG (top) and YAlO₃ (bottom) according to [3].*

7.7.2   Rare-earth ions

The 13 elements following lanthanum (La, atomic number 57) with $N = 58$ (cerium, Ce) to 70 (ytterbium, Yb) are called the *lanthanides* or rare-earth metals.[4]

Being impurity ions, the lanthanides are usually triply ionized, with electron configuration $[Xe]4f^n$ with $1 \leq n \leq 13$ for the nth element after lanthanum. The optical properties of an initially transparent host crystal are determined by the 4f electrons, which are localized within the core of these ions and thus couple only relatively weakly to the lattice of the host crystal.

To a good approximation the electronic states are described by LS coupling and Hund's rules [111]. Because of the large number of electrons that each contribute orbital angular momentum $\ell = 3$, there are in general a multitude of fine-structure states, which lead to the wealth of levels in Fig. 7.20.

Example: Energy levels of the neodymium Nd^{3+} ion

The Nd^{3+} ion has three electrons in the 4f shell. According to Hund's rules, they couple in the ground state to the maximum total spin $S = 3/2$ and total orbital angular momentum $L = 3 + 2 + 1 = 6$. From the ⁴I multiplet the ground state is expected to be at $J = 9/2$ due to the less than half-occupied shell. Unlike the free atom or ion, the magnetic degeneracy in the m quantum number is lifted by anisotropic crystal fields in the local vicinity. The coupling to the lattice oscillations ('phonons') leads eventually to the homogeneously broadened multiplets in Fig. 7.21.

---

[4]The rare-earth metals are not rare at all within the Earth's crust. Since their chemical properties are very similar, it was difficult for quite a long time to produce them with high purity. The element promethium (Pm) cannot be used because of its strong radioactivity.

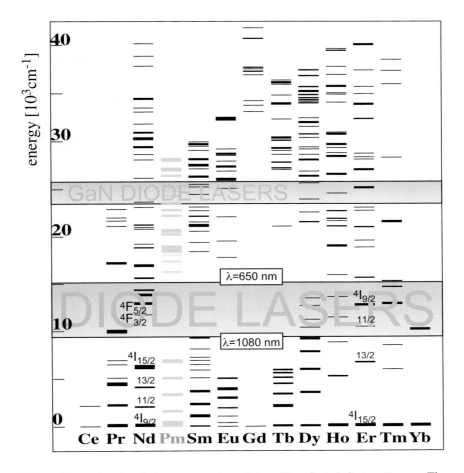

Fig. 7.20: *Energy levels of the rare-earth metals with selected designations. The range accessible by diode laser radiation for optical excitation is marked (according to [43]).*

The rigorous dipole selection rules of the free atom ($\Delta\ell = \pm 1$) are lifted by the (weak) coupling of the electronic states to the vicinity of the electrical crystal field that causes a mixture of $4f^n$ and $4f^{n-1}5d$ states. The energy shift by this interaction is quite small, but for the radiative decay the dipole coupling is now predominant and reduces the lifetimes of the states dramatically to the range of some microseconds. Therefore, intensive absorption and fluorescence of the rare-earth ions can be observed on transitions between the fine-structure levels.

On the other hand, fluorescence cannot be observed from every level since there are competing relaxation processes caused by the coupling of the ionic states to the lattice oscillations, or phonons, of the host lattice, which can lead to fully radiation-

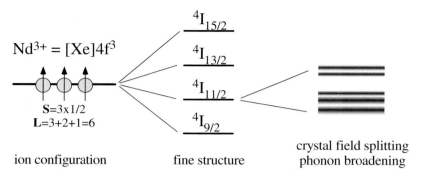

Fig. 7.21: *Energy levels of neodymium Nd³⁺ ions within the solid state. The details of the splitting depend on the host lattice.*

free transitions. Those processes are the more probable, the more the fine-structure levels are lying next to each other. In Fig. 7.19 the fluorescence lines are quite narrow, showing that the atomic character of the ions is largely maintained.

7.8   Selected solid-state lasers

From Fig. 7.20 it can easily be imagined that there are countless laser media containing rare-earth ions [54]. We have selected special solid-state lasers that play an important role as efficient, powerful or low-noise fixed-frequency lasers. Those lasers are used, for example, as pump lasers for the excitation of tunable laser systems or for materials processing that demands intensive laser radiation with good spatial coherence properties. Tunable lasers, which increasingly employ solid-state systems, will also be discussed in a subsequent section about vibronic lasers (Section 7.9).

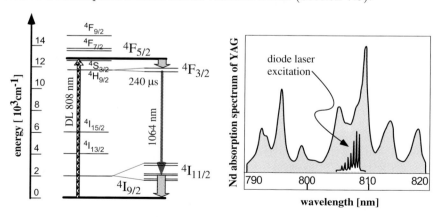

Fig. 7.22: *Laser transitions of neodymium lasers and absorption spectrum.*

7.8.1   The neodymium laser

The neodymium laser is among the systems already developed in the very early days of the laser. It was originally excited with high-pressure noble-gas lamps. Only a small amount of their light energy was absorbed, while the larger part was removed as heat and dissipated. Although the idea to excite neodymium lasers by laser diodes arose quite early, it could not be realized until the advent of reliable high-power laser diodes by the end of the 1980s. Nowadays, neodymium lasers are primarily excited by efficient, and indispensable, laser diodes, and no end to this successful development is in sight even after more than ten years. In Tab. 7.3 we have already presented hosts that have great significance for practical applications.

The neodymium amplifier

The energetic structure of neodymium ions has already been described above (section 7.7.2). We have already mentioned as well that with ions within the solid state a much higher density of excited atoms can be achieved than within the gas laser. In most of the host crystals this is valid for concentrations up to a few per cent. Above this level the ions interact with each other causing detrimental non-radiative relaxations. But there are also special materials, e.g. Nd:LSB (Nd:LaScB), which stoichiometrically contain 25% neodymium. Owing to the extremely high gain density of these materials, remarkably compact, intense laser light sources can be built.

The $^4I_{9/2} \rightarrow \, ^4F_{5/2}$ transition of the Nd^{3+} ion can be excited very advantageously by diode lasers at the wavelength of 808 nm where the upper $^4F_{3/2}$ laser level is populated very rapidly by phonon relaxation.

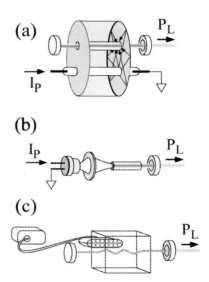

Fig. 7.23: *Configurations of neodymium lasers. (a) Pump lamp and laser bar located at the two foci of an elliptical resonator. (b) Neodymium laser longitudinally pumped with a diode laser. (c) In the slab laser, the pump energy is supplied transversely. The laser beam is guided by total internal reflection within the crystal.*

Since the lower $^4I_{11/2}$ level is emptied just as quickly, the neodymium laser makes an excellent four-level laser system.

Configurations and operating conditions

Because of its widespread application potential, there are numerous technical variants of the neodymium laser. Before diode laser pumps were available with sufficient quality, the crystal within a continuous-wave laser was generally excited by a high-pressure Xe lamp placed at the second focus of an elliptical cavity in order to achieve a high coupling efficiency (Fig. 7.23(a)).

Using diode lasers, life has become much more easy in this respect. In Fig. 7.23(b) such a linear laser rod pumped from the end ('end-pumped laser') is presented. One of its end mirrors is integrated into the laser rod. With this arrangement the pump power is inhomogeneously absorbed so that the gain also varies strongly along the laser beam. That is why a Z-shaped resonator is often used, which allows symmetrical pumping from both sides.

Further enhanced output power can be achieved by using so-called 'slab' geometries in which the pump energy is supplied transversely. With this layout the light of several pumping diode lasers can be used at the same time. It is technically advantageous to operate the heat-producing laser diodes spatially separated from the laser head. The pump light is then transported to the laser amplifier by fibre bundles in a literally flexible way, and optimal geometric pump arrangements can be used. Even with considerable output power of several watts, the laser head itself measures no more than 50x15x15 cm^3. The end of technological developments in this field is still not in sight.

7.8.2   Applications of neodymium lasers

Neodymium lasers have been used in countless applications for a long time, and the more recent advent of efficient diode laser pumps at the wavelength of 808 nm has lent additional stimulus. Here we present two recent examples that symbolize the large range of possible applications: one is the powerful frequency-doubled neodymium laser, which has replaced the expensive argon-ion laser technology more and more, and the other is the extremely frequency-stable monolithic *miser*.

Frequency-doubled neodymium lasers

In Fig. 7.24 we have presented a neodymium laser concept that allows the generation of very intense visible laser radiation at $1064/2 = 532$ nm. The pump energy is applied to the Nd:YVO$_4$ material through fibre bundles, and the Z-shaped resonator offers a convenient geometry to combine the power of several diode lasers and generate very high power at the fundamental wavelength at 1061 nm. In one arm of the laser the light is focused into a nonlinear crystal (here LBO, see Section 12.4), which causes efficient frequency doubling.

In principle, it has been clear for a long time that intense visible laser radiation can be generated with the concept presented here. Before it could be used for producing commercial devices, however, not only technological problems caused by the large power circulating in the resonator had to be solved, but also physical issues such as e.g. the so-called 'greening problem'. This is caused by mode competition [6] leading to very strong intensity fluctuations. It can be solved by operating the laser either at a single frequency or at a large number of simultaneously oscillating frequencies.

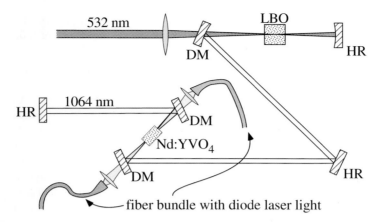

Fig. 7.24: *Powerful frequency-doubled neodymium laser. The light of the laser diodes is supplied through fibre bundles over dichroic mirrors (DM). Within the resonator, 1064 nm light circulates at high intensity. The nonlinear crystal is used for the frequency doubling. HR = high reflector [86].*

The monolithic miniature laser (miser)

The passive frequency stability of any common laser (i.e. in the absence of active control elements) is predominantly determined by the mechanical stability of the resonator, which undergoes length variations due to acoustic disturbances such as environmental vibrations, sound, etc.

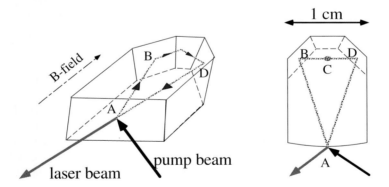

Fig. 7.25: *Monolithic neodymium ring laser. The beam is led out of the plane between B and D. In connection with a magnetic field (see text), this 'out-of-plane' configuration realizes an optical diode and makes unidirectional operation possible.*

It is therefore advantageous to build laser resonators as compact and also as light as possible, since devices with a small mass have higher mechanical resonance frequencies, which can be excited less easily by environmental acoustic noise. In the extreme case, the components of a ring laser (see Section 7.9.2) – laser medium, mirrors, optical diode – can even be integrated into one single crystal. T. Kane and R. Byer [55, 32] realized this concept in 1985 and called it *miser*, a short form of the term 'monolithically integrated laser'.

The miser is pumped by diode laser light. The ring resonator is closed by using total internal reflection at suitably ground and polished crystal planes. An interesting and intrinsic optical diode is also integrated into the device. The so-called 'out-of-plane' configuration of the resonator mode (in Fig. 7.25 the trajectory BCD) causes a rotation of the polarization of the laser field due to the 'slant' reflection angles in analogy to a $\lambda/2$ device. In addition, a magnetic field in the direction of the long axis of the miser causes non-reciprocal Faraday rotation. In one of the directions the rotations compensate each other, whereas in the other direction they add. Since the reflectivity of the exit facet depends on the polarization, one of the two directions is strongly favoured in laser operation.

7.8.3   Erbium lasers

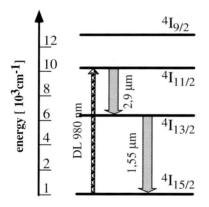

Fig. 7.26: *Part of the energy level scheme of the erbium laser showing the two important laser transitions. The exact transition wavelength depends on the host crystal.*

Erbium (Er) ions can be dissolved in the same host crystals as neodymium ions and are especially interesting for applications at infrared wavelengths near $1.55\,\mu$m. They can be excited by strained quantum well (SQW) laser diodes (see Section 9.3.4) at 980 nm very efficiently. Another long-wavelength laser transition is used mostly for medical applications. A favourable feature is the eye-safe operation at these long wavelengths.

Erbium-doped fibre amplifiers

A significant technological breakthrough was achieved by D. Payne and E. Desurvire [25] in 1989 when they were able to demonstrate amplification at the wavelength of 1550 nm by using Er-doped optical fibres. *Erbium-doped fibre amplifiers* (EDFAs) have very soon become an important amplifier device (gain typically 30–40 dB) for the long-distance transmission of data. It is due to them that today the residual losses of optical fibres in this third telecommunications window (see Fig. 3.10 on p. 81) do not impose any constraints on the achievement of even the highest possible transmission rates over large distances. This breakthrough again was not conceivable without the availability of inexpensive and robust diode lasers for excitation, as we will discuss in more detail in the following subsection about fibre lasers.

7.8.4   Fibre lasers

The total gain of an optical wave in a laser medium is determined by the inversion density (which determines the gain coefficient) and the length of the amplifying medium. E. Snitzer already mentioned in 1961 [96] that optical waveguides or fibres with suitable doping of the core should offer the best qualifications to achieve high total gains.

Although the attractive concept of fibre lasers had already been recognized quite early, the advent of robust and convenient diode lasers was instrumental in making fibre lasers attractive devices. Even the mediocre transverse coherence properties of an array of laser diodes (see Section 9.6) are far superior to conventional lamps as used in the conventional neodymium laser configuration in Fig. 7.23 with regard to focusability and can be used for efficient excitation of the small active fibre volumes.

Fibre lasers are a field of active technological development that is ongoing, and an excellent account of the state of art is given in [26]. We will limit ourselves to the presentation of a few specific concepts since the layout of a fibre laser does not differ basically from other laser types – one might say that it just has a very long and thin amplifying medium [117].

Tab. 7.4: *Elements and wavelengths of selected fibre lasers.*

Wavelength (µm)	Element	Wavelength (µm)	Element
3.40	Er	0.85	Er
2.30	Tm	0.72	Pr
1.55	Er	0.65	Sm
1.38	Ho	0.55	Ho
1.06	Nd	0.48	Tm
1.03-1.12	Yb	0.38	Nd
0.98	Er		

Cladding pumping

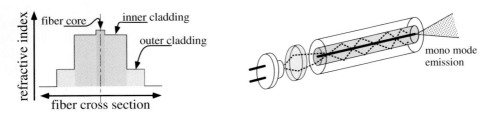

Fig. 7.27: *Cladding pumping in the fibre laser.*

An interesting trick to make the application of the pump energy more efficient has been developed with the so-called 'cladding pumping'. It is quite obvious that monomode fibres with a narrow core should be used as the active medium for the sake of obtaining good-quality transverse laser modes. But then efficient coupling of the pump laser radiation from high-power laser diodes becomes difficult, since direct concentration or focusing of their power to the small active fibre core volume is difficult. This problem can be overcome by using a double fibre cladding that generates a multimode waveguide around the active fibre core. The pump power is coupled into this multimode fibre, and again and again it scatters into the core and is absorbed there. To optimize the scattering, the core is given for instance a slightly star-shaped structure instead of a purely cylindrical one.

Tab. 7.4 contains a number of available wavelengths for fibre lasers widely spread across the infrared and visible spectrum.

With fibre laser media not only very low threshold values for lasing are achieved but also a remarkable output power of several tens of watts. Fibre lasers have not reached the end of their development by any means. Continued interest is also being shown in the development of light sources for blue wavelengths. There are several concepts, e.g. the so-called 'up-conversion' lasers, which can emit blue or even shorter-wave radiation from higher energy levels excited by stepwise absorption of several pump photons.

Fibre Bragg gratings

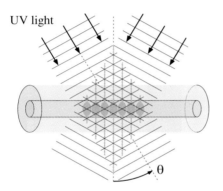

Fig. 7.28: *Production of a Bragg reflector (qualitatively).*

To make fibre lasers practicable, many relevant components for the control of a light beam, e.g. mirrors, output couplers and modulators, have been directly integrated into the fibre. For their detailed discussion, we refer the reader to the specialist literature [116] and restrict ourselves to the example of fibre Bragg gratings (FBGs) [61] used as efficient mirrors and spectral filters.

The Bragg grating is realized by a periodic modulation of the refraction coefficient along the direction of propagation. For this, the Ge-doped fibre core is exposed to two intense UV beams crossing each other at an adjustable angle θ. The UV light induces changes that can be of chemical or photo-refractive nature[5] and are proportional to the local intensity of the standing-wave field. The period Λ of the Bragg mirror can be determined by the choice of the crossing angle and the UV wavelength λ, $\Lambda = \lambda/(2\sin\theta)$.

7.8.5 Ytterbium lasers: Towards higher power with thin disc and fibre lasers

In a recent development the dominance of the Nd lasers is challenged by Yb lasers, which can be dissolved within the host materials of Tab.7.3 as well as rare earth ions. Many physical and technical details about Yb can be found in [26].

---

[5]For the photo-refractive effect, charges are released by illumination and transported within the crystal lattice, which causes a spatial modification of the refractive index up 10^{-3}.

This relatively new laser material offers putative advantages in the strive for ever more output power: Yb ion doping can reach 25% and thus strongly exceeds the 1-2% limit for Nd ions, offering higher gain density. Also, the Yb ions are excited at 940 nm (Nd: 808 nm) while lasing takes place typically between 1030 and 1120 nm, therefore less heat is generated which generally impairs the laser process. Finally, Yb suffers less from excitation into non lasing state and reabsorption of fluorescent light than Nd.

Technological breakthroughs have furthermore supported the advancement of Yb lasers: The thin disc technology and improved performance of fibre lasers which reduce the problems of heat dissipation

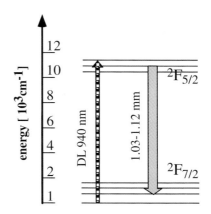

Fig. 7.29: *Part of the energy level scheme of the ytterbium laser showing the pump and the laser transition. The exact transition wavelength depends on the host crystal.*

associated with laser rods in high power applications. Commercially available output power exceeds 4 kW while excellent coherence properties are preserved.

The most important advantage of the thin disc concept (Fig. 7.30) compared to conventional laser rods is the much improved removal of heat generated in the excitation process. The thin disc is mounted on a heat sink and has a favorable surface to volume ratio. Heat gradients occur in the longitudinal rather than in the transverse direction, and hence the Gaussian resonator is much less disturbed by heat gradients, resulting in laser beams with excellent mode purity. As a consequence of the thin disc arrangement single pass absorption of the diode laser pump light is relatively small. However, the multi pass configuration (several 10 times in applications) allows to deliver the pump energy with up to 90% efficiency to the small laser volume.

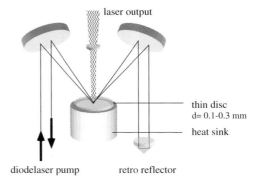

Fig. 7.30: *Cavity and pump geometry for a thin disc laser. The pump is sent through the crystal in multi pass configuration for efficient absorption.*

7.9   Tunable lasers with vibronic states

Even around 1990, the market for tunable laser light sources was dominated by dye lasers due to their convenience and – using multiple chemicals – tunability across the visible spectrum. Since then technical development has favoured solid-state systems, which are particularly interesting if they can be excited by diode lasers.

The tunability of so-called vibronic laser materials comes from the strong coupling of electronic excited states of certain ions (especially 3d elements) to the lattice oscillations. In principle, also semiconductor or diode lasers are among them, and we shall devote the next subsection to them due to their special significance. Even the dye laser can be conceptually assigned to this class since their band-like energy scheme is generated by the oscillations of large molecules. Fig. 7.31 offers an overview about important tunable laser materials.

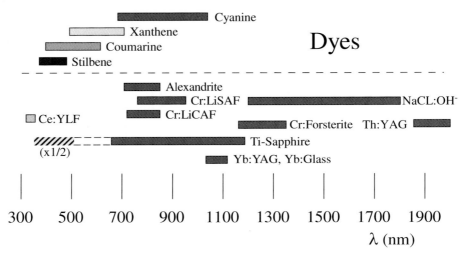

Fig. 7.31: *Tuning ranges of selected laser systems (pulsed and continuous-wave). The frequency-doubled range of the Ti–sapphire laser is hatched.*

7.9.1   Vibronic laser materials

Vibronic laser materials are tunable over large wavelength ranges. Here we present some important systems with their physical properties, and we explain the technical concepts of widely tunable ring lasers in which these laser materials are normally used.

Transition-metal ions

The 3d transition metals lose their outer 4s electron in ionic solid states and additionally some 3d electrons, their configuration being $[Ar]3d^n$. Often the third as well as the fourth ionization state of these ions can be found. Crystal fields have much more effect on the 3d electrons than on the 4f electrons of the rare-earth metals, since those form the outermost shell of electrons. The coupling to the lattice oscillations (which are described by a configuration coordinate Q) leads to a band-like distribution of states. The transitions are called 'vibronic'. On the one hand, these transitions have a large

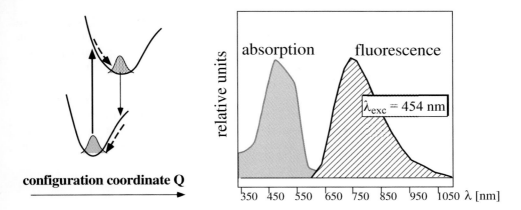

Fig. 7.32: *(a) Vibronic states of solid-state ions. The shaded curves indicate the (quasi-)thermal distributions in the configuration coordinate Q. If the equilibrium positions at the ground and the excited state do not coincide, absorption and emission wavelengths are well separated and offer optimum conditions for a four-level laser system. Relaxation to a thermal or quasi-thermal distribution takes a few picoseconds only. (b) Absorption and fluorescence spectra of a Ti–sapphire crystal. The fluorescence spectrum was excited at a pump wavelength of 454 nm.*

bandwidth, which accounts for broadband absorption as well as for fluorescence. No less important is the very short relaxation time, which leads to the thermal equilibrium position of the vibronic states within picoseconds. The chromium ions and especially the titanium ions have belonged to this important class of laser ions since the first demonstration in the 1980s [75]. The extraordinary position of the Ti–sapphire laser can clearly be recognized in Fig. 7.31, too.

Colour centres

In contrast to the optical impurities of rare-earth and transition metals, colour centres are generated not by impurity atoms but by vacancy lattice sites. They have been investigated for a long time. In an ionic crystal such vacancies have an effective charge relative to the crystal to which electrons or holes can be bound. Different types are collected in Fig. 7.33. Like the transition-metal ions, the electronic excitations have a broadband vibronic structure and are well suitable for the generation of laser radiation.

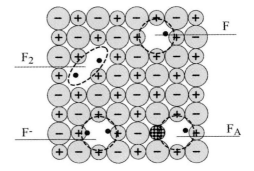

Fig. 7.33: *Models of some colour centres.*

The operation of a colour centre laser is technologically quite costly. They have to be held at the temperature of liquid nitrogen (77 K) and some of them even require an auxiliary light source. By this means colour centres are brought back from parasitic states in which they can drop through spontaneous transitions and which do not take part in the laser cycle.

However, owing to a lack of better tunable alternatives, the colour centre lasers still have continued relevance for near-infrared wavelengths between 1 and 3 μm. They may soon experience replacement because of the improvement of optical parametric oscillators and the application of so-called 'periodically poled' nonlinear crystals (see Section 12).

Dyes

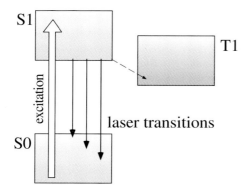

Fig. 7.34: *Laser process in the dye laser (schematically).*

Especially for wavelengths of 550–630 nm the dye laser is still a tunable light source without competition. In this range of the visible spectrum, our color sensual perception changes quickly from green to yellow to red. For this reason the light of dye lasers is superior to all solid-state lasers so far developed with regard to aesthetic and emotional quality.

Dyes are organic molecules with a carbon–carbon double bond, i.e. with a pair of electrons. In Fig. 7.34 the typical energy level scheme of a dye is presented. The paired ground state (S0) consists of a $^1S_0$ state, i.e. orbital angular momentum and total spin vanish. The dye molecules are dissolved (in alcohol or, if they are ejected from a nozzle into free space, in liquids with a higher viscosity such as glycol). The electronic states have a vibration–rotation fine structure that is broadened to continuous bands because of the interaction with the solvent, similar to the vibronic ions. After absorption, the molecules relax rapidly to the upper band edge where the laser emission takes place. Some classes of dye molecules are also shown in Fig. 7.31.

In complete analogy to two-electron atoms like helium, there are a singlet and a triplet system in dye molecules [111], only the transitions between them (intercombination lines) are not as strongly suppressed. The lifetime of the triplet states is very long, however, so that the molecules accumulate there after several absorption–emission cycles and no longer take part in the laser process. Pulsed dye lasers can be operated in an optical cell, which is stirred, but with continuous-wave laser operation, glass cells rapidly alter. Instead, creation of a jet stream expanding freely into air has been successful. The liquid is ejected into a jet stream from a flat 'nozzle' into the focus of a pump laser and laser resonator. The surface of the jet stream has optical quality. One of the most robust dye molecules is rhodamine 6G, which delivers an output power of up to several watts. Furthermore, it can be used for a long time, in contrast to many other dyes, which age rapidly.

7.9.2   Tunable ring lasers

The success of vibronic laser materials is closely related to the success of the ring laser, which allows user-friendly setting of a wavelength or frequency. It is quite remarkable that with this device the fluorescence spectrum of these materials, which has a spectral width of some 10–100 nm or 100 THz, can be narrowed to some megahertz, i.e. up to eight orders just by a few optical components!

In contrast to the linear standing-wave laser, a travelling wave propagates in the ring laser. In the linear laser the so-called 'hole burning' occurs since the amplification has no effect in the nodes of the standing-wave field. For this reason the gain profile is periodically modulated and makes the oscillation of another spectrally adjacent mode possible, which fits to the periodic gain pattern and the resonator. In the ring laser the entire gain volume contributes to a single laser line. Therefore it is the preferred device for spectroscopic applications with high spectral resolution.

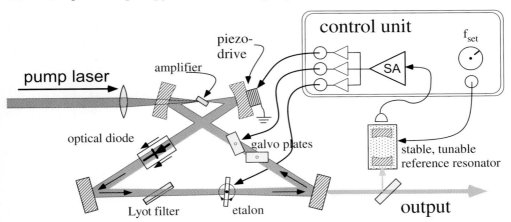

Fig. 7.35: *Ring laser system with optical components for frequency control. SA = servo-amplifier.*

In Fig. 7.35 we show one of numerous proven concepts of a ring laser. This layout is usually called a *bowtie* resonator. The foci of pump laser beam and laser mode are tightly overlapped between two spherical mirrors, which in the rest of the laser generate a Gaussian mode propagating with low divergence, which is also coupled out by one of the partially transparent resonator mirrors. To avoid losses the amplifier (Ti–sapphire rod, colour centre crystal, or dye laser jet) as well as other optical components are inserted into the resonator at the Brewster angle. An optical diode (see Section 3.6.6) allows unidirectional operation.

For wavelength control in general, several optical components with hierarchical spectral resolution (free spectral range) are used, which we list here: a Lyot or bire-fringent filter (p. 95) ensures rough spectral narrowing, one or two etalons (thin and thick etalon, see Section 5.5) with different free spectral ranges select a single res-onator mode. For tuning from the MHz to GHz scale, the resonator length can be varied with different elements: a pair of so-called 'galvo plates' varies the light path by a small synchronous rotation of the glass plates mounted at the Brewster angle, albeit at relatively low scanning speed. More rapid tuning is required for frequency stabilization. For this purpose the resonator length is adjusted by means of a light mirror mounted on a piezo-translator which allows implementation of 100 kHz band-width servo-loops; even higher actuation speed may be achieved by phase modulators (EOMs, see Section 3.6.1) inserted into the resonator.

In experiments a voltage-controlled variation of the laser wavelength is desirable. For this purpose so-called *feed-forward* values are applied to the optical components of the ring laser. Simultaneously, the laser frequency is compared to the also voltage-controlled reference frequency of a passive optical resonator (e.g. according to p. 201) and is servoed to this value by suitable electronic feedback circuitry. With this method, typical continuous tuning ranges of 30 GHz or $1 \, \text{cm}^{-1}$ are achieved, which offer excel-lent conditions for experiments in high-resolution spectroscopy.

Usually the commercially available ring lasers are quite voluminous devices. But it is also possible to build very compact and therefore inherently more stable devices, as C. Zimmermann et al. [119] have demonstrated very successfully with tiny Ti–sapphire lasers not exceeding several centimetres in diameter (though in standing-wave configuration).

8 Laser dynamics

In this chapter we shall take a closer look at the dynamical properties of laser light sources, e.g. the response of a laser system to changes in the operating parameters or to fluctuations of amplitude and phase of its electromagnetic field. For this we first have to investigate theoretically the correlation between the microscopic properties of the laser system and macroscopically measurable quantities like intensity and phase.

8.1 Basic laser theory

In Section 6.2 we studied the response of a simplified polarizable system with only two states to an external driving field. There we found that this polarization can amplify a light field and thus itself become a source of electromagnetic fields.

We know the relation between polarization and electric field already from the wave equation,

$$\left(\nabla^2 - \frac{n^2}{c^2}\frac{\partial^2}{\partial t^2}\right)\mathbf{E} = -\frac{1}{\epsilon_0 c^2}\frac{\partial^2}{\partial t^2}\mathbf{P}. \tag{8.1}$$

Here we have already taken into account that laser radiation is often generated by particles diluted in a host material with refraction coefficient n. The electric field $\mathbf{E}$ contains the dynamics of the laser field, while the polarization $\mathbf{P}$ contains the dynamics of the atoms or other excited particles, which is determined in the simplest approximation according to the Bloch equations (6.28).

8.1.1 The resonator field

In general, multiple eigenfrequencies can be excited in a laser resonator so that we expect a complicated time evolution of field and polarization. This situation though is mostly undesirable for applications. Therefore we concentrate on the special case of only one single mode of the resonator being excited. In many cases this situation is in fact routinely achieved for practical laser operation.

Formally speaking, we decompose the field into its eigenmodes labelled by index k

$$\mathbf{E}(\mathbf{r}, t) = \tfrac{1}{2}\sum_k [E_k(t)\, e^{-i\Omega_k t} + \text{c.c.}]\mathbf{u}_k(\mathbf{r}).$$

Optics, Light and Laser. Dieter Meschede
Copyright © 2004 Wiley-VCH Verlag GmbH & Co. KGaA
ISBN: 3-527-40364-7

The amplitudes $E_k(t)$ correspond to an average of the amplitude in the resonator volume V. The spatial distributions $\mathbf{u}_k$ obey an orthogonality relation,

$$\frac{1}{V} \int_V \mathbf{u}_k \mathbf{u}_k \, dV = \delta_{\mathrm{kl}}, \tag{8.2}$$

and Ω_k is the passive eigenfrequency of the resonator (without a polarizable medium), so that the Helmholtz equation (Eq. (2.11)) is valid:

$$\nabla^2 \mathbf{u}_k(\mathbf{r}) = -\frac{n^2 \Omega_k^2}{c^2} \mathbf{u}_k(\mathbf{r}).$$

The polarization can be expanded within the same set of functions $\mathbf{u}_k(\mathbf{r})$,

$$\mathbf{P}(\mathbf{r}, t) = \tfrac{1}{2} \sum_k [P_k(t) \, e^{-i\Omega_k t} + \mathrm{c.c.}] \mathbf{u}_k(\mathbf{r}).$$

Owing to (8.2) equation (8.1) decomposes into a set of separate equations, of which we use only one but for the very important special case of the *single-mode* or *single-frequency laser*:

$$\left(\Omega^2 + \frac{d^2}{dt^2} \right) E(t) \, e^{-i\omega t} = -\frac{1}{n^2 \epsilon_0} \frac{d^2}{dt^2} P(t) \, e^{-i\omega t}.$$

From this equation, among other things we have to determine the 'true' oscillation frequency of the light field.

Damping of the resonator field

A rigorous theory of the damping of the resonator field cannot be presented here but can be found in e. g. [107] or [35]. As for the Bloch equations, we limit ourselves to a phenomenological approach and assume the energy of the stored field relaxes with rate γ_c. The field amplitude then has to decay with γ_c,

$$\frac{d}{dt} E_n(t) = -\frac{\gamma_\mathrm{c}}{2} E_n(t).$$

Another frequently used measure of resonator damping is the Q-factor (derived from *quality*). For a resonator with eigenfrequency Ω it is given by

$$Q = \Omega/\gamma_\mathrm{c}.$$

Damping of the field is caused not only by the outcoupling of a usable light field E_{out},

$$E_{\mathrm{out}}(t) = \tfrac{1}{2} \gamma_{\mathrm{out}} E_n(t),$$

but also by scattering or absorption losses within the resonator,

$$\gamma_\mathrm{c} = \gamma_{\mathrm{out}} + \gamma_{\mathrm{loss}}. \tag{8.3}$$

We now insert this term into the wave equation (8.1) as well and eliminate the local variation,

$$\left(\Omega^2 + \gamma_\mathrm{c} \frac{d}{dt} + \frac{d^2}{dt^2} \right) \mathbf{E} \, e^{-i\omega t} = -\frac{1}{\epsilon_0 c^2} \frac{d^2}{dt^2} \mathbf{P} \, e^{-i\omega t}.$$

Now we are interested first in the change of the amplitude, which is slow compared to the oscillation with the light frequencies ω or Ω. In the *slowly varying envelope approximation* (SVEA), which has already been used several times, we neglect contributions of the form

$$\left[\frac{d}{dt}E(t),\ \gamma_c E(t)\right] \ll \omega E(t)$$

and obtain

$$(-\Omega^2 + \omega^2)E(t) + 2i\omega\frac{d}{dt}E(t) + i\gamma_c\omega E = -\frac{\omega^2}{n^2\epsilon_0}P(t).$$

In the customary approximation $(-\Omega^2 + \omega^2) \simeq 2\omega(\omega - \Omega)$, we get the *simplified amplitude Maxwell equations*,

$$\frac{d}{dt}E(t) = i\left(\omega - \Omega + i\frac{\gamma_c}{2}\right)E(t) + \frac{i\omega}{2n^2\epsilon_0}P(t). \tag{8.4}$$

In the absence of polarizable matter, i.e. $P(t) = 0$, we recover a field oscillating with the frequency $\omega = \Omega$ which is dampened with rate $\gamma_c/2$ exactly as we expected it. The macroscopic polarization is already known from (6.11), and its dynamics is described through the optical Bloch equations (6.28). There the occupation number difference $w(t)$ occurs, which we replace by the inversion density $\mathcal{N}$ and the total inversion n, respectively, with the definition

$$\mathcal{N}(t) = \frac{n(t)}{V} = \frac{N_{At}}{V}w(t).$$

The entire system of atoms and light field is then described by the *Maxwell–Bloch equations*:

$$\begin{aligned}
\frac{d}{dt}E(t) &= i\left(\omega - \Omega + i\frac{\gamma_c}{2}\right)E(t) + \frac{i\omega}{2n^2\epsilon_0}P(t), \\[2mm]
\frac{d}{dt}P(t) &= (-i\delta - \gamma')P(t) - i\frac{d_{eg}^2}{\hbar}E(t)\mathcal{N}(t), \\[2mm]
\frac{d}{dt}\mathcal{N}(t) &= -\frac{1}{\hbar}\Im\{P(t)^*E(t)\} - \gamma[\mathcal{N}(t) - \mathcal{N}_0].
\end{aligned} \tag{8.5}$$

Lasing can only start if the inversion is maintained by an appropriate pumping process generating the unsaturated inversion density $\mathcal{N}_0 = n_0/V$ (Eq. (6.44)). All in all there are *five* equations since field strength E and polarization P are complex quantities.

Using this system of equations, several important properties of laser dynamics can be understood. Let us introduce another transparent form of the equations that can be obtained when we normalize the intensive quantities field amplitude $E(t)$, polarization density $P(t)$ and inversion density $\mathcal{N}(t)$ to the extensive quantities field strength per photon $a(t)$, number of dipoles $\pi(t)$ and total inversion $n(t)$. For this we use the average 'field strength of a photon' (already introduced in Section 2.1.8) $\langle\mathcal{E}_{ph}\rangle = \sqrt{\hbar\omega/\epsilon\epsilon_0 V_{mod}}$,

$$\begin{aligned}
a(t) &:= E(t)\sqrt{\frac{\epsilon\epsilon_0 V_{mod}}{\hbar\omega}}, \\[2mm]
\pi(t) &:= N(u + iv) = VP(t)/d_{eg}.
\end{aligned} \tag{8.6}$$

Furthermore, for the Rabi frequency Ω_R and the detuning δ (between electric field frequency and the eigenfrequency of the polarized medium) it is advantageous to use normalized quantities,

$$g := -\frac{d_{eg}}{\hbar}\sqrt{\frac{\hbar\omega}{\epsilon\epsilon_0 V}} \qquad \text{and} \qquad \alpha := \frac{(\omega - \omega_0)}{\gamma'} = \frac{\delta}{\gamma'}. \tag{8.7}$$

The coupling factor g describes the rate (or Rabi frequency) with which the internal excitation state of the polarizable medium is changed at a field strength corresponding to just one photon. The alpha parameter α is the detuning normalized to the transverse relaxation rate γ'. It will again be of interest in the section about semiconductor lasers, where it has considerable influence on the linewidth.

With the normalized quantities, Eqs. (8.5) have a new transparent structure

$$\begin{align}
\text{(i)} \qquad \dot{a}(t) &= i(\Omega - \omega + \tfrac{1}{2}i\gamma_c)a(t) + \tfrac{1}{2}ig\pi(t), \\
\text{(ii)} \qquad \dot{\pi}(t) &= -\gamma'(1 + i\alpha)\pi(t) - iga(t)n(t), \\
\text{(iii)} \qquad \dot{n}(t) &= -g\,\mathfrak{Im}\{\pi(t)^*a(t)\} - \gamma[n(t) - n_0].
\end{align} \tag{8.8}$$

The field amplitude $a(t)$, the polarization $\pi(t)$ and the inversion $n(t)$ are coupled by the single-photon Rabi frequency g. At the same time, there is damping with the relaxation time constants γ_c, γ' and γ, respectively. The dynamical properties of the laser system are determined by the ratio of these four parameters, which we have compiled for important laser types in Tab. 8.1.

Tab. 8.1: *Typical time constants of important laser types.*

Laser	Wavelength	Rates			
	λ (µm)	γ_c (s^{-1})	γ (s^{-1})	γ' (s^{-1})	g (s^{-1})
Helium–neon	0.63	10^7	5×10^7	10^9	10^4–10^6
Neodymium	1.06	10^8	10^3–10^4	10^{11}	10^8–10^{10}
Diode	0.85	10^{10}–10^{11}	3–4×10^8	10^{12}	10^8–10^9

Eqs. (8.8) already have great similarity to a quantum theory of the laser field. By analogy, for instance, normalized amplitudes may be simply promoted to field operators, $a(t) \to \hat{a}(t)$, to obtain the correct quantum equations.

8.1.2   Steady-state laser operation

We are now interested in stationary values a^{st}, π^{st} and n^{st}, and begin first by using Eq. (8.8(ii)),

$$\pi^{st} = -i\frac{ga^{st}n^{st}}{\gamma'(1 + i\alpha)} = -i\frac{\kappa n^{st}}{g}(1 - i\alpha)a^{st}. \tag{8.9}$$

Here we have already introduced the quantity

$$\kappa := \frac{g^2}{\gamma'(1 + \alpha^2)}, \tag{8.10}$$

which plays the role of the Einstein B coefficient, as we will see more clearly in the relation with Eq. (8.19). Also, κn^{st} may be interpreted as the rate of stimulated emission.

Saturated gain

We insert the result into (8.8(i)), sort into real and imaginary parts and achieve a very transparent equation with

$$\dot{a}(t) = \{i[\omega - \Omega - \tfrac{1}{2}\kappa n(t)\alpha] - \tfrac{1}{2}[\gamma_c - \kappa n(t)]\}a(t). \tag{8.11}$$

This describes dynamical properties of the amplitude of the resonator field in a good approximation if the damping rate of the polarization γ' dominating in eq. (8.8) is much larger than all the other time constants. In that case $\pi(t)$ can always be replaced by its quasi-stationary value.

For the moment we are only interested in the stationary values for the inversion n^{st} and the amplitude a^{st}:

$$0 = [i(\Omega - \omega - \tfrac{1}{2}\kappa n^{st}\alpha) - \tfrac{1}{2}(\gamma_c - \kappa n^{st})]a^{st}. \tag{8.12}$$

If a laser field already exists ($a^{st} \neq 0$), the real and imaginary parts of eq. (8.12) have to be satisfied separately. Especially the real part clearly illustrates that the rate of stimulated emission κn^{st} corresponds exactly to the gain rate G_S, for it has to compensate exactly the loss rate γ_c,

$$n^{st} = \gamma_c/\kappa \qquad \text{or} \qquad G = \gamma_c = \kappa n^{st}. \tag{8.13}$$

Once the laser oscillation has started, the gain no longer depends on the pumping rate but only on the loss properties of the system. This case is called 'saturated gain' $G = G_S = \gamma_c$. When the laser has not started yet, the (small signal) gain increases linearly with the inversion according to Eq. (6.44), $G = \kappa n_0$. This relation is presented in Fig. 8.1.

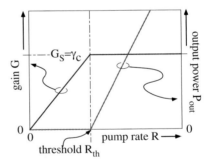

Fig. 8.1: *Saturated gain and laser power.*

Mode pulling

The imaginary part of Eq. (8.11) delivers the 'true' laser frequency ω with which the combined system of resonator and polarized medium oscillates. We replace $\kappa n^{st} = \gamma_c$, use $\alpha = (\omega - \omega_0)/\gamma'$ according to Eq. (8.7) and get the result

$$\omega = \frac{\gamma'\Omega + \gamma_c\omega_0/2}{\gamma' + \gamma_c/2}.$$

According to this, the eigenfrequencies of either component are weighted with the damping rates for the polarization of the respective other part. The true oscillation frequency always lies between the frequencies of the amplifying medium (ω_0) and the resonator (Ω).

Field strength and number of photons in the resonator

According to Eq. (8.6) the photon number and the normalized field strength are connected by $n_{\mathrm{ph}}(t) = |a(t)|^2$. Therefore from the third equation of (8.8) it can be derived that

$$\overline{n}_{\mathrm{ph}} = |a^{\mathrm{st}}|^2 = \frac{\gamma}{\gamma_c}(n_0 - n^{\mathrm{st}}). \tag{8.14}$$

Only when the unsaturated inversion n_0 meets the saturated value n^{st} does laser oscillation start because the photon number must be positive. Below that threshold we here get the result $\overline{n}_{\mathrm{ph}} = 0$. In a refined theoretical description we will see in section 8.3 that even below the threshold stimulated emission leads to an increased photon number in the resonator.

Laser threshold

Above threshold, the unsaturated inversion n_0 must be larger than the inversion at steady-state operation, n^{st}, and thus delivers a value for the pumping power or rate R_{th} at the laser threshold according to $n_0 \geq n^{\mathrm{st}}$ and eq. (6.44). A transparent form is obtained by using the coupling parameter g according to Eqs. (8.7) and (8.13),

$$R_{\mathrm{th}} = \frac{\gamma_c \gamma}{\kappa} \frac{1}{1 - \gamma/\gamma_{\mathrm{dep}}} = \gamma n_{\mathrm{e}}^{\mathrm{st}}. \tag{8.15}$$

In this model at threshold apparently the entire pumping energy is still lost due to spontaneous processes since the laser field has not started lasing yet. Above threshold we can now describe the number of photons within the laser resonator (8.14) by means of the pumping rate,

$$\overline{n}_{\mathrm{ph}} = \frac{1 - \gamma/\gamma_{\mathrm{dep}}}{\gamma_c}(R - R_{\mathrm{th}}) \xrightarrow{\gamma/\gamma_{\mathrm{dep}} \to 0} \frac{1}{\gamma_c}(R - R_{\mathrm{th}}), \tag{8.16}$$

which has a simple form especially for the 'good' four-level laser ($\gamma/\gamma_{\mathrm{dep}} \to 0$).

For the interpretation of Eq. (8.15), it can also be taken into account that most lasers are operated in an open geometry. Then the coupling constant according to (8.7) is connected to the natural decay constant according to (6.41) by $g^2 = \gamma(3\pi c^3/\omega^2 \epsilon V_{\mathrm{mod}})$ with V_{mod} for the mode volume of the resonator field. Using (8.10) one can obtain

$$R_{\mathrm{th}} = \gamma_c \gamma' \frac{1 + \alpha^2}{1 - \gamma/\gamma_{\mathrm{dep}}} \frac{\omega^2 \epsilon V_{\mathrm{mod}}}{3\pi c^3}.$$

It is intuitively clear that a smaller outcoupling (low γ_c, see eq. (8.3)) reduces the laser threshold. According to this relation there are also advantages for small transition strengths (low γ, γ'), fast depopulation rates for the lower laser level (large γ_{dep}) and good correspondence of laser frequency and resonance frequency of the amplifying medium ($\alpha = 0$). The attractive construction of UV lasers suffers among other things from the influence of the transition frequency ω visible here. On the other hand, the concentration of the resonator field onto a small volume V_{mod} is favourable. We are going to follow this path further under the topic *micro-lasers* and *threshold-less lasers* in section 8.3.

Laser power and outcoupling

The outcoupled laser power is directly connected to the number of photons in the resonator according to

$$P_{\text{out}} = h\nu\gamma_{\text{out}}\overline{n}_{\text{ph}} = h\nu\gamma_{\text{out}}\frac{\gamma}{\gamma_{\text{c}}}(n_0^{\text{st}} - n^{\text{st}}). \tag{8.17}$$

It is worth while to consider the influence of outcoupling on resonator damping,

$$P_{\text{out}} = h\nu\gamma_{\text{out}}\left(\frac{R}{\gamma_{\text{out}} + \gamma_{\text{loss}}} - \frac{\gamma}{\kappa}\right). \tag{8.18}$$

For very small outcoupling ($\gamma_{\text{out}} \ll \gamma_{\text{loss}}$), the output power increases with γ_{out}, passes through a maximum, and at $R/(\gamma_{\text{out}}+\gamma_{\text{loss}}) = \gamma/\kappa$ laser oscillation dies out. In order to achieve an output power as high as possible, γ_{out} has to be controlled by the reflectivity of the resonator mirrors. With the example of the helium–neon laser in Fig. 7.8 on p. 199, we have already investigated this question in slightly different terms.

8.2   Laser rate equations

The Maxwell–Bloch equations (8.5) and (8.8) describe the dynamical behaviour of each of the two components of the electric field $E(t)$ and $a(t)$, respectively, and the polarization density $P(t)$ and the dipole number $\pi(t)$, respectively. Furthermore, the inversion has to be taken into account through its density $\mathcal{N}(t)$ or the total inversion $n(t)$. The equations raise the expectation of, in principle, a complicated dynamical behaviour that finds its special expression in the isomorphy of the laser equations with the Lorentz equations of non-linear dynamics that literally lead to 'chaos'.

   However, most conventional lasers behave dynamically in a very well-natured way – or in good approximation according to the stationary description that we just have dealt with intensively. They owe their stability to a fact that also simplifies the mathematical treatment of the Maxwell–Bloch equations enormously. The relaxation rate of the macroscopic phase between laser field and polarization, γ', is typically very much larger than the relaxation rates of inversion (γ) and resonator field (γ_{c}). Under these circumstances the polarization density follows the amplitude of the electric field nearly instantaneously and therefore according to Eq. (8.8) can always be replaced by its instantaneous ratio to field strength $a(t)$ and inversion density $n(t)$,

$$\pi(t) \simeq -\frac{-iga(t)n(t)}{\gamma'(1 + i\alpha)}.$$

Once the polarization density has been 'eliminated adiabatically', it is not worth further investigating the phase dependence of the electric field because it is only interesting in relation to the polarization. Instead of this we investigate the time-varying dynamics of the photon number according to

$$\frac{d}{dt}|a(t)|^2 = a(t)\frac{d}{dt}a^*(t) + a^*(t)\frac{d}{dt}a(t).$$

We obtain the simplified *laser rate equations* where we use the pumping rate $R \simeq n_0/\gamma$ instead of the unsaturated inversion n_0,

$$\text{(i)} \qquad \frac{d}{dt} n_{\mathrm{ph}}(t) \;=\; -\gamma_{\mathrm{c}} n_{\mathrm{ph}}(t) + \kappa n_{\mathrm{ph}}(t) n(t),$$

$$\text{(ii)} \qquad \frac{d}{dt} n(t) \;=\; -\kappa n_{\mathrm{ph}}(t) n(t) - \gamma n(t) + R. \tag{8.19}$$

Unlike common linear differential equations, these equations are connected nonlinearly by the coupling term $\kappa n_{\mathrm{ph}}(t) n(t)$, which is the rate of stimulated emission:

$$R_{\mathrm{stim}} = \kappa n_{\mathrm{ph}}(t) n(t); \tag{8.20}$$

the rate of change of the photon number depends on the number of photons already present.

At first we again study the equilibrium values $\overline{n}_{\mathrm{ph}}$ and n^{st}. Equation (8.19(i)) yields two solutions, the first of which, $\overline{n}_{\mathrm{ph}} = 0$, describes the situation below the laser threshold. There the inversion grows linearly with the pumping rate according to (8.19(ii)) and (6.44) (we again assume the case of a 'good' four-level laser with $\gamma/\gamma_{\mathrm{dep}} \ll 1$):

$$\overline{n}_{\mathrm{ph}} = 0 \qquad \text{and} \qquad n^{\mathrm{st}} = n_0 \simeq R/\gamma.$$

When laser oscillation starts ($\overline{n}_{\mathrm{ph}} > 0$), then according to (8.19(i)) the inversion in equilibrium always has to be clamped at the saturation value n^{st}, and eq. (8.13) can again be found.

As expected we recover the behaviour of Fig. 8.1. The gain only grows until laser threshold is reached and then becomes saturated, i.e. a constant value due to *gain clamping*. At the same time the number of photons increases due to (8.19(ii)) according to

$$\overline{n}_{\mathrm{ph}} = \frac{1}{\gamma_{\mathrm{c}}} \left(R - \frac{\gamma \gamma_{\mathrm{c}}}{\kappa} \right), \tag{8.21}$$

with the value $R_{\mathrm{th}} = \gamma \gamma_{\mathrm{c}}/\kappa$ just corresponding to the pumping power at the threshold. A linear dependence of laser power ($\propto n_{\mathrm{ph}}$) and pumping rate R is predicted, which most frequently occurs with common lasers, as shown for diode lasers, for example, in Fig. 9.12.

8.2.1   Laser spiking and relaxation oscillations

The laser rate equations (8.19) are nonlinear and can in principle only be investigated by numerical analysis. In Fig. 8.2 we present two examples where the laser is switched on suddenly. For $t < 0$ we have $R = 0$ and the switching is instantaneous, at least compared to one of the two relaxation rates γ (inversion density) or γ_{c} (resonator field). The numerical simulation can easily be done with many programs of computer algebra and shows very well the phenomenon of 'laser spiking', being observed, for example, at fast turn-on (nanoseconds or faster) of neodymium or diode lasers.

Relaxation oscillations in a narrower sense occur when the gain (or the loss rate) changes suddenly. Fluctuations of gain are induced by the fluctuations of the pumping

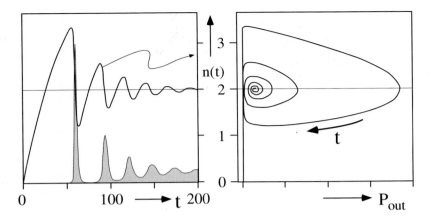

Fig. 8.2: *Numerical evaluation of relaxation oscillations. Left: inversion n( ) and gain, respectively, and output power $P_{\text{out}}(t)$. Right: phase-space representation. The system parameters in Eq. (8.19) are chosen to be: $\kappa = 1$, $\gamma_c = 2$, $\gamma = 0.02$, $R = 0.1$.*

processes, e.g. by switching on or off an optical pump laser. For many purposes, e.g. for the stability analysis of frequency and amplitude of a laser oscillator, it is sufficient to consider small deviations of the photon numbers and the inversion from their equilibrium values:

$$n_{\text{ph}}(t) = \overline{n}_{\text{ph}} + \delta n_{\text{ph}}(t) \qquad \text{and} \qquad n(t) = n^{\text{st}} + \delta n(t).$$

We insert into Eq. (8.19), neglect products of the type $\delta n \delta n_{\text{ph}}$ and obtain the linearized equations

$$(i) \qquad \frac{d}{dt}\delta n_{\text{ph}} \;=\; \kappa \overline{n}_{\text{ph}}\delta n = \left(\frac{\kappa R}{\gamma_c} - \gamma\right)\delta n,$$

$$(ii) \qquad \frac{d}{dt}\delta n \;=\; -(\gamma + \kappa \overline{n}_{\text{ph}})\delta n - \gamma_c \delta n_{\text{ph}}.$$

(8.22)

For simplicity we introduce the normalized pumping rate $\rho = R/R_{\text{th}} = \kappa R/\gamma\gamma_c$, which has the value 1 at threshold, and for both $x = \{\delta n_{\text{ph}}, \delta n\}$ we obtain the usual equation of the damped harmonic oscillator,

$$\ddot{x} + \gamma\rho\dot{x} + \gamma\gamma_c(\rho - 1)x = 0. \tag{8.23}$$

From this we infer without further difficulties that the system can oscillate for

$$(\gamma_c/\gamma)\left[1 - \sqrt{1 - (\gamma/\gamma_c)}\right] < \rho/2 < (\gamma_c/\gamma)\left[1 + \sqrt{1 - (\gamma/\gamma_c)}\right]$$

with normalized frequency

$$\omega_{\text{rel}}/\gamma = \sqrt{(\gamma_c/\gamma)(\rho - 1) - (\rho/2)^2}, \tag{8.24}$$

and is damped with the rate

$$\gamma_{\text{rel}} = \gamma\rho/2 = \gamma R/2R_{\text{th}}. \tag{8.25}$$

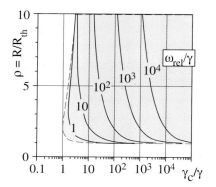

Fig. 8.3: *Relaxation oscillations as a function of γ_c/γ and ρ.*

Especially solid-state lasers typically have long lifetimes in the excited laser level and thus large γ_c/γ ratios, e.g. 10^3–10^4 for semiconductor lasers and 10^4–10^5 for Nd lasers. In Fig. 8.3 it can be seen that in this case relaxation oscillations are triggered immediately above the laser threshold at $\rho = 1$. They can also be driven by external forces, e.g. by modulating the pumping rate appropriately, and they play an important role for the amplitude and frequency stability of laser sources (see Section 8.4), since they are induced by noise sources of all kinds.

Example: Relaxation oscillations in the Nd:YAG laser

We consider the 1064 nm line of the Nd:YAG laser with the following characteristic quantities:

Natural lifetime	$\tau = 240\,\mu s$	$\gamma = 4.2 \times 10^3\,\mathrm{s}^{-1}$,
Resonator storage time	$\tau_c = 20\,\mathrm{ns}$	$\gamma_c = 5.0 \times 10^7\,\mathrm{s}^{-1}$,
Normalized pump rate	$R/R_{th} = 1.0\text{–}1.5$.	

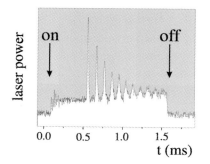
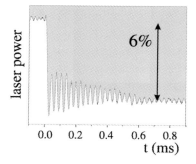

Fig. 8.4: *Spiking and relaxation oscillations in the Nd:YAG laser. The power of the pumping laser diode is modulated by a square signal. Complete modulation (left) causes spiking; partial modulation (right, 6%) causes relaxation oscillations. Compare Fig. 8.2.*

The properties of the relaxation oscillations observed in experiment correspond to the theoretical estimates. For the Nd:YAG parameters the second term in Eq. (8.24) can be neglected in calculating the oscillation frequency due to $\gamma \ll \gamma_c$,

$$\omega_{rel} \simeq \sqrt{\gamma\gamma_c}\sqrt{\rho - 1} \simeq 72\,\mathrm{kHz}\,\sqrt{\rho - 1},$$

and according to (8.25) the damping rate is $\gamma_{rel} \simeq 2 \times 10^3\,\mathrm{s}^{-1}\,R/R_{th}$.

8.3   Threshold-less lasers and microlasers

We have already seen in the section about spontaneous emission that a reflecting environment changes the rate of spontaneous emission. In principle, this effect occurs in every laser resonator, though it is mostly so small that it can be neglected without any problems. The influence is so small because in open resonator geometry (Fig. 8.5) the more or less isotropic spontaneous radiation of an excited medium, e.g. of an atomic gas, is emitted only with a small fraction into that solid angle which is occupied by the electric field modes of the laser resonator.

These changes though can no longer be neglected if the resonator becomes very small, or if, as a result of large steps of the refraction coefficient of the laser medium, the emitted power is more and more confined to the resonator. For this case, the modified effect of the spontaneous emission is often taken into account by the so-called 'spontaneous emission coefficient' β. The β factor indicates which geometrical part of the radiation field couples to the laser mode (rate $\beta\gamma$), and which part is emitted into the remaining volume (rate $(1-\beta)\gamma$).

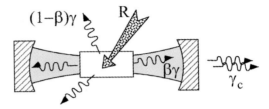

Fig. 8.5: *Relaxation and pumping rates of the laser. The β coefficient of spontaneous emission is a coarse measure for the fraction of spontaneous emission coupled to the laser mode (β, shaded area) and to the other solid angle $(1 - \beta)$.*

Spontaneous emission can be considered as stimulated emission by a single photon, and therefore we set the coupling coefficient $\beta\gamma = \kappa(n_{\rm ph}=1)$. With this trick we can account for spontaneous emission in the laser rate equation (8.19(i)), and by replacing $n_{\rm ph} \to n_{\rm ph} + 1$ we obtain

$$\text{(i)} \qquad \frac{d}{dt}n_{\rm ph}(t) = -\gamma_c n_{\rm ph}(t) + \beta\gamma n_{\rm ph}(t)n(t) + \beta\gamma n(t),$$

$$\text{(ii)} \qquad \frac{d}{dt}n(t) = -\beta\gamma n_{\rm ph}(t)n(t) - \beta\gamma n(t) - (1 - \beta)\gamma n(t) + R.$$

For the steady-state situation the equations can immediately be simplified to

$$\text{(i)} \qquad 0 = -\gamma_c \overline{n}_{\rm ph} + \beta\gamma n^{\rm st}(\overline{n}_{\rm ph} + 1),$$

$$\text{(ii)} \qquad 0 = R - \beta\gamma \overline{n}_{\rm ph} n^{\rm st} - \gamma n^{\rm st},$$

where spontaneous emission is especially prominent through the factor $\overline{n}_{\rm ph} + 1$ in (i). In order to solve this system of equations, it is convenient to express the pumping rate as a function of the photon number in the resonator. We substitute $n^{\rm st}$ in (ii) by means of (i) and obtain

$$\frac{R}{\gamma_c} = \left(\frac{1}{\beta} + \overline{n}_{\rm ph}\right)\frac{\overline{n}_{\rm ph}}{\overline{n}_{\rm ph} + 1}. \qquad (8.26)$$

Far above the laser threshold, i.e. for $\overline{n}_{\rm ph} \gg 1/\beta \geq 1$, the relation between pumping rate and photon number obviously turns again into the result Eq. (8.16), as expected.

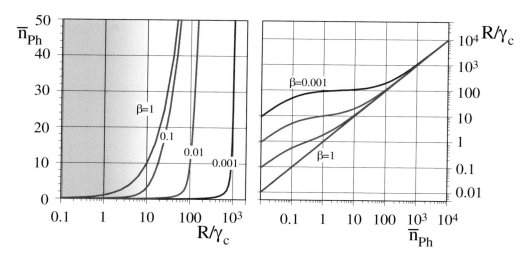

Fig. 8.6: *Threshold behaviour of laser oscillators. Photon number in the resonator as a function of the pumping rate.*

According to the condition (8.26) the laser threshold is reached when the photon number in the resonator meets or exceeds the value $1/\beta$. So in a common laser ($\beta \ll 1$) at threshold there are already very many photons present in the laser mode. To be more exact, there are so many that the rate of stimulated emission into the laser mode precisely equals the total spontaneous decay rate. Above this threshold, additional pumping power is used predominantly to increase the photon number and thus to build up the coherent radiation field.

Excursion: Micro-maser, micro-laser and single-atom laser
Experiments first with the so-called micro-maser, and later on with the micro-laser, have had a strong stimulating effect on the concept of the 'threshold-less' laser. The term 'micro-' does not refer so much to the miniaturized layout but rather to the microscopic character of interaction. The coupling between the field of a micro-laser or micro-maser is so strong that an excited atom does not forego its energy once and forever to the electromagnetic field like in a common laser. For this so-called *strong coupling regime* the rate g of Eq. (8.7) has to be larger than every other time constant (see Tab. 8.1):

$$g \gg \gamma, \gamma_c, \gamma'.$$

Then the resonator field stores the emitted energy and the atom (the polarized matter) can re-absorb the radiation energy. So the energy oscillates between atom and resonator (Fig. 8.7). This situation can be realized already – or even particularly well – with a single atom, and thus the term 'single-atom maser', which was very often used at first.

In order to realize the situation of a micro-maser experimentally, resonators with extremely long radiation storage times have to be used. Since superconducting resonators for microwaves have been available for a longer time, the micro-maser was realized before the micro-laser. The description of the micro-maser requires a joint treatment of atom and

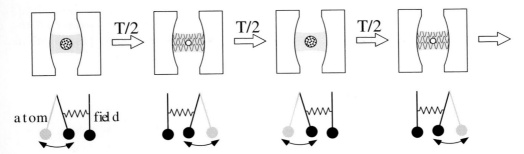

Fig. 8.7: *In a single-atom laser ('micro-laser') the coupling between atom and field is so strong that the oscillation energy oscillates like for two coupled pendula.*

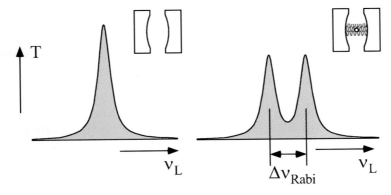

Fig. 8.8: *Transmission spectrum of an empty resonator (left) and a resonator with an excited atoms (right). A single photon is sufficient to cause the so-called 'vacuum Rabi splitting'.*

field according to quantum theory within the framework of the so-called Jaynes–Cummings model, which goes far beyond the scope of this excursion. However, it is intuitively clear that the transmission of the combined system of resonator and atom exhibits a different spectral behaviour from the empty resonator following the ordinary Lorentz curve.

Threshold-less lasers are extraordinarily interesting for applications in integrated optics. For example, semiconductor components may be designed where single electron–hole pairs are directly converted into single photons. Current research follows different routes to construct radiation fields confined to a small mode volumes, with long storage times, and intense coupling to the excited medium. At optical wavelengths a small mode volume also means using miniaturized resonators. For the traditional layouts following the linear resonator, the integration of highly reflective mirrors to achieve large storage densities though is difficult. A solution is offered by the appropriate use of total reflection. Tiny electrical resonators from a monolithic substrate with very high quality have already been realized. At the rim of mushroom- or mesa-shaped semiconductor lasers and dielectric spheres made from silica, circulating field modes, so-called 'whispering-gallery modes', have been prepared and shown to be long-lived. Recently also micro-resonators with an oval geometry have been discussed for micro-laser applications because they allow a particularly strong coupling of laser medium and radiation field (see Section 5.6.4) [80].

8.4   Laser noise

All physical quantities are subject to fluctuations, and the laser light field is no exception: the perfect harmonic wave with fixed amplitude and phase remains a fiction! But the laser light field approaches this ideal of a harmonic oscillator more closely than any other physical phenomenon. According to an old estimate by Schawlow and Townes, the coherent laser light field shows extremely small fluctuations of amplitude and phase. Not the least for this reason it has continued to inspire wide areas of experimental physics to this day. The 'narrow linewidth' (sub-Hertz) has already been introduced in Section 7.3.1 with the example of the He–Ne laser. It promises extremely long coherence times ($>1\,$s) or enormous lengths ($>10^8\,$m), which can be used for high-precision measurements for a wide variety of phenomena.

Usually the so-called 'Schawlow–Townes limit' of the linewidth is hidden by technical and generally much bigger fluctuations. If this fundamental limit is realized, however, it offers information about the physical properties of the laser system. In this section we investigate what physical processes impair ideal oscillator performance.

8.4.1   Amplitude and phase noise

The stationary values of the laser light field (Section 8.1.2) have been determined through the photon number $\bar{n}_{\mathrm{ph}}$ and the true laser frequency ω on p. 231. There we assumed that the phase evolution of the field behaves like a perfect oscillator according to classical electrodynamics:

$$E(t) = \mathfrak{Re}\{E_0 \exp[-i(\omega t + \phi_0)]\}.$$

The coupled system of polarized laser medium and resonator field though is also coupled to its environment, e.g. by the spontaneous emission causing stochastic fluctuations of the field strength and the other system quantities.[1] More realistically we thus introduce noise terms,

$$E(t) \to \mathfrak{Re}\{[E_0 + e_{\mathrm{N}}(t)] \exp[-i(\omega t + \phi_0 + \delta\phi(t))]\},$$

where we assume that we can distinguish contributions to amplitude noise ($e_{\mathrm{N}}(t)$) and to phase noise ($\delta\phi(t)$), although this separation is not unambiguous. Furthermore we assume that the fluctuations are not too fast – i.e. $(de_{\mathrm{N}}/dt)/e_{\mathrm{N}}, d\delta\phi/dt \ll \omega$.

In Fig. 8.9 the effect of white, i.e. frequency-independent, noise of the amplitude and phase, respectively, on the power spectrum of the electromagnetic field (for the

---

[1]In the micro-maser (see p. 238) though the aim is to eliminate exactly this coupling to the environment.

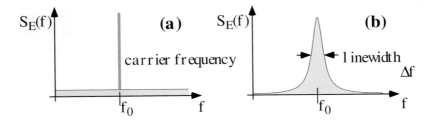

Fig. 8.9: *Field spectrum for (a) white amplitude noise and (b) white phase noise, respectively. The spectral width of the carrier frequency in (a) is limited only by the resolution of the spectrum analyser.*

definition see Appendix A.1) is presented. The exact calculation requires information about the spectral properties of the noise quantities.

We begin with the amplitude fluctuations and first assume perfect phase evolution $(\delta\phi(t) = 0)$. If the fluctuations of the noise amplitude are entirely random, i.e. very 'fast' even during the integration time T of the analyser, they are only correlated at all at delay time $\tau = 0$ ('delta-correlated') and we can describe the correlation function of the noise amplitude using the mean square value $e_{\rm rms}^2 = \langle |e_{\rm N}(t)|^2\rangle$,

$$\langle e_{\rm N}(t)\rangle = 0 \qquad \text{and} \qquad \langle e_{\rm N}(t)e_{\rm N}^*(t+\tau)\rangle = e_{\rm rms}^2 T\delta(\tau).$$

With this information we can calculate the correlation function of an electromagnetic field with amplitude fluctuations where we take advantage of the Poynting theorem (see Appendix A.2),

$$\begin{aligned}
C_E(\tau) &= \langle \mathfrak{Re}\{E(t)\}\,\mathfrak{Re}\{E(t+\tau)\}\rangle \\
&= \tfrac{1}{2}\langle \mathfrak{Re}\{E(t)E^*(t+\tau)\}\rangle \\
&= \frac{1}{2T}\int_{-T/2}^{T/2}\mathfrak{Re}\{[E_0 + e_{\rm N}(t)][E_0^* + e_{\rm N}^*(t+\tau)]\}\,dt \\
&= \tfrac{1}{2}|E_0|^2 + e_{\rm rms}^2 T\delta(\tau).
\end{aligned}$$

The finite integration interval causes errors of magnitude $\mathcal{O}(1/\omega T)$, which can be neglected since at optical frequencies ωT is always very large. Using the Wiener–Khintchin theorem (Eq. (A.9)) the spectrum

$$S_E(f) = \tfrac{1}{2}E_0^2\delta(f) + e_{\rm rms}^2/\Delta f$$

can be obtained. The 'Fourier frequencies' f give the distance to the much larger optical carrier frequency $\omega = 2\pi\nu$. The second contribution causes a 'white noise floor', and we have already replaced $T = 1/\Delta f$ to indicate that in an experiment the filter bandwidth always has to be inserted here. The first contribution represents the carrier frequency like for a perfect harmonic oscillation. The delta function indicates that the entire power in this component can always be found in one channel of the spectrum analyser and so its width is always limited by the filter bandwidth.

In order to study the influence of a fluctuating phase, we follow the presentations of Yariv [115] and Loudon [66] and calculate the correlation function of an electromagnetic field $E(t) = \Re\{E_0 \, e^{-i[\omega t + \theta(t)]}\}$ with a slowly fluctuating phase $\theta(t)$ again using the Poynting theorem (see Appendix A.2),

$$
\begin{aligned}
C_E(\tau) &= \tfrac{1}{2}|E_0|^2 \langle \Re\{e^{i[\omega\tau + \Delta\theta(t,\tau)]}\}\rangle \\
&= \tfrac{1}{2}|E_0|^2 \Re\{e^{i\omega\tau} \langle e^{i\Delta\theta(t,\tau)}\rangle\}.
\end{aligned}
\tag{8.27}
$$

The average extends only over the fluctuating part with $\Delta\theta(t,\tau) = \theta(t + \tau) - \theta(t)$. Though we do not know the exact variation in time (that is just the nature of noise), we assume it to exhibit stationary behaviour so that properties such as the frequency spectrum do not depend on time itself. If the statistical distribution of the average phase deviations $\Delta\theta(\tau)$ is known, we can use the ensemble average over the probability distribution $p(\Delta\theta(\tau))$ instead of the time average to calculate the average in (8.27). For symmetric distributions we have to take only the real part into account,

$$
\langle e^{i\Delta\theta(\tau)}\rangle = \langle \cos\Delta\theta(\tau)\rangle = \int_{-\infty}^{\infty} \cos\Delta\theta \; p(\Delta\theta(\tau)) \, d\Delta\theta.
\tag{8.28}
$$

The wanted probability distribution is completely characterized when $p(\Delta\theta(\tau))$ is explicitly given or, for a known type such as the normal distribution, one of its so-called statistical moments, e.g. the mean square deviation $\Delta\theta_{\mathrm{rms}}^2$, is given.

How do we obtain the required statistical information? From the point of view of experimental physics, one would simply measure the phase fluctuations, for example by heterodyning the laser field under investigation with a stable reference wave to determine the macroscopic phase. Theoretical models are, however, necessary to establish a connection with the microscopic physical properties of the laser system.

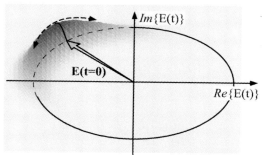

Let us first concentrate on the widely known phase-diffusion model. For this we study the phasor model of the amplitude of the laser field in Fig. 8.10. By the Maxwell–Bloch equations, only the amplitude of the laser field is fixed, but not the phase, since there is no restoring force binding the phase to a certain value. Thus the phase diffuses unobstructedly away from its initial value. We will see that for this – if technical disturbances can be excluded – especially spontaneous emission processes are responsible.

Fig. 8.10: *Phasor model of the laser field. with phase diffusion.*

The phase can change in one dimension only and therefore our model is one-dimensional too. We assume that the phase is subject to small leaps occurring at a rate R still to be determined. The leaps are completely independent of each other, i.e. in every single case the direction of the next step is entirely random. This results in stochastic motion known from the Brownian motion of molecules. Therefore it is also

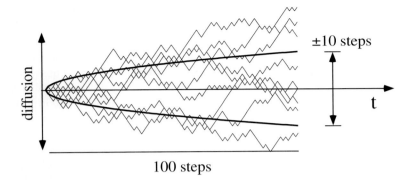

Fig. 8.11: *One-dimensional random walk. Every single step is randomly set in either the +
or − direction. The thick line marks the mean square deviation.*

called a *random walk* and is often compared to the walk of a drunk who is not aware
of his next step.

Using the random-number generator of the computer, such motions can easily
be simulated. In Fig. 8.11 several trajectories are presented and additionally the
time-dependent expectation value of the square fluctuation is shown. The root mean
square deviation of the Gaussian normal distribution after N steps is $(\Delta N_{\mathrm{rms}}^2)^{1/2} =
(\langle N^2 \rangle - \langle N \rangle^2)^{1/2} = \sqrt{N}$, as is well known. Since the number of steps increases in
proportion with time τ, the r.m.s. deviation has to be proportional to $\sqrt{\tau}$. Therefore
we can construct the normal distribution

$$p(\Delta\theta(\tau)) = \frac{\exp(-\Delta\theta^2/2\Delta\theta_{\mathrm{rms}}^2)}{\sqrt{2\pi}\,\Delta\theta_{\mathrm{rms}}} \qquad \text{with} \qquad \int_{-\infty}^{\infty} p(\Delta\theta)\,d\Delta\theta = 1,$$

with mean square value $(\Delta\theta_{\mathrm{rms}})^2 = \theta_0^2 R\tau$ and θ_0 for the length of a single step. Now
we can evaluate the integral of Eq. (8.28), obtaining the simple result

$$\langle \cos\Delta\theta(\tau) \rangle = \exp(-\Delta\theta_{\mathrm{rms}}^2/2) = \exp(-\theta_0^2 R\tau/2).$$

The complete correlation function ($\omega = 2\pi\nu$) reads

$$C_E(\tau) = \tfrac{1}{2}\langle E(0)E^*(\tau) \rangle = \tfrac{1}{2}|E_0|^2\, e^{i2\pi\nu\tau - \theta_0^2 R\tau/2}.$$

The correlation function can also be interpreted as the average projection of the
field vector onto its initial value at the time $\tau = 0$. Its form is identical with the
time dependence of a damped harmonic oscillator. We calculate the spectrum again
according to the Wiener–Khintchin theorem (Eq. (A.9)) and find that white phase
noise (Fig. 8.9) leads to a Lorentz-shaped line with width $\Delta\omega = 2\pi\Delta\nu_{1/2} = \theta_0^2 R$
centred at the carrier frequency ν:

$$S_E(\nu + f) = \frac{|E_0|^2}{T} \frac{\theta_0^2 R/2}{(2\pi f)^2 + (\theta_0^2 R/2)^2}. \qquad (8.29)$$

8.4.2   The microscopic origin of laser noise

The considerations of the previous section are generally valid for oscillators of every kind. We now have to correlate the macroscopically observed properties to the specific microscopic properties of the laser. A rigorous theory (i.e. a consequent theoretical calculation of correlation functions as in Eq. (8.27)) requires a treatment according to quantum electrodynamics, for which we have to refer to the relevant literature. The theories by Haken [39] and by Lax and Louisell [67] respectively are among the important successes in the quantum theory of 'open systems' and were presented shortly after the invention of the laser. We have to limit ourselves here to simplified models, but we can put forward some reflections about the nature of the noise forces.

The fluctuations of laser light field reflect several noise sources. The best-known process is caused by the spontaneous emission out of the amplifying medium into the environment. These radiation processes do not contribute to the laser field but cause stochastic fluctuations of the inversion and the (dielectric) polarization. Since the amplitudes of resonator field and polarization relax back to their steady state, amplitude and phase fluctuations result.[2] Other noise processes are caused because the resonator field also suffers from random losses, or because the pumping process transfers its noise properties to the stimulated emission. It is normally 'incoherent', i.e. the excitation states are produced with a certain rate but with a random, typically Poisson distribution. In a semiconductor laser, electron–hole pairs are injected into the amplification zone. For large current density the charge carriers repel each other, and successive arrival times are more evenly spaced out. It has been shown that this 'regularization' of the pumping process also gives rise to a decrease of the intensity fluctuations [114]!

Many processes can be heuristically interpreted through the 'grainy' structure of the quantized light field. Let us therefore study changes of amplitude and phase of the laser field when 'photons' are added to or taken away from it.

8.4.3   Laser intensity noise

The time evolution of the laser amplitude was investigated in Section 8.2.1 for the system reacting to sudden changes of the pumping rate, for example through deterministic switching events. In our simple model such changes are now caused by small random changes of the photon number $n_{ph}(t) = \overline{n}_{ph} + \delta n_{ph}(t)$ fluctuating around the mean value $\overline{n}_{ph}$.

Quantum limit of the laser amplitude

Let us estimate the mean square deviation $\langle \delta n_{ph}^2 \rangle$ of the photon number, and hence the field amplitude, without ascertaining the distribution more exactly. For this we

---

[2]In another formulation it is often said that spontaneous emission radiates 'into the laser mode'. In this interpretation, polarization and laser field both separately have to relax back again to their equilibrium relation. Since in the theoretical description used here the coupling of resonator field and polarization is already completely included, the interpretation chosen here appears to be physically more conclusive.

rewrite the linearized equation (8.23) for the photon number by inserting the stationary photon number $\overline{n}_{\mathrm{ph}}$ from Eq. (8.21):

$$\frac{d^2}{dt^2}\delta n_{\mathrm{ph}} + (\kappa\overline{n}_{\mathrm{ph}} + \gamma)\frac{d}{dt}\delta n_{\mathrm{ph}} + \gamma_c\kappa\overline{n}_{\mathrm{ph}}\delta n_{\mathrm{ph}} = 0.$$

We multiply this equation by δn_{ph} and arrive at

$$\frac{1}{2}\frac{d^2}{dt^2}\delta n_{\mathrm{ph}}^2 - \frac{1}{2}(\kappa\overline{n}_{\mathrm{ph}} + \gamma)\frac{d}{dt}\delta n_{\mathrm{ph}}^2 - \frac{1}{2}\left(\frac{d}{dt}\delta n_{\mathrm{ph}}\right)^2 + \gamma_c\kappa\overline{n}_{\mathrm{ph}}\delta n_{\mathrm{ph}}^2 = 0. \qquad (8.30)$$

When we search for the steady-state solution of the mean value $\langle\delta n_{\mathrm{ph}}^2\rangle$, we can eliminate the average of the derivatives $\langle(d/dt)\delta n_{\mathrm{ph}}\rangle$ but not that of the square of the fluctuation rate $[(d/dt)\delta n_{\mathrm{ph}}]^2$:

$$-\tfrac{1}{2}\langle[(d/dt)\delta n_{\mathrm{ph}}]^2\rangle + \gamma_c\kappa\overline{n}_{\mathrm{ph}}\langle\delta n_{\mathrm{ph}}^2\rangle = 0. \qquad (8.31)$$

We cannot give a rigorous theoretical description here, but for an intuitive treatment we can use Eq. (8.22(i)), $[(d/dt)\delta n_{\mathrm{ph}}]^2 = (\kappa\overline{n}_{\mathrm{ph}}\delta n)^2$. It is reasonable to assume an inversion undergoing random fluctuations and hence obeying Poisson statistics induced by both spontaneous emission and the stochastic pumping process, and yielding mean square value $\delta n^2 = n^{\mathrm{st}}$. We can now evaluate Eq. (8.30) and find with $\gamma_c = \kappa n^{\mathrm{st}}$ from (8.13)

$$\langle\delta n_{\mathrm{ph}}^2\rangle = \langle(d/dt)\delta n_{\mathrm{ph}}^2\rangle/\gamma_c\kappa\overline{n}_{\mathrm{ph}} = \overline{n}_{\mathrm{ph}}.$$

Most importantly, we find that the number of photons in the resonator fluctuates by an amount proportional to $\sqrt{\overline{n}_{\mathrm{ph}}}$. A more exact analysis shows that the distribution indeed again has the shape of a Poisson distribution (which for large numbers is essentially a Gaussian distribution).

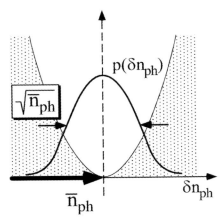

Fig. 8.12: *Distribution of the photon number of the laser field. The fluctuations of the photon number are stabilized by an effective potential (see Eq. (8.31)).*

The investigation of the photon number distribution offers an intuitive picture, which we study in a bit more detail in Fig. 8.13. The total number of photons is proportional to the field energy ($E^2 \propto h\nu\overline{n}_{\mathrm{ph}}$). Removal or addition of one 'photon' changes the field energy by the amount $h\nu$.

Relative intensity noise (RIN)

The fluctuations of the external laser power $P(t) = P_0 + \delta P(t)$ are measured in an experiment. Using Eq. (8.21) the fluctuations of the photon number can be converted into the r.m.s. deviation of the laser power $\delta P_{\mathrm{rms}} = \langle\delta P^2\rangle^{1/2}$. Thus: $\delta P_{\mathrm{rms}} = \sqrt{h\nu\gamma_{\mathrm{out}}}\sqrt{P}$.

Intensity fluctuations of the idealized laser are caused by quantum fluctuations only, the fundamental physical limit. According to the results of the previous section, their relative significance decreases with increasing laser power because $\sqrt{\delta n_{ph}^2}/\overline{n}_{ph} = 1/\sqrt{\overline{n}_{ph}}$. Moreover, many laser types show noise contributions that are not always exactly identified but increasing proportionally to the output power.

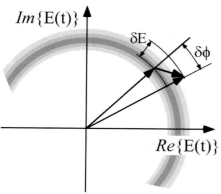

For the quantitative characterization of the amplitude noise, the *relative intensity noise* has been introduced:

$$\text{RIN} := \frac{\delta P_{rms}^2}{P^2},$$

which is a phenomenological quantity and can be straightforwardly measured. For a more exact analysis of the intensity noise, again its spectral distribution has to be determined. In the simplest case of completely random fluctuations, it shows a flat spectrum, i.e. white noise of Fig. 8.9.[3] In the section about semiconductor lasers we will find that in the intensity spectrum, for example, the relaxation oscillations play a role as well.

Fig. 8.13: *Effect of a 'photon' on the time evolution of the laser field. See also Fig. 8.10.*

8.4.4   Schawlow–Townes linewidth

In order to determine the laser linewidth $\Delta\nu$ from Eq. (8.29), we have to know the length of the single step ($\langle\theta_0^2\rangle = 1/2\overline{n}_{ph}$) and to determine the rate of spontaneous emission R_{spont}:

$$
\begin{aligned}
\Delta\nu_{1/2} &= \langle\theta_0^2\rangle R_{spont} \\
&= R_{spont}/2\overline{n}_{ph}.
\end{aligned}
$$

Spontaneous processes occur at a ratio $1 : \overline{n}_{ph}$ to the stimulated processes with regard to the evolution of the resonator field. The rate is proportional to the number of excited particles n_e^{st} so that drawing on Eqs. (8.13) and (8.20) we can write

$$R_{spont} = R_{stim}/\overline{n}_{ph} = \kappa n_e^{st} = \gamma_c n_e^{st}/n^{st}.$$

On the other hand, according to Eq. (8.17), we can connect the photon number with the output power, $\overline{n}_{ph} = P/h\nu\gamma_{out} \simeq P/h\nu\gamma_c$. So finally we arrive at the linewidth

$$\Delta\nu_L = \frac{n_e^{st}}{n^{st}}\frac{\pi h\nu}{P}\gamma_c^2. \tag{8.32}$$

In a 'good' four-level laser the first factor is $n_e^{st}/n^{st} \simeq 1$. This surprising formula was presented by Schawlow and Townes as long ago as 1958, and is called the *Schawlow–Townes linewidth*. As we already calculated in the section about He–Ne lasers, an extremely small linewidth of a few Hertz or less is expected even for conventional lasers.

---

[3]We should be aware of the fact that even 'white noise' has an upper limit frequency – otherwise the r.m.s. value of the fluctuation would be unbounded according to Eq. (A.6)!

Larger linewidths are only observed for small resonators with low mirror reflectance, like e.g. in semiconductor lasers. They are also subject to an additional broadening mechanism caused by amplitude–phase coupling (see Section 9.4.2).

8.5   Pulsed lasers

In Section 8.2.1 on relaxation oscillations, we found (see Fig. 8.2) that switching processes can induce short laser pulses with intensities much higher than average. With pulsed lasers, a large amount of radiation energy, in common systems up to several joules, can be delivered within a short period of time. Its peak power depends on the pulse length.

One important method for generating short and very intense laser pulses is realized by the so-called 'Q-switch' concept. Another method creates a coherent superposition of very many partial waves ('mode locking') resulting in a periodic sequence of extremely short laser pulses.

8.5.1   'Q-switch'

Pulsed neodymium lasers are among the most common systems offering very high peak powers. In such pulsed lasers, the pump energy is supplied through an excitation pulse, e.g. from a flash lamp. The pump pulse (Fig. 8.14) builds up the inversion until the laser threshold is passed. Then stimulated emission starts and the system relaxes to the equilibrium value. In the neodymium laser the amplitude damping occurs so fast that the output power follows the excitation pulse with small relaxation oscillations.

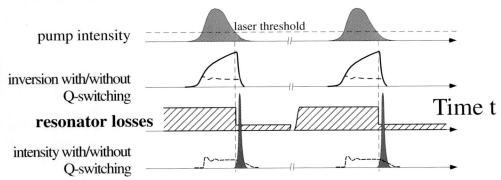

Fig. 8.14: *Time chart of pulsed laser oscillation with and without (dashed lines) Q-switch.*

Alternatively, lasing can initially be suppressed by increasing the resonator losses with a Q-switch. If the accumulation time is short compared to the decay period of the upper laser level (for the neodymium laser, e.g. 0.4 ms), the laser medium acts as an energy storage device and the inversion continues to increase. If the Q-switch triggered by an external impulse is again set to high Q-factor or low loss mode, stimulated emission begins and now, by fast exhaustion of the accumulated energy, a laser pulse is generated that is short compared to the non-switched operation with much higher peak power. The repetition rate of such a laser system usually lies between 10 Hz and 1 kHz.

Technical Q-switches

Q-switches have to fulfil two conditions: in the open state the resonator Q-factor has to be reduced efficiently, whereas in the closed state its insertion loss has to be small compared to other losses. Typical systems for a Q-switch are presented in Fig. 8.15 and described below.

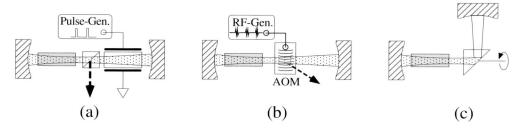

Fig. 8.15: *Q-switch and cavity dumping: (a) electro-optical (Pockels cell); (b) acousto-optical (Bragg cell) and (c) mechanical (rotating prism). See text.*

(a) Pockels cell   The Pockels effect already described in Section 3.6.1 makes a voltage-driven retarder plate available. In combination with a polarizer (Fig. 8.15), the resonator transmission can be modulated very efficiently. The switching time of a Pockels cell is in the nanosecond domain. It is primarily limited by the capacitance of the crystal electrodes and the resistance of the electrical leads.

(b) Acousto-optical modulator (AOM)   In the acousto-optical modulator (see also section 3.6.4), a radio-frequency generator induces an acoustic wave causing a periodic variation of the refraction coefficient. Laser radiation is deflected by diffraction off this grating out of the resonator, and frequency-shifted at the same time. The radio-frequency power can be switched by suitable semiconductor components with nanosecond rise times.

(c) Rotating prism   The Q-switch can also be realized by a mechanical rotating prism, which allows the laser to start only in a narrow acceptance angle range.

Cavity dumping

The Pockels cell and the acousto-optical modulator (AOM) of Fig. 8.15 provide a second output port. This may be used for the so-called *cavity dumping* method. For this, in the closed laser oscillator, a strong oscillation builds up within the resonator first. Through an external pulse triggering the AOM or Pockels cell, this energy is then dumped out of the resonator. The method can also be combined with the mode locking concept of the following section in order to achieve particularly high peak powers.

8.5.2   Mode locking

Even the simplest superposition of two laser beams with different frequencies ω and $\omega + \Omega$ causes periodic swelling up and down, as is well known from amplitude modulation. For equal partial amplitudes with $I_0 = c\epsilon_0|E_0|^2$ we have

$$
\begin{aligned}
E(t) &= E_0\, e^{-i\omega t} + E_0\, e^{-i\omega t}\, e^{-i\Omega t}\, e^{-i\phi}, \\
I(t) &= \tfrac{1}{2} c\epsilon_0 |E(t)|^2 = I_0[1 + \cos(\Omega t + \phi)].
\end{aligned}
$$

When we neglect the dispersive influence of the optical elements, laser resonators of length $n\ell$ (n is refraction coefficient) provide an equidistant frequency spectrum with $\Omega = 2\pi c/2n\ell$ (Eq. (5.18)) that virtually offers itself for synthesis of time-periodic intensity patterns. *Mode locking* establishes a technical procedure to physically realize Fourier time series consisting of many optical waves.

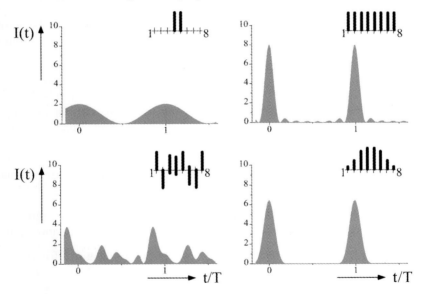

Fig. 8.16: *Intensity variation in time for the superposition of up to eight harmonic waves. The vertical bars indicate the relative strength and phase position of the partial waves. Upper left: Amplitude modulation with two waves for comparison.*

While for two waves the phase ϕ causes only an overall phase shift of the sinusoidal modulation pattern, the pattern originating from superposition of multiple waves also depends on individual phase positions, as we show in Fig. 8.16 with the example of eight superimposed waves. We can calculate the field amplitude in general according to

$$E_{\mathrm{N}}(t) = \frac{E_0}{\sqrt{N}} \, e^{-i\omega t} \, e^{iN\Omega t/2} \sum_{n=1}^{N} \alpha_n \, e^{-in\Omega t} \, e^{-i\phi_n},$$

where important characteristic quantities include

pulse sequence frequency $f = \Omega/2\pi,$
and   pulse period $T = 2\pi/\Omega.$

The mid-frequency is called the 'carrier frequency' $\omega_0 = \omega - N\Omega/2$, and different waves with frequency differences $\Delta f = nf = n\Omega/2\pi$ contribute to the total wave with phases ϕ_n. The partial amplitudes have been chosen in such a way that the intensity $I_0 = (c\epsilon_0/2)E_0^2 \sum_n \alpha_n^2$ is distributed among partial amplitudes with $\alpha_n E_0$ and $\sum_n \alpha_n^2 = 1$. Thus the intensity maxima in Fig. 8.16 are comparable to each other at equal mean power.

In Fig. 8.16 three characteristic situations are presented:

1. In the upper right part all partial waves have identical amplitudes $\alpha_n = \sqrt{1/n}$ and are in phase with $\phi_n = 0$ for all n. For this situation, very sharp periodic maxima with a small peak width $\Delta t \approx 2\pi/(N\Omega) = T/N$ occur. The secondary maxima are characteristic for an amplitude distribution with a sharp boundary.

2. In the lower right part the partial waves are in phase as well, though the amplitudes have been chosen following a Gaussian distribution, which is symmetrical to the carrier frequency ω_0 ($\alpha_n \propto \exp\{-[(2n - N - 1)/2]^2/2\}$). By this distribution, the 'ears' that occur between the maxima in the previous example are suppressed very efficiently and the laser power is concentrated to the maxima. The achievable peak power though is slightly lower. This situation resembles closely the conditions of a real laser resonator. In Fig. 8.17 a frequency spectrum of a periodic train of 27 ps Ti–sapphire laser pulses measured in a Fabry–Perot resonator is shown.

3. In the lower left part of Fig. 8.16 for comparison the situation for random phases ϕ_n of the partial waves is presented, which makes a noisy but periodic pattern.

Let us now study the relation between pulse length and bandwidth (see section 3.4) and therefore consider a periodic series of Gaussian-shaped pulses with $E(t) = \sum_n E_0 \exp\{-[(t - nT)/\Delta t]^2/2\}$. According to the theory of Fourier series, we can obtain the nth Fourier amplitude for $n\Omega$ from

$$\mathcal{E}_n = E_0 \int_{-\tau/2}^{\tau/2} e^{-(t/\Delta t)^2/2} \, e^{-in\Omega t} \, dt \approx E_0 \, e^{-(n\Omega\Delta t)^2/2}.$$

For this, to a good approximation for very sharp pulses, only that single pulse centred at $t = 0$ is taken into account, and the integration limits are extended to $\pm\tau/2 \to \pm\infty$. We define a bandwidth by $2\pi f_{\mathrm{B}} = \Omega_{\mathrm{B}} = 2N\Omega$ with $2N$ now indicating the effective number of the participating laser modes. The contribution of the modes to the total

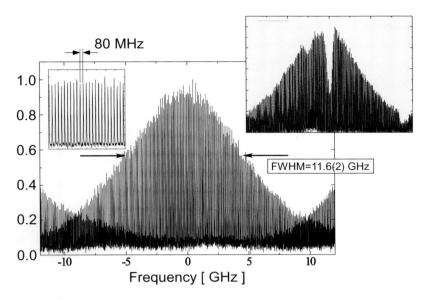

Fig. 8.17: *Frequency spectrum of the 27 ps pulses of a mode-locked Ti–sapphire laser recorded with a Fabry–Perot resonator with 7.5 mm distance between the mirrors or* $\Delta_{\text{FSR}} = 20$ *GHz. The smaller pictures show an enlarged detail and the absorption of a caesium vapour cell in the ray trajectory [10], respectively.*

power drops down to $1/e$ of the central mode at $n = N$. The bandwidth f_B and the pulse length $2\Delta t$ (measured at relative amplitude value $1/e$) are connected to each other and to the peak width Δt_{FWHM} according to

$$\Delta t = \frac{1}{N\Omega} = \frac{\Delta t_{\text{FWHM}}}{\sqrt{8\ln(2)}} = \frac{\Delta t_{\text{FWHM}}}{2.35}.$$

8.5.3   Methods of mode locking

In order to achieve pulses as short as possible, it is first important to use a laser amplifier with a very large bandwidth. For a sufficiently long lifetime of the upper laser level, the excitation can conveniently be generated by a continuous-wave laser. The stored energy is withdrawn from the laser medium by pulses separated typically by 12.5 ns, a time that is short compared to, for example, the lifetime of 4 µs of the upper Ti–sapphire laser level. For other systems like the dye laser, also the so-called 'synchronous pumping' excitation scheme is used. In that case and owing to the short lifetime of the upper laser level, the pumping laser delivers a periodic and exactly synchronized sequence of short pulses. As a certain special case, which we skip here, we just mention the diode laser. By suitable modulation of the injection current (Section 9.4.1) it directly delivers very short pulses down to 10 ps. It has been intensively studied because of its significance for optical communication.

Tab. 8.2 contains important examples of lasers used for the generation of extremely short pulses, and for comparison the limited potential of the helium–neon classic. The typical repetition rate of mode-locked lasers is 80 MHz and 12.5 ns pulse distance, respectively, which is determined by the characteristic construction lengths ℓ setting the repetition rate at $T = 2n\ell/c$.

Tab. 8.2: *Mode locking and bandwidth.*

Laser	Wavelength λ	Bandwidth f_B	Pulse duration $2\Delta t$	Pulse length $\ell_P = 2c\Delta t$
Helium–neon	633 nm	1 GHz	150 ps	–
Nd:YLF laser	1047 nm	0.4 THz	2 ps	0.6 mm
Nd:glass laser	1054 nm	8 THz	60 fs	18 µm
GaAs diode laser	850 nm	2 THz	20 ps	6 mm
Ti–sapphire laser	900 nm	100 THz	6–8 fs	2 µm
NaCl–OH$^-$ laser	1600 nm	400 nm	4 fs	1.5 µm

Mode locking within the laser resonator is achieved by modulation of the resonator losses synchronized to the pulse circulation. In Fig. 8.18 the mode coupler is set to transmission only when the pulse passes and to opaque otherwise. This modulation can be controlled either actively by the Q-switch components of Fig. 8.15, or by passive nonlinear elements. Among them is the so-called 'saturable absorber', which is mainly used for dye lasers. A saturable absorber (optical saturation of an electric dipole transition is treated in Section 11.2.1) has an absorption coefficient that dies away at intensities above the so-called saturation intensity I_{sat},

$$\alpha(I(t)) = \frac{\alpha_0}{1 + I(t)/I_{\text{sat}}}.$$

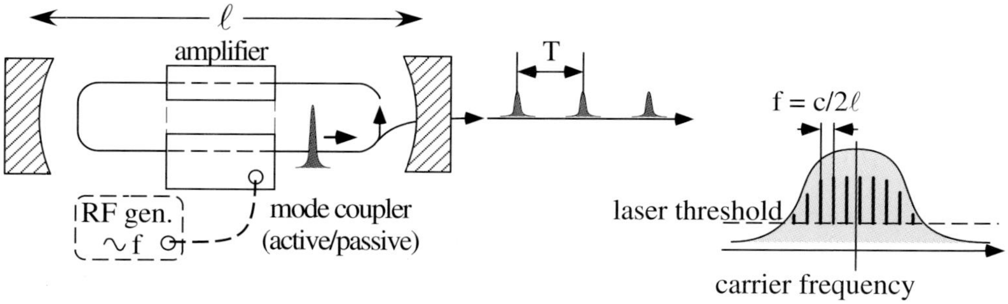

Fig. 8.18: *Laser with mode locking. In the resonator, a spatially well localized light pulse is circulating. Mode locking is achieved actively, e.g. by modulation of the cavity Q-factor, or passively by saturable absorbers or Kerr lens mode locking.*

By means of an intense laser pulse circulating in the resonator (Fig. 8.18), the absorber is easily saturated and hence resonator losses are rapidly reduced during pulse passage. This passive modulation leads to self-locking of the laser modes. A variant of the passive mode locking not studied intensively any more (*colliding pulse mode locking,* CPM laser) uses two pulses circulating in the resonator that hit each other exactly in the saturable absorber.

The most successful method in technical applications at this time is the so-called Kerr lens mode locking (KLM), which causes a time-dependent variation of the resonator geometry due to the intensity dependence of the refraction coefficient

$$n = n_0 + n_2 I(t).$$

Kerr lens mode locking is an example of the application of self-focusing and will be discussed in more detail in the section on nonlinear optics (13.2.1). The dispersive nonlinearity reacts extremely fast, essentially instantaneously, to variations of the intensity, and therefore is advantageous for very short pulses. At the centre of a Gaussian-shaped beam profile (for positive n_2) the refractive index is increased more strongly than in the wings, and hence causes self-focusing, which changes the beam geometry as presented in Fig. 8.19. Since the resonator losses depend on the beam geometry (the alignment of the resonator!), this phenomenon has the same effect as a saturable absorber and can be used for mode locking.

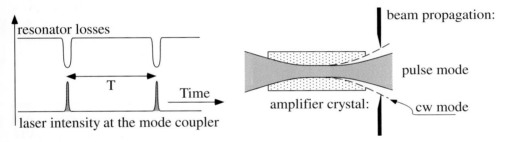

Fig. 8.19: *Time dependence of resonator losses and the influence of a Kerr lens on the beam geometry. The mode locking effect can be supported by the use of an additional aperture.*

Kerr lens mode locking was discovered by W. Sibbett et al. in the Ti–sapphire laser in 1991 [98]. It has led to revolutionary simplifications for the generation of ultrafast pulses due to its particularly simple application, since the nonlinear passive mode locker, the Kerr lens, is intrinsic to the Ti–sapphire amplifier crystal.

The KLM method alone though is not sufficient to generate shorter pulses than about 1 ps. In Section 3.4.1 we investigated the influence of dispersion and group velocity dispersion (GVD) on the shape of propagating light pulses, which naturally play an important role when the shortest light pulses are to be generated in a laser resonator containing several dispersive elements. The GVD can be compensated through the arrangement of prisms of Fig. 13.5 on p. 381. The prism combination is traversed twice per round trip in the resonator. In a ring resonator two pairs of prisms have to be supplied to recombine the beams again. Another technique for dispersion control is offered by dielectric mirrors with specially designed coatings (*chirped mirrors*). Very compact femtosecond oscillators can be built with them.

Here we have considered the mode-locked lasers only in their simplest situation, that is for steady-state conditions. The operation of mode-locked lasers though raises many interesting questions about laser dynamics, for which we refer the reader to the specialized literature. Such questions include, for example, the starting behaviour – How does the passively locked laser get to this state at all? From a naive point of view, we can make, for example, intensity fluctuations responsible for this, which may always be triggered by slight mechanical vibrations.

Another phenomenon is the *amplified spontaneous emission* (ASE), which sometimes causes annoying side effects in experiments. It occurs because, during the pumping phase between the pulses, the amplifier already emits radiation energy, which is intensified in the direction of the desired laser beams due to geometry.

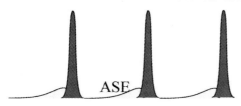

The ASE can be suppressed by, for example, saturable absorbers that transmit light only above a certain threshold intensity or separate it from the laser beam by external spatial filtering since the ASE in principle has much larger divergence.

Fig. 8.20: *Amplified spontaneous emission.*

8.5.4   Measurement of short pulses

The measurement of the temporal properties (especially pulse duration) of short pulses is limited to about 100 ps by common photodiodes and oscilloscopes due to their limited bandwidth (several GHz). On the electronic side, the so-called *streak camera* can be used, a channel plate generating an electron beam that is deflected rapidly similar to an oscilloscope. It leaves a trace on the camera and so converts the time dependence into a local variation. With recent models a time resolution down to 100 fs can be achieved.

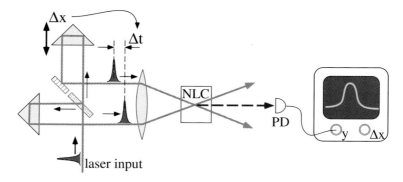

Fig. 8.21: *Autocorrelator for the measurement of the time dependence of very short laser pulses. In the direction of the photodiode (PD), a signal only occurs when the laser pulses are correctly superimposed in both time and space in the nonlinear crystal (NLC).*

A purely optical standard method is offered by the autocorrelator, e.g. according to the layout of Fig. 8.21: a pulsed laser beam is split into two partial beams and superimposed in a nonlinear crystal in such a way that a frequency-doubled signal (details about frequency doubling will be presented in Section 12.4) occurs. On the photodiode a signal is only registered if the partial pulses are superimposed correctly. The voltage signal as a function of the displacement $\Delta x = c\Delta t$ of one arm relative to the other one,

$$I_{\mathrm{PD}}(\Delta t) \propto E(t)E(t + \Delta t),$$

also has pulse shape, but is the result of a convolution of the pulse with itself (therefore autocorrelation), from which the pulse shape has to be deduced by some suitable transformations or models.

8.5.5   Tera- and petawatt lasers

The new possibilities of generating extremely short laser pulses have also opened a window to the generation of extremely intense laser 'flashes', at least for a very short period in time. The field intensities are so large that matter is transferred to completely new states, which at best can be anticipated in special stars.

Even with a 'common' femtosecond oscillator (Ti–sapphire laser, 850 nm, $f = 80\,\mathrm{MHz}$, $\langle P \rangle = 1\,\mathrm{W}$, medium power), using appropriate components for the compensation of group velocity dispersion [102], pulses can be generated with a duration of only $2\Delta t = 10\,\mathrm{fs}$. Even though such pulses only contain small amounts of energy E_{pulse}, they already make available considerable peak power P_{max} and peak field intensities E_{max}:

$$
\begin{aligned}
E_{\mathrm{pulse}} &= 1\,\mathrm{W}/80\,\mathrm{MHz} &&= 12.5\,\mathrm{nJ}, \\
P_{\mathrm{max}} &\approx E_{\mathrm{pulse}}/(2\Delta t) &&\simeq 1\,\mathrm{MW}, \\
E_{\mathrm{max}} &\approx 2P_{\mathrm{max}}/(\pi w_0^2 c\epsilon_0) &&= 7 \times 10^7\,\mathrm{V\,cm^{-1}}.
\end{aligned}
$$

For the calculation of the field intensity, we have assumed the laser power to be concentrated onto a focal spot with a diameter of $10\,\mu\mathrm{m}$. Besides, there, even a $1\,\mathrm{mW}$ He–Ne laser reaches field intensities of about $1\,\mathrm{kV\,cm^{-1}}$! According to this an increase of the pulse energy to $1\,\mathrm{J}$, which can be achieved today using table-top equipment, promises a power of about $100\,\mathrm{TW}$ and even the petawatt range is in sight. For this, field intensities of up to $10^{12}\,\mathrm{V\,cm^{-1}}$ are achieved, about 1000 times the 'atomic field intensity' $E_{\mathrm{at}} = e/4\pi\epsilon_0 a_0^2 = 10^9\,\mathrm{V\,cm^{-1}}$ experienced by an electron in the lowest hydrogen orbit!

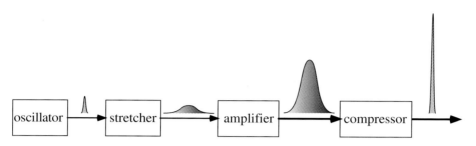

Fig. 8.22: *Chirped pulse amplification. By stretching, the peak power is decreased far enough that amplification without damage becomes possible.*

However, the generation and use of such intense laser pulses are hindered by this highly interesting strong interaction with matter. In common materials (initiated by multiphoton ionization) dielectric optical breakdown occurs and destroys the amplifier. An elegant solution for this situation is offered by the method of *chirped pulse amplification* (CPA, see Fig. 8.22), for which the short pulse is first stretched (in space and time) to decrease the peak power. The stretched pulse is amplified and the stretching is reversed immediately before the application to recover the original pulse shape.

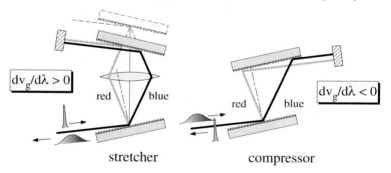

Fig. 8.23: *Grating stretcher and compressor for femtosecond pulses.*

Optical gratings have proven to be very suitable components for achieving stretching as well as compression [99]. The concept of a grating stretcher and compressor is presented in Fig. 8.23. The grating deflects red and blue parts of an incident pulse in different directions. In the stretcher two gratings are combined with 1 : 1 imaging properties. In a completely symmetric layout (dashed upper grating on the left in Fig. 8.23) the upper grating would not change the shape of the impulses at all, only at the drawn position.

8.5.6   White light lasers

Very recently so called *white light lasers* and *super continua* have become an exciting object of research. It seems contradictory at first to speak of white light in this context since the bias from classical optics suggests an absolutely incoherent light source with this term. White light covering the full spectral range of visible colors can, however, be generated from ultrashort pulse lasers. The photograph from the cover of this book shows white laser light dispersed by a two dimensional grating and exhibiting the full range of colors. Furthermore, as shown in Fig. 8.24, this light field shows well modulated interferences, it is thus coherent and truly laser light!

Ultrashort, intense laser pulses are the basis for nonlinear processes transforming their original, relatively narrow spectrum (typically less than 10% of the visible spectrum) into an extremely broadband spectrum which may cover the entire visible spectrum and beyond. The generation of coherent white light is a field of active research and not yet fully understood, but it seems clear that it is essential to provide efficient nonlinear conversion processes with fibres driven in

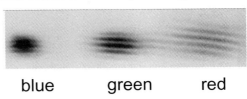

blue   green   red

Fig. 8.24: *Interference pattern observed when a white light laser is superposed with itself. The light beam was dispersed by a prism to demonstrate simultaneous interference of all contributing colors. With permission of Harald Telle and Jörn Stenger.*

the strong guiding limit. Remember that most optical fibres (See Section 1.7) are operated in the *weak guiding limit* where small steps in the index of refraction of order 1% provide guiding but also cause the optical wave to be spread out over a relatively large cross section. In so called *photonic fibres* [87] or *tapered fibres* [106] the strong guiding limit is realized and the optical wave is confined by large index of refraction steps corresponding to the glass-air interface to a cross section with diameter 1-2 μm.

Fig. 8.25 gives an example of the schematic setup for a tapered fibre. It is drawn out to very narrow cross sections from a conventional fibre providing efficient coupling into the tapered section. Several processes of the nature described in chaps. 12 and 13 on nonlinear optics are responsi-

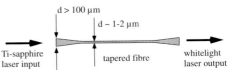

Fig. 8.25: *White light generation with a tapered optical fibre.*

ble for the spectral broadening occuring during propagation of the short light pulse through the tapered section.

9 Semiconductor lasers

Immediately after the demonstration of the ruby laser (1960) and the helium–neon laser (1962), the lasing of diodes, or 'semiconductor lasers', was also predicted and a little bit later was realized experimentally.[1] However, it took more than 20 years for those components to become commercially successful products, since numerous technological problems had to be overcome. The first laser diodes, for example, could operate only at cryogenic temperatures, while applications in general require operating temperatures close to room temperature. Moreover, GaAs was the first relevant material for the manufacture of laser diodes, and not silicon, which, then as now, otherwise dominates semiconductor technologies.

Today, laser diodes belong to the most important 'opto-electronic' devices because they allow the direct transformation of electrical current into (coherent!) light. Therefore there are countless physical, technical and economic reasons to dedicate a chapter of its own to these components and related laser devices.

9.1 Semiconductors

For a detailed description of the physical properties of semiconducting materials, we refer the reader to the known literature [45, 57]. Here we summarize properties of importance for the interaction with optical radiation.

9.1.1 Electrons and holes

In Fig. 9.1 the valence and conduction bands of a semiconducting material are presented. Electrons carry the current in the conduction band (CB), whereas holes[2] do so in the valence band. The distribution of the electrons into the existing states is described by the Fermi function $f(E)$

$$f_{el}(E, \varepsilon_F) = [1 + e^{(E-\varepsilon_{CB})/kT}]^{-1}, \tag{9.1}$$

which is determined by the Fermi energy for electrons of the conduction band $\varepsilon_F = \varepsilon_{CB}$ and temperature T. Especially at $T = 0$ all energy states below the Fermi energy are

---

[1] John v. Neumann (1903-1957) carried out the first documented theoretical consideration of a semiconductor laser in 1953. This unpublished manuscript was reproduced in [105]

[2] It should not be forgotten that the term *hole* is only an – albeit very successful – abbreviation for a basically very complex physical many-particle system. Most of the physical properties (conductivity, Hall effect, etc.) of the electrons of the valence band can be very well described as if there were free particles with a positive charge and a well-defined effective mass.

Optics, Light and Laser. Dieter Meschede
Copyright © 2004 Wiley-VCH Verlag GmbH & Co. KGaA
ISBN: 3-527-40364-7

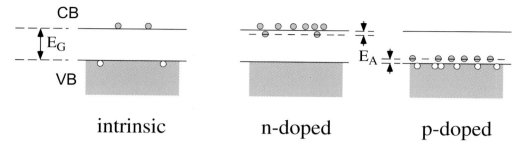

intrinsic   n-doped   p-doped

Fig. 9.1: *The band model for semiconductors. Electrons and holes can move freely and independently from each other. CB = conduction band; VB = valence band; E_g = bandgap energy; E_A = excitation energy of the dopant impurities.*

completely filled, and above it completely empty. The distribution is described in analogy by

$$f_h = 1 - f_{el} = [1 + e^{(\varepsilon_{VB} - E)/kT}]^{-1}. \tag{9.2}$$

In equilibrium the occupation numbers of electrons and holes are characterized by a common Fermi energy $\varepsilon_{CB} = \varepsilon_{VB}$. In forward-biased operation at a pn junction a non-equilibrium situation relevant for laser operation arises with different Fermi energies for electrons and holes, $\varepsilon_{CB} \neq \varepsilon_{VB}$.

Some important situations of the Fermi distribution in semiconductors are presented in Fig. 9.2. At $T = 0$ the Fermi energy gives exactly the energy up to which the energy levels are occupied.

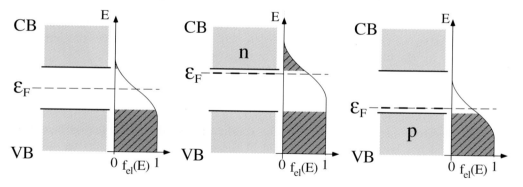

Fig. 9.2: *Fermi distribution in intrinsic, p- and n-doped semiconductors, respectively.*

9.1.2   Doped semiconductors

An *intrinsic* semiconductor consists of a pure crystal, e.g. the technologically most important material Si from main group IV in the periodic table or the III–V compound GaAs. In such a material the Fermi energy is found close to the middle of the bandgap. The occupation probability of the states can then approximately be described according to Boltzmann's formula

$$f_{\mathrm{el}}(T) \simeq e^{-E_{\mathrm{g}}/kT}.$$

The bandgap energy E_{g} depends on the material and is of the order of a few eV; therefore at room temperature ($kT \simeq 1/40\,eV$) there are only very few electrons in the conduction band. The revolutionary significance of semiconductors arises in principle from the possibility to increase the conductivity dramatically via doping (e.g. in Si, with impurity ions from main groups III or V) and even via different concentrations for holes and electrons (Fig. 9.1). The deficit or excess of electrons of the impurity atoms generate energy states near the band edges which are easy to excite at thermal energies. Electron charge carriers are generated in this way in an *n-doped* semiconductor, and holes in a *p-doped* system, respectively. The Fermi energy lies in this case near the acceptor (p doping) or the donor (n doping) level. Already at room temperature such a doped semiconductor exhibits a large conductivity caused by electrons in n-type and holes in p-type material.

9.1.3   pn junctions

If electrons and holes collide with each other, they can 'recombine', emitting dipole radiation. Such processes are facilitated by having an interface between p- and n-doped semiconducting material (a pn junction), which is the heart of every semiconductor diode. Fig. 9.3 presents the essential properties of a pn junction.

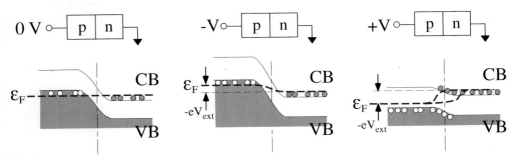

Fig. 9.3: *Free charge carriers at a pn junction. Left: Zero-bias equilibrium. At the interface, electrons diffuse into the p-doped area and holes into the n-doped one, where they can recombine. At the barrier, a layer is depleted of charge carriers and an electrical field is generated, counteracting any further diffusion. Middle: At reverse bias, the depletion zone is enlarged. Right: At forward bias, a current flows through the junction, electrons and holes flood the barrier layer and cause recombination radiation. Within the conduction and valence bands, there is thermal equilibrium characterized by two different Fermi energies for electrons and holes.*

9.2   Optical properties of semiconductors

9.2.1   Semiconductors for opto-electronics

From the opto-electronic point of view, the energy gap at the band edge is the most important physical quantity, since it determines the wavelength of the recombination radiation. It is presented in Fig. 9.4 for some important opto-electronic semiconductors as a function of the lattice constants, which have technological meaning for the formation of compound crystals. A particular gift of Nature for this is the extremely small difference of the lattice constants of GaAs and AlAs. Because of the excellent lattice match, the bandgap can be controlled over a wide range by the mixing ratio x in $(Al_xGa_{1-x})As$ compound crystals (Fig. 9.5).

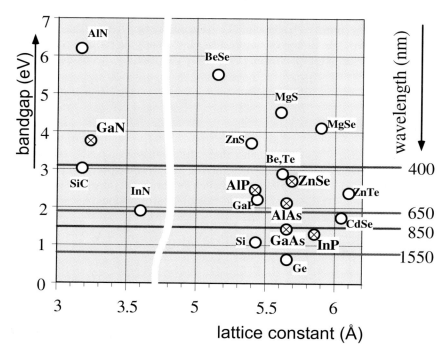

Fig. 9.4: *Bandgap energy of some important semiconducting materials. Materials for which lasing has been realized already are marked with a cross. At the right-hand side, some technically relevant laser wavelengths are given.*

Other compound crystals have been in use as well for quite a long time. Especially, the wavelength of 1.55 µm that is most important for optical telecommunications can be obtained from a quaternary InGaAsP crystal. Silicon, the economically most significant semiconducting material, does not play any role, since it does not have a direct bandgap but only an indirect one (see Section 9.2.4).

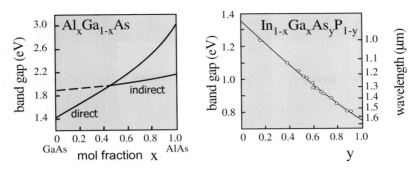

Fig. 9.5: *Bandgap energy in AlGaAs and InGaAsP as a function of the mixing ratio.*

Excursion: Blue luminescent gallium nitride, a scientific fairytale

The development of laser diodes experienced rapid progress in the 1980s and 1990s, but 1996 will go down in history as a very special year. In that year, Shuji Nakamura, with the Japanese company Nichia Chemical Industries Ltd, was able to present the world's first blue laser diodes to an astonished audience. He had made the devices based on GaN, which had been considered completely unsuitable for opto-electronics! This research was supported neither commercially nor academically, and success would not have been possible without the confidence of his boss, Nobuo Ogawa. With no experience at all on this topic in his company and not very much in touch with semiconductor lasers, since 1989 he had allowed the then 36-year-old engineer to pursue a research programme that went against all established opinions about the potential of gallium nitride [77].

In fact, there had been commercial interest in blue luminescence long before the interest in blue laser radiation, since only with blue light sources was there the hope to produce fully coloured screens based on semiconductors. Worldwide large sums had been invested in research on ZnSe, which was supposed to have the biggest chance of success. In textbooks, it could be read that GaN was unsuitable in spite of its well-known and attractive physical properties (direct bandgap of 1.95–6.2 eV for (Al,Ga,In)N), since it could not be p-doped. This assertion though could not be maintained any more after 1988, when Akasashi et al. were successful with the preparation of such crystals, though at first with a costly electron-beam technique. S. Nakamura succeeded crucially in the thermal treatment of GaN samples by replacing the NH_3 atmosphere by N_2. He found that the ammonia atmosphere dissociated and the released hydrogen atoms passivated the acceptors in GaN.

With this, though by far not all the problems were solved, the gate to the blue laser diode had been widely opened. Less than 10 years after this discovery, blue laser diodes could be bought in spite of all predictions – an event from the scientific book of fairytales.

9.2.2   Absorption and emission of light

In a semiconductor, electrons are excited from the valence band to the conduction band on absorption of light with a wavelength

$$\lambda < E_{\mathrm{g}}/hc,$$

so that electron–hole pairs are generated. Under certain conditions, e.g. at very low temperatures, absorption of light can be observed already below the band edge. During this process no freely mobile charge carriers are generated, rather pairs bound in 'excitonic' states with a total energy slightly below the edge of the conduction band. Excitons, which resemble atoms made from pairs of electrons and holes, will however not play any role in our considerations.

If free electrons and holes are available, they can recombine under emission of light that again has a wavelength corresponding roughly to the band edge due to energy conservation. The 'recombination radiation' though has furthermore to fulfil momentum conservation[3] for the electron–hole pair $(\hbar\mathbf{k}_{\mathrm{el}}, \hbar\mathbf{k}_{\mathrm{h}})$ as well as for the emitted photon $(\hbar\mathbf{k}_{\mathrm{ph}})$:

$$
\begin{aligned}
\text{energy:} \quad & E_{\mathrm{el}}(\mathbf{k}_{\mathrm{el}}) = E_{\mathrm{h}}(\mathbf{k}_{\mathrm{h}}) + \hbar\omega, \\
\text{momentum:} \quad & \hbar\mathbf{k}_{\mathrm{el}} = \hbar\mathbf{k}_{\mathrm{h}} + \hbar\mathbf{k}_{\mathrm{ph}}.
\end{aligned}
\tag{9.3}
$$

The $\mathbf{k}$-vectors of the charge carriers are of magnitude π/a_0 with a_0 indicating the lattice constant and therefore very much larger than $2\pi/\lambda$. That is why optical transitions only take place if the lowest-lying electronic states in the E–$\mathbf{k}$ diagram (the 'dispersion relation') are directly above the highest-lying hole states.

In Fig. 9.6 the situation for two particularly important semiconductors is schematically presented. In the so-called 'direct' semiconductor GaAs, at $\mathbf{k} = 0$, a conduction band edge with 'light' electrons meets a valence band edge with 'heavy' holes (the effective mass of the charge carriers is inversely proportional to the curvature of the bands); there direct optical transitions are possible. Silicon, on the other hand, is an indirect semiconductor. The band edge of the electrons occurs at large k_{el} values, that of the holes at $k = 0$; thus silicon cannot radiate! There are however weaker and more complex processes, e.g. with the participation of a phonon which supplements a large $\mathbf{k}$ contribution and thus ensures momentum conservation in Eq. (9.3) at negligible energy expense.

The recombination radiation is caused by an optical dipole transition with a spontaneous lifetime τ_{rec} of typically

$$
\text{recombination time } \tau_{\mathrm{rec}} \simeq 4 \times 10^{-9} \text{ s.}
$$

The recombination rate is also called the 'inter-band' decay rate and is very slow compared to the collision time T' of the charge carriers with defects and phonons within the conduction and valence bands. This 'intra-band' scattering takes place on the picosecond time scale

$$
\text{relaxation time } T' \simeq 10^{-12} \text{ s,}
$$

and ensures that, owing to relaxation within each of the bands, there is an equilibrium state determined by the crystal temperature.

9.2.3   Inversion in the laser diode

In a semiconductor, coherent light is generated by stimulated recombination radiation. In the beginning the pn junctions had to be very deeply cooled down to the

---

[3]In a crystal it is more exact to speak about *quasi-momentum conservation*.

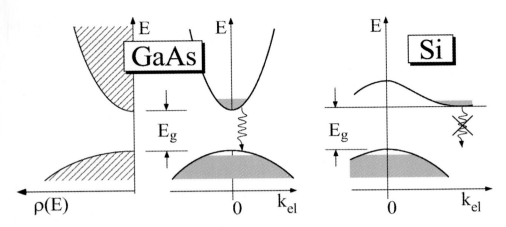

Fig. 9.6: *Left: Electronic density of states and simplified dispersion relation for direct semiconductors (GaAs). The different curvatures of the bands are the origin of the different effective masses (see Eq. (9.4)). In equilibrium there are charge carriers only at the band edges (shadowed areas indicate states filled with electrons). Optical transitions start and end with hardly any change of the **k**-vector since the momentum of the photons is not visible on this scale. They can only take place if electrons of the conduction band meet an unoccupied state, a hole in the valence band. Right: For indirect semiconductors (Si), direct optical transitions are suppressed.*

temperature of liquid helium in order to suppress loss processes competing with luminescence and to generate an adequate inversion density for lasing. The development of the heterostructure laser, which we are going to discuss a little later, has overcome this problem and contributed decisively to the still growing success of semiconductor lasers.

The amplification is determined among other things by the number of charge carriers that can emit recombination radiation at a certain energy difference. For this, their density of states has to be calculated according to the methods of Appendix B.3. Close to the band edges the E–$\mathbf{k}$ dispersion relation is quadratic like for free particles. Its curvature is proportional to the inverse effective mass m^* (Fig. 9.6), which e.g. in GaAs yields light electrons with $m^*_{\mathrm{el}} = 0.067 m_{\mathrm{el}}$ and heavy holes with $m^*_{\mathrm{h}} = 0.55 m_{\mathrm{h}}$:

$$E_{\mathrm{el,h}} = E_{\mathrm{CB,VB}} + \frac{\hbar^2 k_x^2}{2m^*_{\mathrm{el,h}}} + \frac{\hbar^2 k_y^2}{2m^*_{\mathrm{el,h}}} + \frac{\hbar^2 k_z^2}{2m^*_{\mathrm{el,h}}}. \tag{9.4}$$

By $E_{\mathrm{CB,VB}}$, the lower edge of the conduction and valence band respectively is meant. In the three-dimensional volume $k_x^2 + k_y^2 + k_z^2 = k^2$ and using $\rho_{\mathrm{el,h}}(k)\,dk = k^2\,dk/2\pi^2$ the density of states for electrons and holes are separately calculated according to

$$\rho_{\mathrm{el,h}}(E)\,dE = \frac{1}{2\pi^2}\left(\frac{2m^*_{\mathrm{el,h}}}{\hbar^2}\right)^{3/2}(E - E_{\mathrm{CB,VB}})^{1/2}\,dE,$$

with E for electrons and holes counted from each band edge $E_{\mathrm{CB,VB}}$. With this we can also determine the density of charge carriers for electrons and holes. We introduce the two quantities $\alpha_{\mathrm{el}} = (E_{\mathrm{CB}} - \varepsilon_{\mathrm{CB}})/kT$ and $\alpha_{\mathrm{h}} = (\varepsilon_{\mathrm{VB}} - E_{\mathrm{VB}})/kT$ and replace the integration variable by $x = (E - E_{\mathrm{CB}})/kT$ and $x = (E_{\mathrm{VB}} - E)/kT$, respectively,

$$
\begin{aligned}
n_{\mathrm{el,h}} &= \int_{E_{\mathrm{CB,VB}}}^{\infty} \rho_{\mathrm{el,h}} f_{\mathrm{el,h}}(E, \varepsilon_{\mathrm{CB,VB}}) \, dE \\
&= \frac{1}{2\pi^2} \left(\frac{2m_{\mathrm{el,h}}^* kT}{\hbar^2} \right) \int_0^{\infty} \frac{\exp(-\alpha_{\mathrm{el,h}}) \sqrt{x} \, dx}{\exp(x) + \exp(-\alpha_{\mathrm{el,h}})}.
\end{aligned}
$$

Estimates can be obtained easily by inserting the characteristic effective masses for GaAs. We obtain after a short calculation for $T = 300\,\mathrm{K}$:

$$
\left\{ \begin{array}{c} n_{\mathrm{el}} \\ n_{\mathrm{h}} \end{array} \right\} = \left\{ \begin{array}{c} 4.7 \times 10^{17}\,\mathrm{cm}^{-3} \\ 1.1 \times 10^{19}\,\mathrm{cm}^{-3} \end{array} \right\} e^{-\alpha_{\mathrm{el,h}}} \int_0^{\infty} \frac{\sqrt{x}\,dx}{e^x + e^{-\alpha_{\mathrm{el,h}}}}. \tag{9.5}
$$

In a laser diode the density of charge carriers is maintained by the injection current (see p. 268). With the help of the implicit equation (9.5) the Fermi energies for the conduction and valence bands can be determined numerically. We usually expect the same concentration for electrons and holes.

Example: Charge carrier densities in GaAs

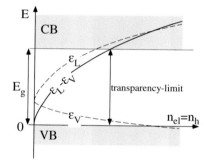

Fig. 9.7: *Density of charge carriers and Fermi energies.*

A case of special interest occurs at $\alpha_{\mathrm{el,h}} = 0$ since there the Fermi energy just reaches the edges of the valence and conduction bands. This case can even be resolved analytically:

$$
\begin{aligned}
n_{\mathrm{el}} &= 4.7 \times 10^{17}\,\mathrm{cm}^{-3} \int_0^{\infty} \frac{\sqrt{x}\,dx}{e^x + 1} \\
&= 3.2 \times 10^{17}\,\mathrm{cm}^{-3}.
\end{aligned}
$$

In general, Eq. (9.5) has to be evaluated by numerical methods. The result of such an evaluation is presented in Fig. 9.7.

Owing to the smaller effective masses, the electron concentration makes the Fermi energy $\varepsilon_{\mathrm{CB}}$ increase faster than the hole concentration $\varepsilon_{\mathrm{VB}}$, and it reaches the band edge first. Through strong p-doping, though, the Fermi energy in the currentless state (i.e. free of charge carriers) is shifted closer to the valence band, so that $\varepsilon_{\mathrm{VB}}$ gets to the valence band edge at a lower density of charge carriers.

As we will see later (Eq. (9.9)) it is already sufficient for inversion if the difference of the Fermi energies $\varepsilon_{\mathrm{CB}} - \varepsilon_{\mathrm{VB}}$ is larger than the bandgap energy E_{g}. There the so-called 'transparency limit' is reached since the radiation field is no longer absorbed but amplified.

In terms of lasing we are more interested in which states can contribute to a transition with energy $E = \hbar\omega > E_{\mathrm{g}} = E_{\mathrm{CB}} - E_{\mathrm{VB}}$ or where we can expect inversion. We obtain the rate of stimulated emission from the Einstein B-coefficient. If the occupation probabilities f_{el} at the energy difference of the direct transition $\hbar\omega$ in valence and conduction band, respectively, are taken into account, the rates of stimulated emission $(R_{\mathrm{CV}}(\mathbf{k}))$ and absorption $(R_{\mathrm{CV}}(\mathbf{k}))$ can be determined for a given electron $\mathbf{k}$-vector and at the energy density of the radiation field $U(\omega(\mathbf{k}))$ with the Einstein coefficient $B_{\mathrm{CV,VC}}$:

$$\text{emission:} \qquad R_{\mathrm{CV}}(\mathbf{k}) = B_{\mathrm{CV}}U(\omega(\mathbf{k}))[f_{\mathrm{el}}^{\mathrm{CB}}(1 - f_{\mathrm{el}}^{\mathrm{VB}})],$$

$$\text{absorption:} \qquad R_{\mathrm{VC}}(\mathbf{k}) = B_{\mathrm{VC}}U(\omega(\mathbf{k}))[f_{\mathrm{el}}^{\mathrm{VB}}(1 - f_{\mathrm{el}}^{\mathrm{CB}})].$$

The number of possible transitions at frequency ω has to be determined from the sum of the dispersion relations for electrons and holes. For this we use the so-called reduced densities of states. With $\mu^{-1} = m_{\mathrm{el}}^{*-1} + m_{\mathrm{h}}^{*-1}$ and $\rho(\omega) = \hbar\rho(E)$, we have

$$\rho_{\mathrm{red}}(\omega) = \frac{1}{2\pi^2}\left(\frac{2\mu}{\hbar}\right)^{3/2}(\omega - E_{\mathrm{g}}/\hbar)^{1/2}. \tag{9.6}$$

Then the difference of emission and absorption rates can be calculated with $B_{\mathrm{CV}} = B_{\mathrm{VC}}$ from

$$\begin{aligned}
R_{\mathrm{CV}} - R_{\mathrm{VC}} &= B_{\mathrm{CV}}U(\omega)[f_{\mathrm{el}}^{\mathrm{CB}}(1 - f_{\mathrm{el}}^{\mathrm{VB}}) - f_{\mathrm{el}}^{\mathrm{VB}}(1 - f_{\mathrm{el}}^{\mathrm{CB}})]\rho_{\mathrm{red}} \\
&= B_{\mathrm{CV}}U(\omega)[f^{\mathrm{CB}} - f^{\mathrm{VB}}]\rho_{\mathrm{red}}.
\end{aligned} \tag{9.7}$$

The role of inversion, which in conventional lasers is given by the occupation number difference of the excited (N_e) and lower state (N_g) of the laser transition, is now taken over by the product

$$(N_e - N_g) \quad \rightarrow \quad (f^{\mathrm{CB}} - f^{\mathrm{VB}})\rho_{\mathrm{red}}[(E_{\mathrm{CB}} - E_{\mathrm{VB}})/\hbar],$$

with the first factor controlled by the injection current.

9.2.4 Small signal gain

Consider a pulse of light propagating in the z direction with group velocity v_{g} and spectral intensity $I(\omega) = v_{\mathrm{g}}U(\omega)$. The change of the intensity by absorption and emission respectively is described according to Eq. (9.7). After a short travel length $\Delta z = v_{\mathrm{g}}\Delta t$ we can thus write $\Delta I = (R_{\mathrm{CV}} - R_{\mathrm{VC}})\hbar\omega\Delta z$. Then the absorption and emission coefficients respectively are determined according to Eq. (6.22)

$$\alpha(\omega) = \frac{\Delta I}{I\Delta z} = \frac{(-R_{\mathrm{VC}} + R_{\mathrm{CV}})\hbar\omega}{v_{\mathrm{g}}U(\omega)}.$$

We use the identity $B_{\mathrm{CV}} = A_{\mathrm{CV}}/[\hbar\omega(\omega^2/\pi^2c^3)] = A_{\mathrm{CV}}/[\hbar\omega\rho_{\mathrm{ph}}(\omega)]$ according to Eq. (6.42) to relate the Einstein coefficient to the microscopic properties of the semiconductor. Then we can write

$$\alpha(\omega) = \frac{1}{v_{\mathrm{g}}\tau}\frac{\rho_{\mathrm{red}}(\omega)}{\rho_{\mathrm{ph}}(\omega)}(f^{\mathrm{CB}} - f^{\mathrm{VB}}) = \alpha_0(f^{\mathrm{CB}} - f^{\mathrm{VB}}), \tag{9.8}$$

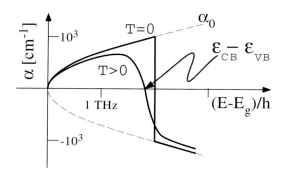

Fig. 9.8: *Absorption and (small signal) gain at a pn junction for a given density of charge carriers at $T = 0\,K$ and at elevated temperature.*

where we have introduced the maximum absorption coefficient $\alpha_0(\omega)$, which is proportional to the reduced density of states.

For an estimate we use specifications characteristic for GaAs lasers: wavelength $\lambda_L = 850\,\text{nm}$; reduced effective mass $\mu = 0.06 m_{\text{el}}$; recombination time $\tau_{\text{rec}} = 4 \times 10^{-9}\,\text{s}^{-1}$; group velocity $v_g \simeq c/3.5$. For typical separations of the laser frequency from the band edge of $1\,\text{THz} = 10^{12}\,\text{Hz}$, corresponding to $2\,\text{nm}$ in terms of wavelengths, one can calculate

$$\alpha_0 = 6.8 \times 10^3\,\text{cm}^{-1}\,\sqrt{(\nu_L - E_g/\hbar)/\text{THz}}.$$

The very large gain factors α_0 are somewhat reduced in a room-temperature laser by the Fermi factor from Eq. (9.8).

Like in the gas laser, amplification in the laser diode is achieved when stimulated emission overcomes the losses caused by outcoupling, scattering and absorption. In Fig. 9.8 we have calculated the gain and loss profile for an example. At $T = 0$ the Fermi distributions are step-like and therefore the value of the absorption coefficient is exactly at $\alpha_0(\omega)$. Moreover, it becomes immediately clear that an inversion of the charge carriers can occur only if there are *different* Fermi energies in the conduction and the valence band,

$$\varepsilon_L - \varepsilon_V > h\nu > E_g. \tag{9.9}$$

The charge carrier distribution corresponds to a dynamic equilibrium that can only be sustained for forward biassed operation of the diode. The more exact calculation of the semiconductor gain is an elaborate matter since it depends on the details of the technical layout which is much more complex as we are going to see later on.

Example: Threshold current of the semiconductor laser
The threshold current density required can easily be determined when the critical density of charge carriers $n_{\text{el}} \geq 10^{18}$ is known. The density of charge carriers is trans-

ported by the injection current to the pn junction and recombines there spontaneously with rate $\tau_{\text{rec}}^{-1} = 2.5 \times 10^8 \, \text{s}^{-1}$:

$$\frac{dn_{\text{el}}}{dt} = -\frac{n_{\text{el}}}{\tau_{\text{rec}}} + \frac{j}{ed}.$$

Without difficulties we derive the stationary current density for a width of the space charge zone $d = 1 \, \mu\text{m}$ of the pn junction

$$j = \frac{n_{\text{el}} ed}{\tau_{\text{rec}}} \geq 4 \, \text{kA cm}^{-2}.$$

For an active zone with a typical area of $0.3 \times 0.001 \, \text{mm}^2$, this current density already corresponds to $12 \, \text{mA}$, which has to be concentrated exactly onto this small volume. It is obvious that it is worth while to technically reduce the natural width of the diffusion zone of the charge carriers in order to lower the threshold current density. This concept is precisely pursued by heterostructure and quantum-film lasers.

9.2.5 Homo- and heterostructures

Although the basic concept for the operation of a semiconductor laser originates from the early days of the laser, it was initially mandatory to cool the pn junction to cryogenic temperatures to obtain lasing at all. The light mobile electrons have a large diffusion length ($\geq 0.5 \, \mu\text{m}$), so that large threshold currents were required, and at room temperature the gain could not overcome the losses caused especially by non-radiative recombination and reabsorption. In the 1970s, however, this problem was solved by the concept of 'heterostructures', and ever since laser diodes have continued

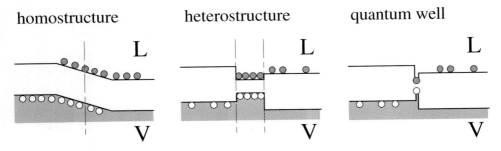

Fig. 9.9: *Band structure for electrons and holes: homostructure, heterostructure and quantum films. The quantum film limit is usually entered at thickness below 200 Å.*

their triumphant route as sources for coherent light. In a heterostructure two different materials (e.g. with different composition and different bandgaps) are adjacent to each other. The interface creates potential steps which inhibit the diffusion of charge carriers across the barrier. For laser materials the bandgap is chosen in such a way that electrons and holes can be locked between two layers with a larger bandgap in a zone with a smaller bandgap ('double heterostructure'). Otherwise the light generated at the centre would be absorbed again in the outer areas of the amplification zone.

This advantage of heterostructures compared to simple homostructures is schematically presented in Fig. 9.9. The strongly simplified potential scheme indicates that the motion of the charge carriers is now limited to a narrow layer ($\simeq 0.1\,\mu$m) in order to realize an accordingly high gain density by their strong confinement. Furthermore, when the refractive index in this area is higher than in the adjacent layers, a favourable waveguide effect is obtained, which in this case is called 'index guiding'. Also, the spatial variation of the charge carriers causes changes of the refractive index and waveguiding again, which is called 'gain guiding' in this case. With further miniaturization of the active layer we get to the realm of quantum film systems, which are not just simply smaller but also show qualitatively novel properties (see Section 9.3.4).

9.3   The heterostructure laser

The most important material for the manufacture of opto-electronic semiconductors until now has been GaAs. As a direct semiconductor, not only does it offer the necessary microscopic properties, but also, by variation of the $Ga_xAl_{1-x}As$ compound crystal composition, it offers widespread technical potential to adjust the bandgap and the refractive index to the requirements for applications. The characteristic wavelength at 850 nm has technological significance as well, because it lies in one of three spectral windows (850, 1310, 1550 nm) suitable for the construction of optical networks. Today, the concepts of the AlGaAs laser have been transferred to other systems as well, like, for example, InAlP.

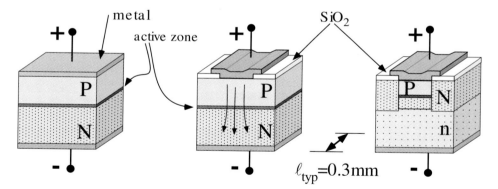

Fig. 9.10: *Layer systems for laser diodes. Left: Plain homostructure. Middle: The current flux is narrowed by insulating oxide layers and causes a concentration of the inversion density. The inhomogeneous amplification, or charge carrier density, furthermore generates a waveguide, which leads the light field along the gain zone ('gain guiding'). Right: Double heterostructures generate a precisely controlled amplification zone as well as an optical waveguide for the light field.*

9.3.1   Construction

Laser crystal

Laser crystals are produced through epitaxial growth.[4] The composition of these layers can be controlled along the growth direction by regulating the precursor flux. The vertical double heterostructure (DH) is controlled by such growth. The lateral structuring on the micrometre scale is engineered through methods known from microelectronics, e.g. optical lithography processes.

Owing to the construction, the laser field propagates along the surface of the crystals, and outcoupling takes place at the edge of a cleavage face. Therefore this type is called an 'edge emitter', in contrast to an alternative layout where light is emitted perpendicularly to the surface. This type will be introduced shortly in Section 9.5.2.

Laser crystals with a length of about 0.2–1 mm are produced by simply cleaving them from a larger epitaxially grown wafer. They can basically be inserted into a suitable package (Fig. 9.11) without any further treatment and be contacted with standard techniques to facilitate handling. The transverse geometric properties of the laser field are determined by the shape of the amplification zone. In the far field an elliptical beam profile is generally observed caused by the diffraction off the heterostructure and the transverse waveguiding. The light of edge-emitting laser diodes thus has to be collimated, which is quite costly for the purposes of application, and is one reason for the development of surface emitters, which offer a circular beam profile from the beginning.

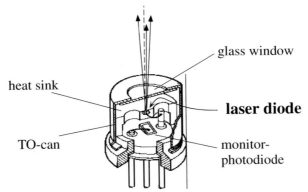

heat sink

glass window

laser diode

TO-can

monitor-
photodiode

Fig. 9.11: *Standard package for laser diodes. The semiconductor device itself is hardly visible and has typical dimensions of 0.3 mm edge length. This type is called an edge emitter.*

9.3.2   Laser operation

In the most frequent and simplest case, the cleavage faces of the crystal already form a laser resonator. At a refractive index $n = 3.5$ the intrinsic reflectivity of a GaAs

---

[4]During epitaxial growth, thin monolayers of the (semiconducting) material are homogeneously deposited on a monocrystalline substrate from molecular beams.

crystal is 30% and is often sufficient to support lasing due to the large gain coefficients of semiconductors. In other cases the reflectivity of the cleavage faces can be modified by suitable coatings. In Fig. 9.12 the output power of a semiconductor laser as a function of the injection current is presented.

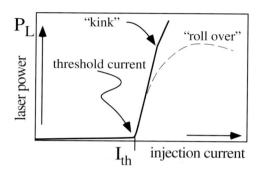

Fig. 9.12: *Current–power diagram of a laser diode. At large currents a 'roll-over' can occur due to heating of the pn junction.*

For many applications, e.g. spectroscopy or optical communications, the use of single-mode lasers (both transverse and longitudinal) is important. The homogeneous gain profile of the laser diode offers excellent preconditions to implement single-mode operation even though the free spectral range of semiconductor lasers at $\ell_{\mathrm{typ}} = 0.3\,\mathrm{mm}$ is, in spite of the substantial $\Delta\nu_{\mathrm{FSR}} = 150\,\mathrm{GHz}$, still very small compared to the gain bandwidth of 10 THz and more. In fact parasitic laser oscillations ('sidebands') are very efficiently suppressed in many components.

The threshold currents of a laser diode vary depending on the layout, but the aim is always a laser threshold as small as possible. It has to be kept in mind that large current densities of $100\,\mathrm{kA\,cm^{-2}}$ and more occur, causing strong local heating and thus leading to damage of the heterostructures. For the same reason the threshold current grows with temperature. In high-power lasers the so-called 'roll-over' occurs, for which increase of the injection current no longer leads to increase of the output power but on the contrary reduces it due to the heating of the pn junction! The relation between threshold current I_{th} and temperature follows an empirical law with a characteristic temperature T_0 and current I_0,

$$I_{\mathrm{th}} = I_0 \exp\left(\frac{T - T_0}{T_0}\right). \tag{9.10}$$

In conventional heterostructure lasers the characteristic temperature has values of about $T_0 = 60\,\mathrm{K}$ but in other layouts such as VCSEL or quantum-film lasers (see section 9.3.4) these values are increased in a favourable direction up to 200–400 K so that the temperature sensitivity of the components is significantly reduced. A qualitative semiconductor laser output power vs. pump curve is presented in Fig. 9.12 and reflects features of the idealized laser of Fig. 8.1. From the slope of the power, the differential quantum efficiency can be obtained, which is typically 30% or more:

$$\text{differential quantum efficiency} = \frac{e}{h\nu}\frac{dP}{dI}.$$

Sometimes there are so-called 'kinks' in the power–current diagram. They are an indication of a modification of the laser mode, e.g. caused by a charge carrier profile switching geometrically from one spatial mode to another at this particular current.

9.3.3   Spectral properties

Emission wavelength and mode profile

The emission wavelength of a semiconductor laser is determined by the combined effect of gain profile and laser resonator as it is for other laser types. We first consider the wavelength selection of the 'freely operating' laser diode without any additional optical elements.

Single-mode operation occurs in many types of laser diodes. It is favoured by a gain profile that is homogeneously broadened as a result of a large intra-band relaxation rate. So at the gain maximum the mode starts lasing by it-self. However, the detailed geometry of the often complex multilayer laser crystal can also allow multimode laser operation, and even in components explicitly called 'single-mode laser' usually further parasitic modes may only be suppressed by a certain finite factor (typically ×100, or 20 dB).

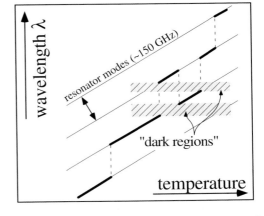

Fig. 9.13: *Mode jumps of diode lasers caused by temperature variations.*

Although the construction length of the resonator is generally very short (0.3–0.5 mm, $n \simeq 3.5$) and already for conventional components delivers a free spectral range of 80–160 GHz (which can be much larger for VCSEL lasers), there are still many resonator modes in the gain profile at a typical spectral width of some 10 nm or some THz.

The refractive index determining the resonator frequency depends sensitively on the temperature as well as on the charge carrier density and the injection current respectively, so that the exact laser frequency ν_L can be tuned over considerable ranges by controlling these parameters:

1. By increasing the temperature of an external heat sink (e.g. a Peltier cooler), we typically find a rate of frequency change at $d\nu_\mathrm{L}/dT = -30\,\mathrm{GHz\,K^{-1}}$, i.e. a redshift.

2. Variation of the injection current causes a shift $d\nu_\mathrm{L}/dI = \eta_\mathrm{th} + \eta_n$. The shift is due to temperature changes within the heterostructure ($\eta_\mathrm{th} \simeq -3\,\mathrm{GHz\,mA^{-1}}$) and also modifications of the charge carrier density ($\eta_n \simeq 0.1\,\mathrm{GHz\,mA^{-1}}$). For slow current variations, the frequency change is dominated by the thermal redshift, but for modulation frequencies exceeding $f_\mathrm{mod} \geq 30\,\mathrm{kHz}$, the influence of the charge carrier density dominates (see Section 9.4.1).

Unfortunately tuning by temperature and current at the pn junction only does not usually allow generation of every frequency within the gain profile. It is impaired by 'dark' zones (Fig. 9.13) since the gain profile and the mode structure of the resonator do not vary synchronously with each other. External optical elements can, however, also be used to access those forbidden domains (see Section 9.5.1).

Electronic wavelength control

When the exact frequency or wavelength of the laser radiation is important like, for example, in spectroscopic applications, then the temperature at the laser diode junction and the injection current have to be controlled very precisely. The high sensitivity to temperature and current fluctuations sets high technical demands on the electronic control devices. If technically caused frequency fluctuations are to be kept lower than the typical 5 MHz caused by intrinsic physical processes (see Section 9.4.2), then according to the variation rates given in the preceding section obviously a temperature stability $\delta T_{\mathrm{rms}} \leq 1\,\mathrm{mK}$ and a current stability $\delta I_{\mathrm{rms}} \leq 1\,\mu\mathrm{A}$ has to be achieved with appropriate servo-controllers.

Considering it more exactly, the spectral properties of the servo-controllers have to be investigated, but this would by far exceed the scope of this book. However, it is quite easy to see that the temperature control cannot have a large servo bandwidth due to its large thermal masses. The bandwidth of the current control is basically limited only by the capacitance of the laser diode itself, but it is advisable in terms of servo-control methods to limit the constant-current source to a small internal bandwidth in order to reduce the current noise and instead of this to provide some additional fast high-impedance modulation inputs like, for example, in Fig. 9.14.

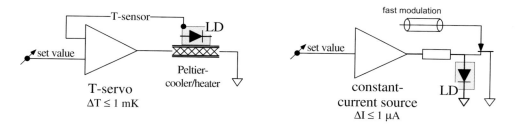

Fig. 9.14: *Left: Temperature control for laser diodes may use thermistors as temperature sensors. Right: Current control will typically inject a well-stabilized current that resists rapid variations. Fast modulation may then be realized by directly injecting additional small currents.*

For wavelength stabilization the devices described here have a merely passive effect – they warrant tight control of operational parameters of the laser diode but do not interrogate the wavelength itself. For many applications, e.g. optical wavelength standards, still better absolute stabilities are required and deviations from a desirable wavelength must be directly sensed, for instance through a spectroscopic signal, and corrected through suitable servo-controls.

9.3.4   Quantum films, quantum wires, quantum dots

Conventional heterostructures serve to hinder the diffusion of electrons and holes and to concentrate the gain into a small zone. The charge carriers though move in a well with dimensions of about 100 nm like more or less classical point-like particles. By further miniaturization (see Fig. 9.15) we reach the realm of quantized electronic motion in which the dynamics of the charge carriers in the vertical direction orthogonal to the layer system is characterized by discrete energy levels according to quantum mechanics.

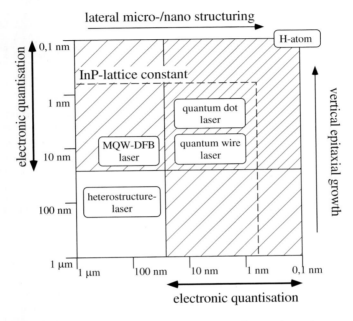

Fig. 9.15: *Semiconductor miniaturization and semiconductor laser types with reduced dimensionality.*

Once miniaturization reaches the quantum border in one dimension, a 'two-dimensional electron gas' is created, which we shall call a 'quantum film' here. In the literature there are also other terms used, e.g. quantum well (QW) lasers. Structures with reduced dimensionality offer lower threshold currents, larger gain and lower temperature sensitivity than conventional DH lasers, advantages already essentially acknowledged since the early 1980s.

Inversion in the quantum film

The two-dimensional character of the charge carrier gas causes a change in the density of states (DOS, see Appendix B.3), the fundamental origin of improved operation characteristics like, for example, low threshold current and lower temperature sensitivity.

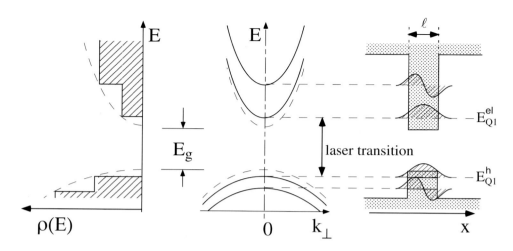

Fig. 9.16: *Band structure (middle) and density of states (left) in a quantum film. On the left the dashed curves show the corresponding density of states for the bulk material. The hatched curves on the right indicate the wavefunction of the stored electrons and holes in the 1D transverse potential.*

In addition to the kinetic energy of the transverse quantum state E_{Qi}, there are two continuous degrees of freedom with momentum components $k_{\perp i}$. For the electrons and holes in the ith sub-band of the quantum film we find ($k_{\perp i}^2 = k_{\perp ix}^2 + k_{\perp iy}^2$):

$$E_i = E_{V,L} + E_{Qi} + \frac{\hbar^2 k_{\perp i}^2}{2m_{el,h}^*}.$$

Among the interesting properties of quantum film lasers is the possibility to control the transition wavelength by choosing the film thickness ℓ, which determines the energy separation of the quantum states in the electronic and hole-like state. According to quantum mechanics we have $E_{Q1} \simeq \hbar^2/2m_{el}^* \ell^2$.

The density of states in the k-plane is $\rho_{el,h}(k)\, dk = k\, dk / 2\pi^2$ and can be converted into an energy density with $dE = \hbar^2 k / m_{el,h}^* \, dk$. In the transverse direction each quantum state (energy $E_{Qi}^{el,h}$, quantum number i) contributes with the density π/ℓ,

$$\rho_{el,h}(E)\, dE = \sum_i \frac{m_{el,h}^{*i}}{\hbar^2 \ell} \Theta(E - E_{Qi}^{el,h}).$$

The theta function has the values $\Theta(x) = 1$ for $x > 0$ and $\Theta(x) = 0$ for $x \leq 0$. Also the effective masses $m_{el,h}^{*i}$ may depend on the quantum number. The density of state grows step-like (see Fig. 9.16) in a quantum film every time the energy reaches a new transverse quantum state. There it has exactly the value corresponding to the volume material (dashed line in Fig. 9.16).

The advantage of the QW laser becomes evident when we determine the dependence of the Fermi energy on the charge carrier concentration as we did on p. 266. With terms similar to the 3D case, e.g. $\alpha_{el}^i = E_L + E_{Qi}^{el} - \varepsilon_L$, we obtain

$$n_{el,h} = \sum_i \frac{m_{el,h}^{*i} kT}{\hbar^2 \ell} e^{-\alpha_{el,h}^i} \int_0^\infty \frac{dx}{e^x + e^{-\alpha_{el,h}^i}}.$$

This integral can be analytically evaluated. Using the parameters of GaAs at $T = 300\,\text{K}$ and for a quantum film with a thickness of $\ell = 100\,\text{Å}$, we find the relation

$$n_{el} = 33 \times 10^{15}\,\text{cm}^{-3}\, \ln\left(1 + e^{-\alpha_{el,h}^i}\right),$$

from which the Fermi energy can be obtained. The value of the first factor is two orders of magnitude smaller than for the volume material (Eq. (9.5))! This indicates that in the QW laser inversion can be expected already at considerably smaller charge carrier concentrations and thus smaller threshold current densities than in conventional DH lasers.

Multiple quantum well (MQW) lasers

For a fair comparison with conventional DH lasers it has to be taken into account that the total gain of a quantum film is smaller than that of a DH laser simply due to the smaller volume. This disadvantage can be largely compensated by introducing multiple identical quantum films in the volume of the laser light field.

In Fig. 9.17 a multiple quantum well (MQW) structure is schematically presented. The charge carriers are to be 'caught' in the potential wells but the relaxation rate, e.g. through collision with a phonon, is quite small due to the small film thickness. To increase the concentration of the charge carriers in the vicinity of the quantum films, an additional heterostructure is provided – the separate confinement heterostructure (SCH) in Fig. 9.17. This structure also acts as a waveguide for the resonator field and focuses the light intensity onto this zone, which is usually much smaller than an optical wavelength. Today MQW lasers have become a standard product of the opto-electronics industry.

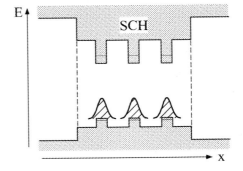

Fig. 9.17: *Concept of a multiple quantum well (MQW) structure consisting of three quantum films. SCH: separate confinement heterostructure.*

Another interesting innovation has been introduced with the 'strained quantum well'. They offer additional technical advantages since the effective masses are increased as a result of mechanical strain in the crystal lattices by a factor 2. Thereby the density of states as well as the threshold current density decrease again.

Let us once again summarize the advantages of quantum film lasers compared with conventional double-heterostructure lasers:

1. The modified density of states causes lower threshold currents since fewer states per charge carrier are available, which can consequently be filled with lower currents. Typically threshold current densities of 50–100 A cm^{-2} are achieved. The lower threshold indirectly improves again the temperature sensitivity since there is less excess heat generated in the heterostructures.

2. The differential gain is larger than for the DH lasers since the electrically dissipated power growing with the current causes a lower reduction of the gain.

3. The threshold condition depends less strongly on the temperature. For conventional DH lasers the transparency threshold grows with $T^{3/2}$, in quantum film lasers only in proportion to T. The characteristic temperatures according to Eq. (9.10) are about 200 K.

Quantum wires and quantum dots

The reduced dimensionality of semiconductor structures can be extended through construction: two-dimensional quantum films become quasi-one-dimensional quantum wires and even zero-dimensional quantum dots when suitable methods of lateral microstructuring are chosen. In Fig. 9.18 this evolution with its effect on the density of states is presented.

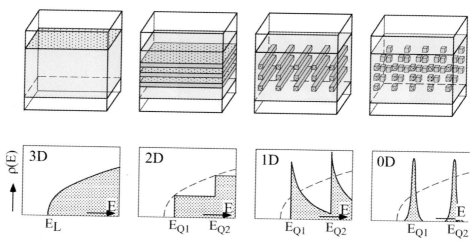

Fig. 9.18: *Evolution from the double-heterostructure laser over quantum films and wires to quantum dots.*

The properties of the density of states continue the tendency of the quantum film laser to realize overall gain already at small current densities. While the layer stack of the quantum film laser can simply be manufactured by controlling the growth processes (in Fig. 9.18 in the vertical direction), the lateral properties have to be manufactured in general by a completely different process. On the one hand, there is no longer a big

difference between manufacturing quantum wires and quantum dots from the technological point of view; on the other hand, the necessary lateral structural dimensions
of 0.1–0.2 nm are not easily achievable with standard methods of optical lithography.
Also, the strictly periodic formation of quantum dots shown in Fig. 9.18 has been
difficult to realize up to now, but on the other hand it is not necessary for the laser
process either. Multiple quantum dots have been produced using self-organization of
a heterogeneous growth process[109, 38].

9.4 Dynamic properties of semiconductor lasers

Among the technically more attractive properties of the laser diode is the possibility
to modulate it directly by varying the injection current. For instance, the speed of
switching the laser on and off determines the rate for generating digital signals and thus
transmitting information. A rate equation approach makes an excellent approximation
since the transverse relaxation is dominated by the fast rate of intra-band scattering
$\gamma'^{-1} = T_2 \simeq 1\,\mathrm{ps}$ and hence the polarization effects remain in equilibrium with the
field amplitude (Section 8.2). To understand the dynamics of laser diodes, we use the
amplitude equation (8.11) and the rate equation (8.19(ii)),

$$
\begin{aligned}
\dot{E}(t) &= \{i[\Omega - \omega - \tfrac{1}{2}\kappa\alpha n(t)] + \tfrac{1}{2}[\kappa n(t) - \gamma_c]\}E(t), \\
\dot{n}(t) &= -\kappa n(t)n_{\mathrm{ph}}(t) - \gamma n(t) + R.
\end{aligned}
\tag{9.11}
$$

Note that here we use $n(t)$ for the charge carrier density. The current density j 'feeds'
the dynamics with $R = j/ed$, we replace $|E(t)|^2 \to n_{\mathrm{ph}}(t)$, and furthermore we now
use the photon lifetime $\gamma_c \to 1/\tau_{\mathrm{ph}}$ and the recombination time $\gamma \to 1/\tau_{\mathrm{rec}}$ instead of
the damping rates. At steady state ($\dot{n} = \dot{n}_{\mathrm{ph}} = 0$) we find

$$
n^{\mathrm{st}} = \frac{1}{\kappa\tau_{\mathrm{ph}}} \qquad \text{and} \qquad \overline{n}_{\mathrm{ph}} = \frac{1}{\kappa\tau_{\mathrm{rec}}}\left(\frac{j}{j_{\mathrm{th}}} - 1\right),
$$

where $j_{\mathrm{th}} = ed/\kappa\tau_{\mathrm{rec}}\tau_{\mathrm{ph}}$. Mostly we are interested in small deviations from the stationary state. Then we can linearize,

$$
n(t) = n^{\mathrm{st}} + \delta n(t) \qquad \text{and} \qquad n_{\mathrm{ph}} = \overline{n}_{\mathrm{ph}} + \delta n_{\mathrm{ph}},
$$

and find the equations of motion, in which we set $j_0/j_{\mathrm{th}} = I_0/I_{\mathrm{th}}$,

$$
\begin{aligned}
\dot{\delta n}_{\mathrm{ph}}(t) &= \frac{1}{\tau_{\mathrm{rec}}}\left(\frac{I_0}{I_{\mathrm{th}}} - 1\right)\delta n(t), \\
\dot{\delta n}(t) &= \frac{j_{\mathrm{mod}}}{ed} - \frac{1}{\tau_{\mathrm{rec}}}\frac{I_0}{I_{\mathrm{th}}}\delta n(t) - \frac{1}{\tau_{\mathrm{ph}}}\delta n_{\mathrm{ph}}.
\end{aligned}
\tag{9.12}
$$

9.4.1 Modulation properties

We consider the effect of small harmonic modulations of the injection current $j_{\mathrm{mod}} = j_0 + j_m\,e^{-i\omega t}$ on the amplitude and the phase of the laser light field.

Amplitude modulation

The modulation of the number of photons is equivalent to the variation of the output power. Therefore we use $\delta n_{\mathrm{ph}}(t) = \delta n_{\mathrm{ph0}}\, e^{-i\omega t}$ and $\delta n(t) = \delta n_0\, e^{-i\omega t}$ and we can replace $\delta n_{\mathrm{ph0}} = -(I_0/I_{\mathrm{th}} - 1)\delta n_0/i\omega\tau_{\mathrm{rec}}$. After a short calculation we get

$$\delta n_{\mathrm{ph0}} = -\frac{\tau_{\mathrm{ph}} j_m}{ed}\,\frac{I_0/I_{\mathrm{th}} - 1}{\omega^2 \tau_{\mathrm{rec}}\tau_{\mathrm{ph}} - (I_0/I_{\mathrm{th}} - 1) + i(I_0/I_{\mathrm{th}})\omega\tau_{\mathrm{ph}}}. \tag{9.13}$$

We are interested in the value of the resulting amplitude modulation according to eq. (9.13)

$$|\delta n_{\mathrm{ph0}}| = \frac{\tau_{\mathrm{ph}} j_m}{ed}\,\frac{I_0/I_{\mathrm{th}} - 1}{\sqrt{[\omega^2 \tau_{\mathrm{rec}}\tau_{\mathrm{ph}} - (I_0/I_{\mathrm{th}} - 1)]^2 + \omega^2 \tau_{\mathrm{ph}}^2 (I_0/I_{\mathrm{th}})^2}}. \tag{9.14}$$

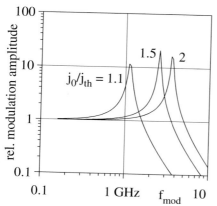

In Fig. 9.19 we present the response of a typical laser diode to a current modulation with frequency $f_{\mathrm{mod}} = \omega/2\pi$ calculated from Eq. (9.14). We have used a spontaneous recombination time $\tau_{\mathrm{rec}} = 2 \times 10^{-9}\,\mathrm{s}$ and a photon lifetime of $\tau_{\mathrm{ph}} = 10^{-12}\,\mathrm{s}$. The frequency of the relaxation resonance grows with injection current as expected according to Eq. (8.24). Experimental data are well represented by this function.

Fig. 9.19: *Amplitude modulation of a diode laser as a function of the modulation frequency.*

For applications, e.g. in optical communications, a large modulation bandwidth is important. In addition to this, the frequency response is to stay flat up to frequencies as high as possible, and furthermore there should not be any major phase rotations (δn_{ph0} is a complex quantity!). Today in compact VCSEL components modulation bandwidths of 40 GHz and more are achieved, and an end of this development is not yet in sight.

Phase modulation

Next we investigate the evolution of the phase $\Phi(t)$ separating off the steady state in Eq. (9.11) with

$$E(t) \to \mathcal{E}\exp[i(\Omega - \omega - \alpha\gamma_{\mathrm{c}}/2)]\exp[i\Phi(t)]$$

and find the coupling

$$\dot\Phi(t) = \tfrac{1}{2}\alpha\kappa\,\delta n(t)$$

of charge carrier dynamics and phase evolution. We again expect a harmonic dependence $\Phi(t) = \Phi_0\, e^{-i\omega t}$, which after a short calculation we can also express by the modulation amplitude of the photon number n_{ph} and so obtain the very transparent result where $\alpha = (\omega - \omega_0)/\gamma$, Eq. (8.7):

$$\Phi(t) = \Phi_0\, e^{-i\omega t} = \frac{\alpha}{2}\frac{\delta n_{\mathrm{ph0}}}{\overline{n}_{\mathrm{ph}}}\, e^{-i\omega t}.$$

The result shows that the factor α describes the coupling of the phase change to the amplitude change. In laser diodes it has typical values of 1.5–6. [It usually vanishes in gas lasers since those oscillate very close to the atomic or molecular resonance lines ($\alpha \simeq 0$).] It also plays a significant role for the linewidth of the laser diode, as we will see in the next subsection.

Up to now we understood the amplitude as well as the phase modulation only as a consequence of the dynamic charge carrier density. The modulation current moreover causes a periodic heating of the heterostructure, which modifies the optical length of the laser diode resonator as well and even dominates the

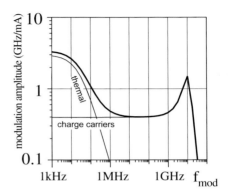

Fig. 9.20: *Phase modulation of semiconductor lasers. The modulation index consists of thermal and charge carrier density contributions.*

modulation depth up to typical critical frequencies of some $10\,\mathrm{kHz}$. Both the temperature and charge carrier density modulation, which we have already identified on p. 273 as the origin for the detuning of laser wavelength with injection current, contribute to the low-frequency limit of the phase modulation amplitude.

9.4.2 Linewidth of the semiconductor laser

When the linewidth of a laser diode is calculated according to the Schawlow–Townes formula (Eq. (8.32)), a higher value than e.g. for the He–Ne laser is already expected from the beginning due to the large linewidth of the empty resonator $\gamma_{\mathrm{c}} \simeq 10^{12}$. In experiments, still larger linewidths of $10\text{–}100\,\mathrm{MHz}$ are observed for a typical $1\,\mathrm{mW}$ laser diode and set in relation to the 'pure' Schawlow–Townes limit $\Delta\nu_{\mathrm{ST}}$. This broadening is described by the so-called α *parameter* which was already introduced in our simple laser theory describing the amplitude–phase coupling (eq. (8.7)),

$$\Delta\nu'_{\mathrm{ST}} = (1 + \alpha^2)\Delta\nu_{\mathrm{ST}}.$$

With semiconductor lasers this is often called Henry's α parameter because C. Henry discovered that, albeit known from the early days of laser physics, it plays a much more significant role for diode lasers than for gas lasers [44].

The α factor was initially introduced as an 'abbreviation' for the normalized detuning in Eq. (8.7). A more detailed analysis shows that it gives the differential ratio of real and imaginary parts of the susceptibility or the refraction coefficient as well,

$$\alpha = \Delta n'/\Delta n''.$$

It can only be calculated with elaborate methods and detailed knowledge of the diode laser construction and is thus preferably obtained from experiment.

Example: 'Pure' Schawlow–Townes linewidth of a GaAs laser

We determine the linewidth according to Eq. (8.32) of a GaAs laser for $1\,\mathrm{mW}$ output power and at a laser frequency of $\nu_{\mathrm{L}} = 350\,\mathrm{THz}$ at $857\,\mathrm{nm}$. The small Fabry–Perot resonator with a length of $0.3\,\mathrm{mm}$ and a refraction coefficient 3.5 leads, for mirror reflectivities of $R = 0.3$, to a linewidth and to decay rates of $\Delta\nu = \gamma_{\mathrm{c}}/2\pi = 3 \times 10^{10}$ which are much larger than for a typical GaAs laser and cause a very much larger Schawlow–Townes linewidth:

$$\Delta\nu_{\mathrm{ST}} \simeq \frac{\pi h \times 350\,\mathrm{THz}\,(2\pi \times 50\,\mathrm{GHz})^2}{1\,\mathrm{mW}} = 1.5\,\mathrm{MHz}.$$

In practice $\Delta\nu'_{\mathrm{ST}} = (1+\alpha^2)\Delta\nu_{\mathrm{ST}}$ with α ranging from 1.5 and 6 is found for enhanced linewidths.

9.4.3   Injection locking

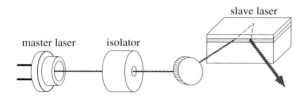

Fig. 9.21: *Injection locking. The coherent light field of the 'master laser' is injected into a 'slave laser' and leaves its coherence properties on it. The isolator serves to decouple the master laser from any radiation emitted by the slave laser.*

In a conventional laser the oscillation of the light field starts by itself from the noise. We are going to study now how a laser oscillator reacts if it is already irradiated by an external monochromatic light field. The considerations are in principle valid for almost all types of lasers but they are particularly important for the applications of laser diodes since in this way the preparation of a light field with good coherence length and high output power in functionally separated components (in a so-called 'master–slave' arrangement) can be achieved.

In Fig. 9.21 we have schematically presented a situation typical for laser diodes. In the 'master laser' a laser light field with well-controlled coherence properties is prepared. Its light is injected into a 'slave laser' and determines the dynamical properties of the latter under conditions we are going to investigate here. The slave laser itself may generally have less advantageous coherence properties as long as it makes high output power available, e.g. in Fig. 9.21 from a broad stripe laser or a tapered amplifier.

Let us insert the coupling to the laser light field to an external field in eq. (9.11) in a heuristic way. The coupling term must have the same structure like the outcoupling term (i.e. $\dot{E} \propto (\gamma_{\mathrm{ext}}/2)E_{\mathrm{ext}}$ where γ_{ext} is the damping rate due to the outcoupling mirror) but the external field oscillates with its own frequency ω_{ext}. We replace $\kappa n \rightarrow G$ and write

$$\dot{E}(t) = [i(\omega - \Omega - \tfrac{1}{2}\alpha G) + \tfrac{1}{2}(G - \gamma_{\text{c}})]E(t) + \tfrac{1}{2}\gamma_{\text{ext}} E_{\text{ext}}\, e^{-i(\omega_{\text{ext}} - \omega)t + i\varphi}.$$

Then we find equations for the equilibrium which we separate into real and imaginary parts,

$$
\begin{aligned}
\text{(i)} \qquad\qquad & \tfrac{1}{2}(G - \gamma_{\text{c}}) + \frac{\gamma_{\text{ext}}}{2}\frac{E_{\text{ext}}}{E}\cos\varphi = 0, \\[4pt]
\text{(ii)} \qquad (\omega_{\text{ext}} - \Omega - \tfrac{1}{2}\alpha G) & + \frac{\gamma_{\text{ext}}}{2}\frac{E_{\text{ext}}}{E}\sin\varphi = 0,
\end{aligned}
\qquad (9.15)
$$

which describe the amplitude (i) and the phase (ii), respectively. If we limit ourselves to the case of small coupling, the modifications of the field amplitude can be neglected. Then we can use the modified saturated gain,

$$G = \gamma_{\text{c}} - 2\Delta_M\cos\varphi,$$

from Eq. (9.15(i)) by introducing the frequency

$$\Delta_{\text{M}} := \frac{\gamma_{\text{ext}}}{2}\frac{E_{\text{ext}}}{E} = \frac{\gamma_{\text{ext}}}{2}\sqrt{\frac{I_{\text{ext}}}{I}}.$$

From Eq. (9.15(ii)) we obtain the relation

$$\omega_{\text{ext}} - (\Omega + \alpha\gamma_{\text{c}}/2) + \alpha\Delta_{\text{M}}\cos\varphi = \Delta_{\text{M}}\sin\varphi.$$

The result can be presented even more conveniently with $\tan\varphi_0 = \alpha$ and $\omega_{\text{free}} := \Omega + \alpha\gamma_{\text{c}}/2$, which is the laser oscillation frequency in the absence of an injected field. For $\alpha = 0$ it is known as the *Adler equation*:

$$\omega_{\text{ext}} - \omega_{\text{free}} = \Delta_{\text{M}}\sqrt{1 + \alpha^2}\,\sin(\varphi - \varphi_0). \qquad (9.16)$$

Then we can derive immediately the limiting conditions for the so-called capture or 'locking range':

$$-1 \le \frac{\omega_{\text{ext}} - \omega_{\text{free}}}{\Delta_{\text{M}}\sqrt{1 + \alpha^2}} \le 1.$$

We find that the slave oscillator locks to the frequency of the external field. The locking range $2\Delta_{\text{M}}$ is larger when more power is injected and when the coupling is stronger, i.e. when the reflectivity of the resonator is lower. According to our analysis for a laser diode, which has typically a low reflectivity, the locking is furthermore supported by the phase–amplitude coupling described by the factor $\sqrt{1 + \alpha^2}$.

The phase condition shows that the locking is made possible through a suitable adjustment of the phase angle φ between master and slave oscillator. A more detailed analysis of the stability, which we skip here, shows that only one of the two adjustment solutions is stable according to Eq. (9.16).

Outside the capture range the locking condition cannot be fulfilled but the external field there causes a phase modulation as well which already leads to a frequency shift of the slave oscillator. The theoretical analysis is a bit more costly but it shows among other things that close to the locking range additional sidebands are generated from the master and the slave light field due to nonlinear mixing processes.

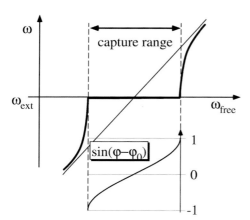

Fig. 9.22: *Frequency characteristic and phase position of a slave laser on injection locking.*

9.4.4   Optical feedback and self-injection locking

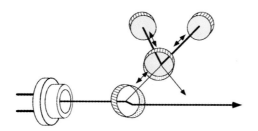

Fig. 9.23: *Optical feedback from a folded resonator. Feedback can only take place in the case of resonance of laser frequency and resonator.*

The coherence properties of laser diodes are extraordinarily sensitive to backscattering from outside. Every randomly caused reflection can trigger considerable and uncontrollable frequency fluctuations. For critical applications, e.g. in spectroscopy, therefore, optical isolators with a high extinction ratio have to ensure that the backscattering occurring at every optical element is suppressed.

The feedback from external components can be described as a form of 'self-injection locking' in immediate analogy to the injection locking described in section 9.4.3,

$$\dot{E}(t) = [i(\omega - \Omega - \tfrac{1}{2}\alpha G) + \tfrac{1}{2}(G - \gamma_c)]E(t) + r(\omega)E(t)\,e^{-i\omega\tau},$$

with $\tau := 2\ell/c$ giving the delay time needed by the light to travel from the laser source to the scattering position at a distance ℓ and back again. The reflection coefficient $r(\omega)$ of the optical element may also depend on the frequency, like e.g. for the resonator in Fig. 9.22 according to Eq. (5.13).

In direct analogy to the case of normal injection locking, the analysis leads again to a characteristic equation for the frequency which now depends critically on the return phase $\omega\tau$:

$$\omega - \omega_{\text{free}} = r(\omega) = \sqrt{1 + \alpha^2}\,\sin[\omega(\tau - \tau_0)].$$

An overview can be most simply obtained graphically. In Fig. 9.24 we present the situation for a simple mirror (left) and a Fabry–Perot resonator (right). It is evident

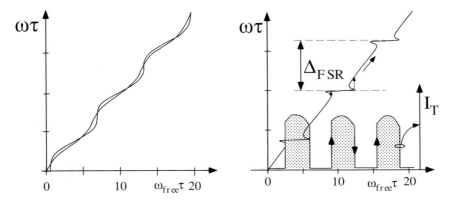

Fig. 9.24: *The effect of feedback on the oscillator frequency of a laser diode. Left: A simple mirror with two different reflected waves. Right: Folded resonator according to Fig. 9.23). The shadowed curve shows the expected transmission of the resonator at tuning of the laser frequency in the positive direction.*

from Fig. 9.24 that back-reflections from acoustically vibrating setups changing the return phase will always cause frequency fluctuations. A stable resonator, however, forces the laser frequency to oscillate at its eigenfrequency if the right conditions are chosen. Thus coupling to the external resonator results in improved coherence properties – the resonator works like a passive flywheel counteracting the phase fluctuations of the active oscillator.

9.5 Laser diodes, diode lasers, laser systems

A laser diode emits coherent light as soon as the injection current exceeds the threshold current through the semiconductor diode. Specific applications, however, set different demands for the wavelength and the coherence properties of the laser radiation. In order to control these properties, the laser diode is used in different optical layouts and is integrated into 'systems' that we are going to call 'diode laser' to distinguish it from the opto-electronic 'laser diode' component.

Owing to the microscopic dimensions of the laser crystal, additional devices like filters may be immediately integrated during manufacture. Such concepts are realized with the so-called DFB, DBR and VCSEL lasers. Another possibility is to achieve frequency control by coupling the laser light back into the resonator as described in the previous section.

9.5.1 Tunable diode lasers (grating tuned lasers)

Among the most unwanted properties of laser diodes are the mode hops that prevent continuous tuning along the entire gain profile as shown in Fig. 9.13. This problem can be solved by using an anti-reflection coating on the laser diode facets and installing it as

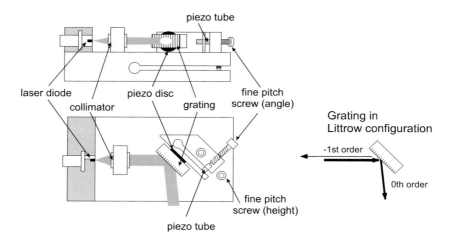

Fig. 9.25: *Construction of a diode laser system according to the Littrow principle. The −1st order of the grating is retro-reflected into the laser diode. The zeroth order makes light available for applications.*

an amplifying medium in an outer resonator with suitable mirrors and filter elements. This 'extended cavity' concept, though, gives up many advantages of the semiconductor laser, such as e.g. the compact layout. Therefore the 'external cavity' method is preferred. The external grating is mounted in the so-called Littrow arrangement for feedback (Fig. 9.25). There the grating reflects about 5–15% of the power exactly back in −1st order into the light source while the rest is reflected away for applications. It thus causes frequency-selective feedback and a corresponding modulation of the gain profile. So by turning the grating most laser diodes can now be tuned to almost every wavelength within their gain profile without any further modifications of their facet reflectivities.

9.5.2   DFB, DBR, VCSEL lasers

The integration of periodic elements for frequency selection has not only been studied with semiconductor lasers but for them it is an attractive choice because the methods of microlithography are required for manufacturing anyway. The concepts of the DFB laser (distributed feedback) and DBR laser (distributed Bragg reflector) are implemented with lateral structures on a suitable substrate (edge emitters), while the VCSEL laser (vertical cavity surface-emitting laser) is realized by a vertical layer stack.

We already know the function of the integrated Bragg end mirrors from the fibre laser (section 7.8.4), and both edge-emitting types differ only in the layout of the Bragg reflector. For the DBR laser it is set aside from the active zone as a selective mirror (and some parameters, such as the centre wavelength, can possibly be controlled, e.g. by injecting a current for refractive index control). For the DFB laser the active zone and the Bragg grating (which is a phase grating in general) are integrated in one element.

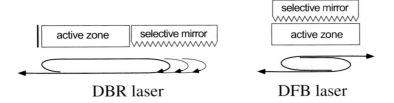

Fig. 9.26: *Principal elements of DBR and DFB lasers.*

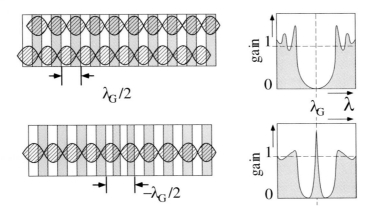

Fig. 9.27: *Grating with uniform period and with $\lambda/4$ shift in the DFB laser. The effective spectral gain profile is qualitatively drawn on the right-hand side.*

Because of simpler and more reliable manufacturing methods, today the DFB laser is in more widespread use than the DBR variant among the edge emitters. Studying the spectral properties of the periodic DFB structure in more detail, one finds that light wave propagation is strongly suppressed in a region centred at the wavelength corresponding to the periodicity of the grating [115, 101]. The cause of this can be seen qualitatively in Fig. 9.27. We can define two stationary waves, which experience a lower average refraction coefficient $n_- = n - \delta n$ at one position and a higher one $n_+ = n + \delta n$ at another position, so that for the same wavelength two frequencies $\nu_\pm = n_\pm c/\lambda$ at the same separation from the centre wavelength $\nu_0 = nc/\lambda$ are allowed. A gain maximum is, however, generated exactly at this position if the so-called $\lambda/4$ shift of the period is inserted at the centre of the DFB structure.

A conceptual example of a VCSEL [19, 52] is presented in Fig. 9.28. The layer structures are epitaxially grown. The active zone has a length of just one wavelength within the material, i.e. $\lambda/n \simeq 250$ nm at an emission wavelength of 850 nm. It is host to several closely adjacent quantum films with a typical thickness of 8 nm. Since the gain length is extremely short, the Bragg mirrors have to have a very high reflectivity of 99.5%. For this, typically 20–40 $Al_xGa_{1-x}As/Al_yGa_{1-y}As$ layer stacks are required with a refraction coefficient contrast as high as possible.

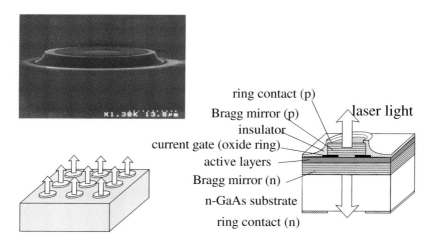

Fig. 9.28: *Concept of VCSEL lasers. (The electron microscope photograph was provided by Dr. Michalzik, University of Ulm [52].)*

For the VCSELs the concentration of the injection current onto the desired cross-sectional area of the laser field is a huge technical challenge. In today's solutions, for example, the resistance of certain layers is strongly increased by proton bombardment, though this causes disadvantageous crystal damage in adjacent material. Using another method the upper Bragg stack is structured into round mesa-like mirrors and finally a thin $Al_{0.97}Ga_{0.03}As$ layer is chemically transformed into an insulating oxide, thus creating current apertures with an inner diameter of only a few micrometres.

9.6   High-power laser diodes

The direct conversion of electrical energy into coherent light with an excellent efficiency promises a wealth of applications. Coherent light provides this energy for e.g. cutting and welding in materials processing with a very high 'quality', so to speak, because its application can be controlled with a very good spatial and time resolution. So interest in increasing the output power of laser diodes up to range of 1 kW and more was quite natural from the beginning.

The 'quality' of a laser beam for machining applications depends on the total power available but at the same time depends crucially on its spatial properties, i.e. the transverse coherence. For practical evaluation, it is customary to use the beam parameter product of beam waist w_0 and divergence angle θ_{div} (see p. 40). When this is normalized to the corresponding product for a perfect TEM_{00} Gaussian beam, it is called the M^2 factor [101]:

$$M^2 = \frac{w_0\theta_{\mathrm{div}}(\text{measured})}{w_0\theta_{\mathrm{div}}(\text{perfect})}. \tag{9.17}$$

It is a measure of the performance of beam cross-section and divergence, and gives an estimate of the fraction of laser light propagating within the dominant Gaussian mode, for only this can be focused to the optimum, i.e. limited by diffraction, or transmitted through a spatial filter (see Fig. 2.11 on p. 47). The M^2 factor grows with decreasing beam quality and should differ as little as possible from unity.

As already indicated in Fig. 9.12, the power increase just by increasing the injection current is seriously limited. On the one hand, owing to the excess heat, the 'roll-over' effect occurs; on the other hand, the light intensity becomes so high that the emitting facets suffer spontaneous damage, often leading to the total loss of the device. These problems are particularly severe for layers containing Al. For this reason in high-power lasers mostly Al-free quantum films are used at least for the gain zone. It is generally observed that the output power for a conventional single laser diode stripe with a facet of about $1 \times 3 \, \mu m^2$ is limited to not more than some $100 \, mW$. Therefore the output power of semiconductor devices can in principle only be increased by spreading the gain over facets as large as possible or over many facets and the volumes connected to them. Today the output power is increased by using laser arrays, broad-area and tapered amplifier lasers as schematically presented in Fig. 9.29.

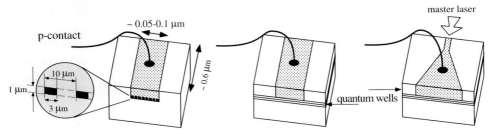

Fig. 9.29: *Concepts for high-power laser diodes: laser diode arrays, broad-area laser and tapered amplifier.*

1. Several laser diode stripes can be placed on a single substrate without any problem. If the separations of the single stripes are not too big, the fields of adjacent modes overlap slightly and are coupled through their phase evolution, i.e. the output power of all individual stripes is coherently coupled, or 'capable of interfering'. The far field of a laser array depends on the relative phase positions of the single stripes. In Fig. 9.30 we show calculations of the idealized field distribution of two and four identical Gaussian emitters, considering all possible combinations of relative phase positions. A realistic laser array often shows a far field with two 'ears'. Their origin becomes evident from this consideration.

2. In a broad-area laser diode a wide diode volume is used for amplification, as the name indicates. However, the control of the transverse field distribution becomes more and more difficult with increasing power, so that broad-area lasers are limited to quite low powers (sub-watts).

3. Tapered amplifiers of trapezoidal shape are used to boost laser light from lower power to high power while transferring high spatial and longitudinal coherence.

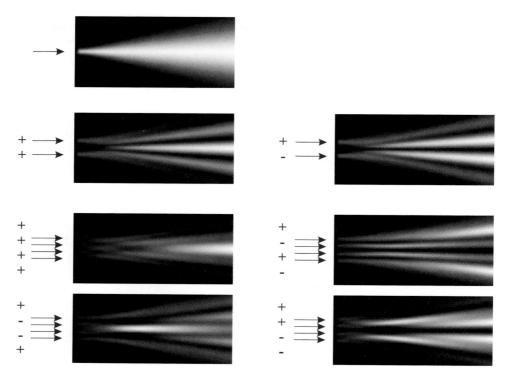

Fig. 9.30: *Beam shapes of laser arrays: (left) the symmetric phase positions of the individual stripes; (right) the antisymmetric phase positions.*

For this concept, the term MOPA (master oscillator power amplifier) has become established. The trapezoidal form has been chosen here to realize maximum gain, but to keep the power density low at the same time to avoid damage. With this concept M^2 factors of 1.05 at output powers of some watts are achieved.

10 Sensors for light

The application of optical instruments depends essentially on how sensitively light can be detected by means of suitable devices. Here we are rather blessed by the human eye, which is – despite all the weaknesses of its imaging optics – an enormously sensitive and versatile receiver.

From the historical point of view, above all, we find light-sensitive plates at the beginning of the development of optical sensors. Photographic emulsions in which light causes a permanent chemical change have been developed to high sensitivity, high resolution and countless applications in more than a century of intensive work.

However, in a physical experiment or in a technical application, when the intensity of a light beam has to be detected and evaluated, then solid-state detectors (and among them especially semiconductor detectors) have out-performed film for quite a long time. They deliver an electrical signal that not only can be saved and recorded without a slow sequence of chemical processes but also are advantageous with regard to linearity.

Until recently, films used to be unbeatable for taking high-contrast pictures with high resolution. With the culture-driving development of semiconductor technology, and the opportunity of processing larger and larger (electronic) data streams faster and faster, that field of application runs into danger of being replaced by opto-electronic components, too. We report about this in the section about image sensors.

Optical sensors generally consist of physical materials that can be coarsely divided up into two classes according to the effect of the incident light beam:

1. **Thermal detectors.** Ideal thermal detectors are *blackbodies*. This means that they absorb all incident light. The energy current of the incident light leads to a temperature increase compared to the environment, which is measured and converted into an electrical signal.

   Among the thermal detectors are thermopiles, bolometers and pyroelectric detectors. The strengths of thermal detectors are their broad spectral sensitivity and their robust layout. Their most significant disadvantage is a slow rise time.

2. **Quantum sensors.** In a quantum detector, a light beam is converted into free charge carriers using the internal or external photo-effect. The current or the charges are directly measured. The often used picture, according to which in a photodiode photons are simply converted into electrons and counted, has to be taken with a pinch of salt. However, a more strict theoretical description of the photon counter is beyond the scope of this text [84, 70].

Optics, Light and Laser. Dieter Meschede
Copyright © 2004 Wiley-VCH Verlag GmbH & Co. KGaA
ISBN: 3-527-40364-7

Among the quantum detectors are photomultipliers, on the one hand, and photoconductors and photodiodes, on the other. The historical development from electron tube to semiconductor technology can also be observed with these components. As suggested by the name, by means of quantum detectors single electrons can be recorded. Their rise time is rarely more than $1\,\mu$s, but often they have to be cooled and are subject to stronger spectral limitations than thermal detectors. In principle, also the emulsions of photographic films belong to the class of quantum sensors, since an individual photon is necessary to reduce each AgBr molecule and thus to cause blackening.

When an optical sensor has to be chosen for a certain application, from the physical point of view it is of interest, for example, whether the detector has a sufficient sensitivity and a fairly short rise time to dynamically record the desired quantity. These properties can be found from the manufacturers' data sheets. For more insight we first have to strike out a bit further and talk about the noise properties of detector signals.

10.1   Characteristics of optical detectors

10.1.1   Sensitivity

In an optical sensor, light pulses are ultimately converted into electric signal voltages $U(t)$ or signal currents $I(t)$. Since all electronic quantities are subject to similar procedures when measured, we use subscripts V_U and V_I for their identification. The *responsivity* $\mathcal{R}$ describes the general response of the detector to the incident light power P_L without taking details like wavelength, absorption probability, circuit wiring, etc., into account:

$$\text{sensitivity} = \frac{(V_U, V_I)}{P_L} \tag{10.1}$$

The physical unit of responsivity is usually $\mathrm{V\,W^{-1}}$ (especially for thermal detectors) or $\mathrm{A\,W^{-1}}$.

10.1.2   Quantum efficiency

In a quantum detector photons are converted into electrons. Even single electrons may be amplified in such a way that their pulses can be registered and counted. Not every incident photon triggers an electron since the absorption probability is lower than unity, or because other processes compete with the photo-effect. The probability for registering an event for each incident photon is called the *quantum efficiency* η. The rate of arrival of photons r_{ph} at a detector with area A can easily be determined according to

$$r_{\mathrm{ph}} = \frac{1}{h\nu} \int_A dx\, dy\, I(x, y). \tag{10.2}$$

If the entire radiation power is absorbed, Eq. (10.2) simplifies to $r_{\mathrm{ph}} = P_L/h\nu$. In an ideal quantum detector, this should become the photo-current $I = e r_{\mathrm{ph}}$, but in physical reality there are competing processes reducing the quantum efficiency. According to Eq. (10.1) the responsivity can be expressed in terms of these elementary quantities:

$$\mathcal{R} = \frac{r_{\text{el}}}{r_{\text{ph}}} \frac{e}{h\nu} = \eta \frac{e}{h\nu}. \qquad (10.3)$$

A practical rule of thumb can be obtained by using the wavelength $\lambda = c/\nu$ in µm instead of the frequency:

$$\mathcal{R} = \eta \frac{\lambda/\text{µm}}{1.24} \; [\text{A W}^{-1}],$$

from which the quantum efficiency can be determined for a known responsivity.

10.1.3 Signal-to-noise ratio

A quantity can only be recognized if it emerges 'from the noise', i.e. if it is larger than the intrinsic noise of the detector. Formally the quantitative concept of 'signal-to-noise ratio' (SNR) has been introduced,

$$\text{SNR} = \frac{\text{signal power}}{\text{noise power}}.$$

In this case we use a generalized concept of power $\mathcal{P}_V(f)$ for an arbitrary physical quantity $V(t) = \mathcal{V}(f) \cos(2\pi f t)$. The average power is

$$\mathcal{P}_V(f) = \tfrac{1}{2} \mathcal{V}^2(f). \qquad (10.4)$$

The physical unit of these powers are A^2, V^2, ..., depending on the basic value. A fluctuating quantity like noise current or voltage is determined not only by one amplitude at one frequency but also by contributions at many frequencies within the bandwidth Δf of the detector. The average power in a frequency interval δf can be measured with a filter of this bandwidth and with mid-frequency f. Therefore we define the power spectral density

$$v_{\text{n}}^2(f) = \frac{\delta V^2(f)}{\delta f},$$

so e.g. $i_{\text{n}}^2(f)$ in $A^2\,\text{Hz}^{-1}$ for current noise, and $e_{\text{n}}^2(f)$ in $V^2\,\text{Hz}^{-1}$ for voltage noise. Since the contributions are not correlated, the square sum of the power contributions in small frequency intervals can be summed to give the average of the noise power (see Appendix A.1):

$$P_V = \int_{\Delta f} v_{\text{n}}^2(f)\, df. \qquad (10.5)$$

If the noise in the bandwidth Δf is constant, the value of the noise power simplifies to $P_V = v_{\text{n}}^2 \Delta f$. For example, the r.m.s. value of the noise current $I_{\text{rms}} = \sqrt{P_I}$ of a photodiode–amplifier combination reads $I_{\text{rms}} = (i_{\text{n}}^2 \Delta f)^{1/2}$, with i_{n}^2 the constant value of the current noise spectral density.

Often the unphysical noise amplitude is given instead of the noise power,

$$\text{noise amplitude} = (\text{noise power spectral density})^{1/2},$$

given in e.g. $A\,\text{Hz}^{-1/2}$ or $V\,\text{Hz}^{-1/2}$. Very generally the noise contribution can be reduced by limiting the bandwidth of the detector. This advantage has to be traded in for reduced dynamic properties – faster signal variations can no longer be registered.

10.1.4   Noise equivalent power (NEP)

The *noise equivalent power* (NEP) is the radiation power that is necessary to exactly compensate the noise power at the detector or to obtain a signal-to-noise ratio of exactly unity. The lower the designed bandwidth of the detector, the lower is the minimum detectable power, but again at the expense of the bandwidth. The minimum detectable radiation power is therefore referred to 1 Hz bandwidth and is given by the unphysical noise amplitude density,

$$\text{NEP} = \frac{(\text{noise power spectral density})^{1/2}}{\text{responsivity}}.$$

Its physical unit is $\mathrm{W\,Hz^{-1/2}}$. The manufacturer of a detector prefers to quote the spectral maximum of the responsivity; though it has to be taken into account that the value depends on the optical wavelength λ as well as on the electrical signal frequency f.

10.1.5   Detectivity 'D-star'

For the sake of completeness, we mention the concept of 'detectivity' D and D^* introduced to make different detector types comparable with each other. First, just the complement of the noise equivalent power $D = \text{NEP}^{-1}$ was introduced as the detectivity. The variant of the 'specific detectivity' called 'D-star' (D^*) has found widespread use since the responsivity of many detectors is proportional to the square root of the detector area $A^{1/2}$:

$$D^* = \frac{\sqrt{A}}{\text{NEP}}. \tag{10.6}$$

The reason for this is the limitation of the detection sensitivity by the thermal background radiation, especially for infrared detectors: the larger the detector area, the more blackbody radiation is absorbed. The physical unit of D^* is $\mathrm{cm/(W/Hz^{1/2})} = 1$ Jones, where the name of the inventor of the detectivity is used for abbreviation. D^* is a measure for the signal-to-noise ratio in a bandwidth of 1 Hz when a detector with an area of diameter 1 cm is illuminated with a radiation power of 1 W .

10.1.6   Rise time

Often very fast events are to be recorded by means of optical detectors, which means that the detector has to react very rapidly to variations of the incident radiant flux. The 'rise time' τ is the time during which the current or voltage changes of the detector reaches $(1-1/e)$ or 63% of the final value when the light source is switched on suddenly. In analogy to that, the 'fall time' is also defined. They depend on the layout of the detector and can be influenced within the physical limits. Thermal detectors are inert and react with delay times of many milliseconds. The 'rise time' of semiconductor detectors is generally limited by the capacitance of the pn junction and is only a few picoseconds in special cases. The finite response time of a detector can be taken into account by, for example, adding the time or frequency dependence to the responsivity of Eq. (10.3),

$$\mathcal{R}(f) = \frac{\mathcal{R}(0)}{1 + (2\pi f \tau)^2}.$$

The charge impulse of a photomultiplier can as well be shorter than 1 ns, though here the longer time of travel through the dynodes layout has to be taken into account. The times of travel in cable connections also have to be taken into account in servo-control applications, since they limit their bandwidth.

10.1.7 Linearity and dynamic range

A linear relation of detector input power and output voltage or current provides optimal conditions for a critical analysis of the quantity to be measured. However, there is always an upper limit – ultimately due to the strong temperature load at high light power – at which deviations from linearity can be observed. The lower limit is mostly given by the noise equivalent power. A quantitative measure for the *dynamic range* can be given according to

$$\text{dynamic range} = \frac{\text{saturation power}}{\text{NEP}}.$$

The dynamic range between these limits can be, for example for photodiodes, an impressive six magnitudes or more.

10.2 Fluctuating opto-electronic quantities

In this section we collect physically different contributions of electrical noise generated in opto-electronic detectors. Besides the intrinsic contributions of the receiver–amplifier combination, like dark current noise and amplifier noise, above all there is the photon noise of the light source.

10.2.1 Dark current noise

A detector generates a fluctuating signal $V_n(t)$ even when there is no incident light signal at all. In fact, the detector sensitivity is decreased not by the average of the background – this can be straightforwardly subtracted – but by its fluctuations. In a thermal detector, spontaneous temperature fluctuations cause the dark noise. In a quantum detector, generally charge carriers spontaneously generated, e.g. by thermionic emission, are responsible for this. In the simplest case, the noise power density of the dark current I_D is $i_D^2 = 2eI_D$ according to the Schottky formula (eq. (A.13) in the Appendix). A proven but sometimes costly method for reducing the dark noise is to apply cryogenic cooling of the detector.

10.2.2   Intrinsic amplifier noise

Photomultipliers and avalanche photodiodes (APDs) have an internal amplification mechanism that multiplies the charge of a photo-electron by several orders of magnitude. The amplification factor G though is subject to fluctuations contributing to noise as well. The *excess noise* factor F_e is calculated according to

$$F_e^2 = \frac{\langle G^2 \rangle}{\langle G \rangle^2} = 1 + \frac{\sigma_G^2}{\langle G \rangle^2}, \tag{10.7}$$

and can also be expressed through the variance of the amplification, $\sigma_G^2 = \langle G^2 \rangle - \langle G \rangle^2$. It affects dark and photo-currents in indistinguishable ways.

10.2.3   Measuring amplifier noise

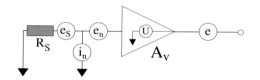

Fig. 10.1: *Noise sources of an idealized amplifier. Voltage sources (e_i) have negligible, current sources (i_n) infinite inner resistance. R_S is the resistance of the detector, A_V the amplification.*

Depending on the application, a detector may operate as a voltage or a current source characterized by its internal resistance R_S. At the input of an idealized test amplifier, we find the voltage noise amplitude e_i, which consists of the uncorrelated contributions of the detector, e_S^2, with source resistance R_S, and the contributions of current and voltage noise of the amplifier (i_n^2 and e_n^2, respectively):

$$e_i^2 = e_S^2 + e_n^2 + i_n^2 R_S^2.$$

The noise voltage at the exit of the amplifier is then $e = A_V e_i$. The noise amplitude of the detector consists of the contributions of the dark current, the parallel resistance of detector and amplifier input, and the photon current i_{ph}^2,

$$e_S^2 = R_S^2 \left(i_{ph}^2 + i_D^2 + \frac{4kT}{R_S} \right).$$

The last contribution takes the thermal or *Johnson* noise of the detector resistance into account. For optical detection, the most desirable situation is obtained when the noise of the photoelectrons generated by the signal source (i_{ph}^2) dominates all intrinsic amplifier contributions,

$$i_{ph}^2 > i_D^2 + \frac{4kT}{R_S} + \frac{e_n^2}{R_S^2} + i_n^2. \tag{10.8}$$

In practical applications it has to be taken into account that all the quantities mentioned previously depend on frequency. If possible the frequency of the signal can be selected to minimize background noise. Advantageous conditions are generally found at high frequencies since all devices at frequencies below a certain *corner frequency* f_c show the so-called $1/f$ or *flicker noise*, which approximately increases with $1/f$

towards low frequencies. The typical spectral behaviour of the amplifier noise is presented in Fig. 10.2.

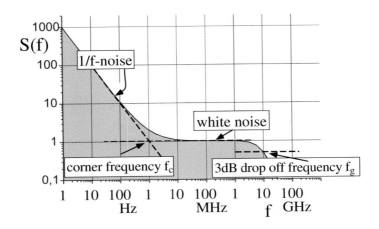

Fig. 10.2: *Spectral properties of typical amplifier noise, schematically.*

10.3   Photon noise and detectivity limits

By conversion of light into photo-electrons in an opto-electronic circuit, something like a copy of the photon current emerges. It is obvious that the fluctuations of the photon current are mapped onto the electron current, too. Now, for the rigorous description of the processes occurring during conversion of light into photo-electrons, a quantum theory of the electromagnetic field is required, but quantum electrodynamics does not offer an intuitive approach and is hence omitted here.

Instead, we assume that the probability of observing an event in a short time interval is proportional to Δt by taking into account the arrival rate of the photons (eq. (10.2)) and the quantum efficiency η,

$$p(1, \Delta t) = \eta r_{\mathrm{ph}}(t)\Delta t. \tag{10.9}$$

Furthermore we assume that, for sufficiently small Δt, no double events occur and that the probabilities in sequential time intervals are statistically independent. The last assumption means that the photo-emission process does not have any after-effects in the detector; this is not necessarily the case any more at high charge carrier density because they repel each other due to Coulomb forces.

These conditions lead to *Poisson statistics* of the counting events. The probability of finding K events in an arbitrary time interval τ is

$$p_\tau(K) = p(K, t, t+\tau) = \frac{\overline{K}_\tau^K}{K!}\, e^{-\overline{K}}.$$

The average $\overline{K}_\tau$ is according to Eq. (10.3)

$$\overline{K}_\tau = \eta r_{\mathrm{ph}} \tau. \tag{10.10}$$

Random conversion of photons into photo-electrons leads to fluctuations of the photo-electron current. In addition, also the light intensity $P_{\mathrm{L}}(t)/A$ can vary. If it happens in a deterministic way, i.e. predictably, we can define the power W_τ integrated in the interval τ,

$$W_\tau(t) = \int_t^{t+\tau} P_{\mathrm{L}}(t')\,dt',$$

and with the abbreviation $\alpha = \eta/h\nu$ we obtain the probability distribution

$$p_\tau(K) = \frac{(\alpha W_\tau)^K}{K!}\, e^{-\alpha W_\tau}. \tag{10.11}$$

The properties of the light source are reflected in the statistics of the photo-electrons, and therefore we consider the light field of a laser and a thermal light source as important examples.

10.3.1   Photon statistics of coherent light fields

The average power of a laser is constant; therefore the arrival rate of photons r_{ph} is constant as well, and we can directly take over the average from Eq. (10.10). The statistical distribution is characterized by the variance

$$\sigma_{K_\tau}^2 = \overline{(K^2 - \overline{K}_\tau^2)},$$

which has the known value for Poisson statistics

$$\sigma_{K_\tau}^2 = \overline{K}_\tau. \tag{10.12}$$

From this relation it is also clear that relative fluctuations decrease with increasing number of events,

$$\frac{\sigma_{K_\tau}}{\overline{K}_\tau} = \frac{1}{\sqrt{\overline{K}_\tau}}, \tag{10.13}$$

and become very small for large $\overline{K}_\tau$. The noise caused by the grainy particle structure of the current is called *shot noise*. It also sounds very loud like the audible drubbing caused by raindrops falling on a tin roof.

We expect a random sequence of charge pulses for the photo-electron current. The spectrum of the current noise depends on the frequency and can be obtained directly from the Schottky formula (Eq. (A.13) in the Appendix),

$$i_{\mathrm{coh}}^2 = 2e\overline{I}_{\mathrm{ph}}. \tag{10.14}$$

This noise current also accounts for the contribution caused by the random conversion of photons into photo-electrons when the quantum efficiency is lower than 100%. We can furthermore interpret the Schottky formula by identifying the r.m.s. value of the counting statistics $\sigma_{K_\tau}^2$ in the time interval τ with the variance of the number of charge carriers, $\sigma_{K_\tau}^2 = \frac{1}{2}I_{\mathrm{rms}}^2/e^2$, which again leads to the result of Eq. (10.14) (the factor $\frac{1}{2}$ occurs because the power spectral density is defined for positive Fourier frequencies only; see Appendix A.1).

A coherent light field generates the photo-current with the lowest possible noise, and therefore it comes very close to our idea of a classical wave with constant amplitude and frequency. We may interpret the noise as a consequence of the 'granularity' of the photo-current, and of its Poisson statistics. Though it has to be pointed out that we did not derive this result here but rather have put it in from the beginning.

10.3.2 Photon statistics in thermal light fields

A thermal light field generates an average photo-current as well, though the intensity is not constant like in a coherent laser beam but subject to strong random fluctuations. Therefore, for the integrated power W_τ we can also give only probabilities $p_{\tau,W}(W_\tau)$ with $\int dW_\tau\, p_{W_\tau}(W_\tau) = 1$ here. The additional fluctuation of the amplitude results in *Mandel's formula*, which is formally similar to a Poisson transformation of the probability density p_{W_τ} (see Eq. (10.11)):

$$p(K) = \int_0^\infty \frac{\alpha W_\tau}{K!}\, e^{-\alpha W_\tau} p_{\tau,W}(W_\tau)\, dW_\tau. \tag{10.15}$$

This contribution has double Poisson character, so to speak. It can be shown that the average of the counting events is

$$\overline{K}_\tau = \alpha \overline{W}_\tau$$

as before, and the variance is

$$\sigma_{K_\tau}^2 = \overline{K}_\tau + \alpha^2 \sigma_{W_\tau}^2. \tag{10.16}$$

Thus the variance of a fluctuating field, like, for example, blackbody radiation (see the excursion on p.184), is in principle larger than for a coherent field. We will see, however, that detection of these strong fluctuations is possible at very short time scales only and hence beyond the dynamic properties of most photodetectors.

We can interpret the relation (10.16). The first term is caused by the random conversion of photons into photo-electrons and is a microscopic property of the light–matter interaction which cannot be removed. The second term represents the fluctuations of the recorded light field and also occurs without the randomness of the photo-electron generation process.

The calculation of σ_W^2 in (10.16) is not a trivial problem at all. We consider the cases of extremely short and very long integration intervals τ. A thermal light field is characterized by random amplitude fluctuations. For very short time intervals, even shorter than the very short remaining coherence time τ_c of the light source of about 1 ps, we can assume a constant intensity so that $W_\tau = P_L\tau$. The intensity itself is randomly distributed and thus follows a negative exponential distribution,

$$p_{\tau,W} = e^{-W/\overline{W}_\tau}/\overline{W}_\tau.$$

By insertion into Eq. (10.15) and integration, the *Bose–Einstein distribution* of quantum statistics is obtained,

$$p_\tau(K) = \frac{1}{1 + \overline{K}_\tau} \left(\frac{\overline{K}_\tau}{1 + \overline{K}_\tau} \right)^K . \tag{10.17}$$

The variance of this field is

$$\sigma_K^2 = \overline{K}_\tau + \overline{K}_\tau^2,$$

and can be interpreted like (10.16) before. Its relative value always remains close to unity:

$$\frac{\sigma_K}{\overline{K}_\tau} = \sqrt{\frac{\overline{K}_\tau}{1 + \overline{K}_\tau}}.$$

The distribution from Eq. (10.17) is well known for a light field when K is replaced by n and $\overline{K}_\tau$ by the mean thermal number of photons,

$$\overline{n}_{\mathrm{ph}} = \frac{1}{e^{h\nu/kT} - 1} . \tag{10.18}$$

The coherence time of a thermal light source though is so short that there exist hardly any detectors with appropriately short response and integration times. The more important limiting case for the thermal light field thus occurs for integration times $\tau \gg \tau_{\mathrm{c}}$. For this case it can be shown [70] that the variance σ_K is well approximated by

$$\sigma_K^2 = \overline{K}_\tau \left(1 + \frac{\overline{K}_\tau \tau_{\mathrm{c}}}{\tau} \right) . \tag{10.19}$$

Thus for most cases $\sigma_K^2 \simeq \overline{K}_\tau$ as for the thermal light field (cf. Eq. (10.12)). Incidentally, these noise properties cannot tell us about the properties of the light field, coherent or thermal! The second term in Eq. (10.19) can be interpreted as the number of photons reaching the detector during a coherence interval. Only when this number becomes larger than unity can a significant increase of the fluctuations be expected.

Ambient radiation of a light source mostly corresponds to the spectrum of the blackbody radiation at 300 K. Its maximum lies at a wavelength of 10 μm and decreases rapidly towards the visible spectral range. Unavoidably at least part of this radiation also enters the detector. Especially for infrared detectors the sensitivity is in general limited by the background radiation. For thermal radiation it is still valid that the coherence time is very short, so that the variance of the photo-electron noise of the thermal radiation can be calculated according to (10.19).

In order to determine the emission rate of photo-electrons r_{el}, we have to multiply the average photon number $\overline{n}_{\mathrm{ph}}$ of Eq. (10.18) with the density of oscillator modes $\rho(\nu) = 8\pi\nu^2/c^3$ at frequency ν, to integrate over the detector area A. Accounting for the quantum efficiency $\eta(\nu)$ and in addition the radiative flux from half the solid angle 2π, we arrive at

$$r_{\mathrm{el}} = A \int_0^\infty d\nu\, \eta(\nu) \frac{2\pi\nu^2}{c^3} \frac{1}{e^{h\nu/kT} - 1} .$$

The spectrum of the charge carrier fluctuations is proportional to the variance of the arrival rate, which we can now calculate according to Eq. (10.19); as for the coherent light field, we obtain a white shot noise spectrum. Since the photo-emission vanishes below a certain critical frequency ν_g or a critical wavelength $\lambda_g = c/\nu_g$, the noise spectrum for a detector with the bandgap $E_g = h\nu_g$ can be calculated according to

$$i_n^2 = 2e^2 r_{el} = 2e^2 A \int_{\nu_g}^{\infty} d\nu\, \eta(\nu)\, \frac{2\pi\nu^2}{c^2}\, \frac{1}{e^{h\nu/kT}-1}.$$

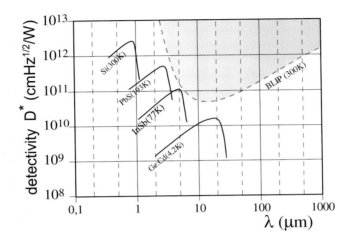

Fig. 10.3: *Specific detectivity for some important semiconductor detectors.*

When we additionally assume the quantum efficiency has maximum value $\eta(\nu) = 1$ everywhere, then according to Eq. (10.6) we obtain the maximum specific detectivity $D^*(\lambda_g, T)$ of an ideal *background-limited photodetector* (BLIP detector), which depends on the environmental temperature T and the critical wavelength λ_g,

$$D^*(\lambda_g, T) = \frac{\lambda_g}{hc}\left(2\int_{c/\lambda_g}^{\infty} d\nu\, \frac{2\pi\nu^2}{c^2}\, \frac{1}{e^{h\nu/kT}-1}\right)^{-1/2}.$$

This reaches a minimum at $\lambda = 14\,\mu\text{m}$, Fig. 10.3. For large wavelengths D^* has to increase linearly since the thermal radiation power does not change any more.

10.3.3   Shot noise limit and 'square-law' detectors

According to Eq. (10.14) the photo-electron noise generated by detection of a coherent laser beam is proportional to P_L. This most favourable case is mainly realized with photodiodes. If the power is chosen large enough according to

$$P_L \geq \frac{h\nu}{\eta}\frac{1}{e^2}\left(2e\bar{I}_D + \frac{4kT}{R_S} + \frac{e_n^2}{R_S^2} + i_n^2\right) = \frac{h\nu}{\eta} r_{th}, \tag{10.20}$$

then the photon noise of the light beam dominates all other contributions in eq. (10.8), which do not depend on the light power. This case is called the 'shot-noise-limited' detection. Incidentally, the term within the large brackets in Eq. (10.20) can be interpreted as the rate r_{th} at which the detector–amplifier combination randomly generates charge carriers. Defining the minimum light power by the value where the same number of charge carriers is generated (SNR $\approx$ 1), we find

$$P_{min} = \frac{h\nu}{\eta} \sqrt{r_{th}\Delta f}$$

in a bandwidth Δf. For sufficiently long integration times (or correspondingly small bandwidth), in principle, arbitrarily small power may be registered. In practice, this potential though is impaired by the dynamics of the signal and slow drifts of the detector–amplifier properties.

Quantum detectors are also called 'square-law' detectors since the trigger probability of a photo-electron is proportional to the square value of the field strength $|E(t)|^2 = 2P_L(t)/c\epsilon_0 A$ of the radiation field illuminating the detector area A. This is especially important for applying so-called heterodyne detection. For this method the field of a local oscillator $\mathcal{E}_{LO} e^{-i\omega t}$ (see Section 7.3.2) is superimposed with a signal field $\mathcal{E}_S e^{-i(\omega+\omega_S)t}$ on the receiver. In general one chooses $P_{LO} \gg P_S$. The photo-current will thus experience a variation in time

$$I_{ph} \simeq \frac{e\eta}{h\nu} \left(P_{LO} + 2\sqrt{P_S P_{LO}} \cos \omega_S t \right).$$

If LO and signal fields oscillate with the same frequency ω, it is called a 'homodyne' detection; otherwise ($\omega_S \neq 0$) it is a 'heterodyne' detection. Superposition of optical fields on a square-law detector generates products of local oscillator and signal fields oscillating at difference frequencies, and thus it acts as an optical mixer.

The detection of a signal at a higher frequency is usually an advantage since it occurs at a lower noise power spectral density (Fig. 10.2). When the LO power is increased until its shot noise density $i_{LO}^2 = 2e^2\eta P_{LO}/h\nu$ dominates all other contributions, the minimum detectable signal power no longer depends on the thermal noise properties of the detector. One has $I_S = 2e\eta\sqrt{P_{min}P_{LO}}/h\nu$ and the minimum power I_S^2 has to be larger than the noise power $i_{LO}^2\Delta f$ in the bandwidth of measurement, Δf,

$$P_{min} = h\nu\Delta f/\eta.$$

In other words, within the time resolution Δf^{-1} of the detector, the signal light has to release at least one photo-electron to make detection possible.

10.4   Thermal detectors

Thermal detectors consist of a temperature sensor coated with an absorber material, e.g. special metal oxides known from illumination technologies. Over wide wavelength ranges they have very 'flat' spectral dependences and are therefore very sought-after for calibration purposes.

In order to achieve a high sensitivity, i.e. a large temperature increase ΔT, the sensor should have a low heat capacity K as well as a low heat loss rate V to the environment caused by heat conduction due to the construction, convection and radiation. The temperature change of the probe follows the differential equation

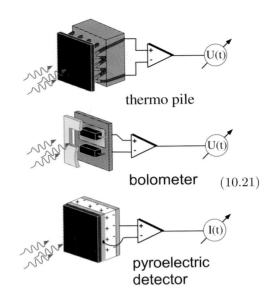

thermo pile

bolometer   (10.21)

pyroelectric detector

$$\frac{d}{dt}\Delta T = \frac{P_{\mathrm{L}}}{K} - \frac{V}{K}\Delta T,$$

from which it can be seen immediately that a thermal detector integrates the incident light power for short times. In equilibrium the obtained temperature increase is $\Delta T = P_{\mathrm{L}}/V$, from which the responsivity R_{th} is determined with the voltage–temperature coefficient of the thermal probe, C_{TV},

Fig. 10.4: *Thermal detectors.*

$$R_{\mathrm{th}} = C_{\mathrm{TV}}/V.$$

However, compromises are necessary since the rise time is determined by the coefficient $\tau = K/V$ according to Eq. (10.21). In the ideal case the minimum detectable power of a thermal detector is caused by unavoidable spontaneous temperature fluctuations, the spectral power density $t^2 = 4k_{\mathrm{B}}T^2V/[V^2 + (2\pi K f)^2]$ of which determines the theoretical responsivity limit (k_{B} = Boltzmann constant). For signal frequencies f far above the detector bandwidth $\Delta f = 1/2\pi\tau$, the idealized noise equivalent power can be given:

$$\mathrm{NEP}_{\mathrm{th}} = T\sqrt{2k_{\mathrm{B}}V}.$$

Obviously it is profitable to lower the environmental temperature – a method that is used in particular for bolometer receivers.

10.4.1   Thermopiles

In these, the light energy is absorbed by a thin blackened absorber plate in close thermal contact with a thin-layer pile of thermocouples made, for example, of copper–constantan. Since the voltage difference of a single element is very small, some 10–100 of them are connected in series with the 'hot' ends receiving the radiation field to be detected and the 'cold' ends kept at ambient temperature. The voltage of the thermopile is proportional to the temperature increase and thus to the power uptake of the absorber.

Thermopiles are mainly used in optics to determine the intensity of high-power light sources, especially laser beams. Owing to their integrating character, they are also capable of determining the average power of pulsed light sources.

10.4.2   Bolometers

The temperature increase by illumination can also be measured by means of a resistor with a large temperature coefficient. This is called a 'bolometer'. For this application, especially semiconductor resistors called thermistors are of interest.

Bolometers are mainly used in a bridge circuit. Only one of two identical thermistors in the same environment is exposed to radiation so that fluctuations of the environmental temperature are already compensated. Very high sensitivities are obtained with bolometers operated at cryogenic temperatures when the heat capacity of the thermistor is very low.

10.4.3   Pyroelectric detectors

In pyroelectric sensors a crystal is used with an electrical polarity that depends on temperature, e.g. $LiTaO_3$. The crystal is inserted into a capacitor. When the temperature changes, a charge is induced on the metallized faces, generating a transient current. The sensitivity for a crystal with pyroelectric coefficient p, heat capacity K and distance d between the capacitor electrodes is

$$\mathcal{R} = p/Kd. \tag{10.22}$$

A pyroelectric detector registers only changes of the incident light power. According to eq. (10.22) its sensitivity is significantly enhanced by thin-layer technology. Therefore the thickness of the crystal is only some $10\,\mu m$, which allows fast rise times. Wide spectral applicability of these detectors is achieved by using an appropriate broadband absorber.

Pyroelectric detectors are cheap and robust, and are often used, for example, in the manufacture of motion sensors.

10.4.4   The Golay cell

An unusual thermal detector is the radiation sensor called a *Golay cell* after its inventor. It is often used as a result of of its high responsivity. The temperature increase by light absorption causes a pressure increase in a small container filled with xenon. On one side the container is closed by a membrane that bulges due to the pressure increase. The small mechanical motion of the surface can be read out very sensitively by means of a 'cat's eye technique'.

10.5   Quantum sensors I: photomultiplier tubes

It may be somewhat surprising that Albert Einstein got his Nobel Prize for physics in 1921 for his 1905 light quanta hypothesis of the photo-electric effect and not for

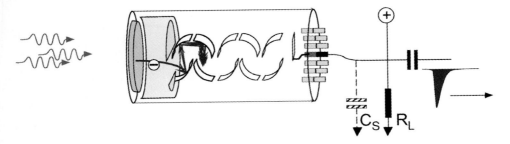

Fig. 10.5: *Layout of a photomultiplier tube (PMT) with transparent dynode. The circuitry operates the PMT in the counting mode.*

any other of his scientific triumphs. He not only used Planck's hypothesis that light energy could be absorbed only in light quanta with fixed value $E_{photon} = h\nu$ but also expanded it by attributing quantum nature to the light field itself. According to Einstein's simple concept the maximum kinetic energy E_{max} of an electron that is emitted from the surface of a material with work function W is

$$E_{max} = h\nu - W. \tag{10.23}$$

In general, though, only a few emitted electrons reach the maximum energy E_{max}. More importantly, the photo-electric effect vanishes completely for frequencies $\nu \leq W/h$, the cut-off frequency or wavelength, which depends on the work function W of the used material.

Photo-cathodes

Common metals mostly have very high values for the work function between 4 and 5 eV, corresponding to cut-off wavelengths between about 310 and 250 nm according to Einstein's equation (10.23). In vacuum it is also possible to use caesium, which immediately corrodes under atmosphere conditions but has the lowest work function of all metals with $W_{Cs} = 1.92$ eV. By coating a dynode with caesium, a photo-cathode becomes sensitized for light frequencies extending nearly across the entire visible spectral range ($\lambda < 647$ nm).

The probability of triggering a photo-electron by the absorption of a photon, the quantum efficiency (QE), is generally lower than unity. Owing to its high QE, approaching 30%, the semiconductor $CsSb_3$ is very often used for photo-cathode coating. It is inserted into vacuum tubes made from different glasses with differing transparencies. Such combinations have led to the classification of the spectral responsivity using the term *S-X cathode* (X = 1, 2, ...). The tri-alkali cathode S-20 (Na_2KCsSb) has been used for a long time now, and 1% QE is achieved even at 850 nm.

Cs-activated GaAs offers 1% QE even further into the infrared at 910 nm wavelength. Still farther in the infrared spectral range, the InGaAs photo-cathode does not exceed 1% QE at any wavelength, but still has 0.1% QE at 1000 nm. In this spectral range, though, the internal photo-effect in semiconductors has a very high quantum efficiency; therefore here the photomultiplier tubes compete with the avalanche photodi-

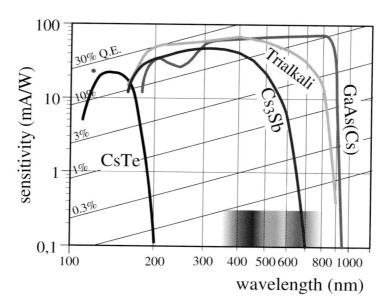

Fig. 10.6: *Spectral responsivity of several important photo-cathodes.*

ode discussed below, which can be considered as semiconductor-based photomultiplier tubes.

Conversely there are situations in which a light detector ought to be sensitive only to UV wavelengths since then visible light such as daylight no longer contributes to the signal background and its noise. For this purpose so-called *solar-blind* cathodes are used made of, for example, Cs_2Te or CsI.

Amplification

The success of photomultiplier tubes (PMT) is not conceivable at all without the enormous amplification obtained with a secondary electron multiplier (SEM), which is connected to the photo-cathode. In a SEM, electrons are accelerated and cause multiple *secondary* electrons to be ejected from the anode. The multiplication factor for a layout with n dynodes at applied voltage U_{PMT} is $\delta = c[U_{PMT}/(n+1)]^\alpha$. The up to 15 steps cause an avalanche-like amplification of the photo-current $I_{ph} = GI_{el}$,

$$G = \text{constant} \times U_{PMT}^{\alpha n}, \tag{10.24}$$

with geometry and dynode material causing a slight attenuation of the theoretical amplification factor of a single step by a factor $\alpha = 0.7$–0.8. At the end of a cascade subjected to a total voltage of about 1–$3\,\text{kV}$, a charge pulse with 10^5 to 10^8 electrons is available. The high intrinsic gain G leads to extreme sensitivity, which reaches values of $R_{PMT} \simeq 10^4$–$10^7\,\text{A W}^{-1}$ according to

$$R_{PMT} = \frac{\eta G e}{h\nu},$$

depending on layout and circuit wiring. Since the gain depends sensitively on the applied voltage due to (10.24), the voltage supply has to be stable and low-noise.

Counting mode and current mode

The input channels of any electronic measuring amplifier usually expect a voltage at the input. Thus the current of the photomultiplier tube has to be converted into a voltage by a load resistance R_L. Especially for low currents the PMT works like an ideal current source, and thus R_L can be chosen arbitrarily large. In practice, however, the rise time is limited by the load resistance and the stray capacitance of the anode to the layout

$$\tau = R_L C_S.$$

In addition, large load resistances cause the anode to discharge slowly. Thus the voltage of the last dynode stage is decreased and therefore also the efficiency of the anode for collecting secondary electrons: the characteristic curve becomes nonlinear and the photomultiplier tube *saturates* for a certain light power.

For the circuitry wiring, the *counting mode* and the *current mode* are distinguished. The counting mode is suitable for very low light powers. The gain G is chosen very high and R_L so low that for a standard $50\,\Omega$ impedance typical voltage pulses of some $10\,\text{mV}$ and some nanoseconds width are observed. These pulses can be processed directly using commercial counting electronics. They cause the 'clicks' of the *photon counter*. Because of the similarity to a *Geiger–Müller tube* for α-rays, this is also called the *Geiger mode*. Of course, a statistical distribution of impulses with different heights and widths is generated, from which signal photon pulses are selected by electronic discriminators.

The current mode is used for larger light intensities, with lower gain G and a load resistance adjusted to the desired bandwidth. The resistance should be chosen high enough to approach as close as possible an ideal current source.

Noise properties of PMTs

A small current flows through a photomultiplier tube even when the tube is operated in total darkness. It is called the 'dark current' I_D and is mainly caused by thermionic emission of electrons from the photo-cathode, which are amplified indistinguishably from photo-electrons.

In the counting mode of the photomultiplier tube, we can directly use the Schottky formula (Eq. (A.13) in the Appendix) if we insert the effective average charge $\langle Ge \rangle$ of a single photo-electron to determine the power density of the shot noise of the dark counting rate R_D. In this case the noise equivalent power is calculated as:

$$\text{NEP}_{\text{count}} = \frac{\sqrt{2R_D}}{\eta} h\nu, \tag{10.25}$$

where we have used the average gain $\langle G \rangle$.

If a photomultiplier tube is used in the current mode, the fluctuations of the gain also cause noise: the noise power density of the current is then $i_n^2 = \langle 2GeI_D \rangle =$

$2e\langle G^2\rangle\langle I_\mathrm{D}\rangle/\langle G\rangle$, since the instantaneous gain is strictly related to the instantaneous current I_D. In current mode the result of (10.25) is increased by the *excess noise* factor $F_\mathrm{e} = \langle G^2\rangle/\langle G\rangle^2$ of eq. (10.7):

$$\mathrm{NEP}_\mathrm{current} = F_\mathrm{e}\frac{\sqrt{2I_\mathrm{D}/\langle eG\rangle}}{\eta\langle G\rangle}h\nu.$$

Their enormous sensitivity has led to numerous applications for photomultiplier tubes, and in addition has caused the development of many specialized types. The most common models are the so-called *side-on* PMTs, in which the photo-electron is ejected from an opaque photo-cathode and first counterpropagates the light beam. The *head-on* models are equipped with a transparent photo-cathode. From their rear the photo-electrons are sent into the secondary emission multiplier. They are advantageous when photo-cathodes with a large area are required, for example in scintillation detectors. For applications in servo-control devices, though, photomultiplier tubes have certain disadvantages when not only the rise time but also the delay time (caused by the travel time within the detector) play a role.

Microchannel plates and channeltrons

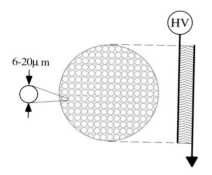

Microchannel plates (MCPs) are actually a variant of the secondary electron multiplier. A single microchannel consists of a glass capillary tube with a diameter of 6–20 µm. The wall is coated with a semi-conducting material (e.g. NiCr) with relatively low conductivity. The ends of the tube are coated with a metal and operate as photo-cathode and anode, respectively; a high voltage drops along the walls and generates a 'continuous dynode'. This type of secondary emission multiplier with a single channel is also known as a *channeltron*. Using an appropriate coating of the input facet, they can be converted into very compact photomultiplier devices. Their disadvantage is the saturation behaviour, which generally begins at lower currents than in photomultiplier tubes because of the high wall resistance.

Fig. 10.7: *Microchannel plate (MCP), schematic.*

A microchannel plate consists of several thousand densely packed capillary tubes subjected in parallel to a high-voltage source and working like an array of SEM tubes. As *MCP-PMTs* they have advantages due to their excellent time resolution and their low sensitivity to magnetic fields (which influence the amplification behaviour of every SEM. Moreover, they allow the detection of very low light intensities with spatial resolution and are therefore used to build the image intensifiers discussed in Section 10.7.3.

10.6   Quantum sensors II: semiconductor sensors

In semiconductors, the photo-electrons do not have to be knocked out of the material but can internally generate free charge carriers. The *internal photo-effect* is used in two different types of photodetectors: *photoconductors* and *photodiodes*. In photoconductors the photoelectric change of the conductivity is measured, while photodiodes are sources of photo-current.

10.6.1   Photoconductors

For the excitation of intrinsic photo-electrons, often a much lower energy is necessary than for the ejection of an electron out of a material. Photoconductors mostly manufactured using thin-layer technology therefore display their strength as infrared receivers.

In an *intrinsic* semiconductor, charge carriers can be generated by thermal motion or absorption of a photon. In this case the cut-off wavelength λ_g is determined by the energy of the bandgap according to Eq. (10.23). In Ge, for instance, it is $0.67\,\text{eV}$, corresponding to a cut-off wavelength of $1.85\,\mu\text{m}$.

Tab. 10.1: *Bandgap energy of selected semiconductors*

Material	E_g (eV) at $300\,\text{K}$	λ_g (μm)
CdTe	1.60	0.78
GaAs	1.42	0.88
Si	1.12	1.11
Ge	0.67	1.85
InSb	0.16	7.77

Tab. 10.2: *Activation energy in some doped semiconductors*

Material	E_A (eV) at $300\,\text{K}$	λ_A (μm)
Ge:Hg	0.088	14
Si:B	0.044	28
Ge:Cu	0.041	30
Ge:Zn	0.033	38

The spectral sensitivity can be extended to even larger wavelengths by using extrinsic (doped) semiconductors. The cut-off wavelength then increases with the activation energy E_A of the donor atoms. Ge is used particularly often since its cut-off wavelength is extended for example by Hg dopants up to the $32\,\mu\text{m}$ limit.

Sensitivity

In a photoconductor the optically induced change of the conductivity is measured. Thus roles are played not only by the rate of charge carrier generation r_L – which behaves like the response of all quantum sensors – but also by the relaxation rate τ_{rec}^{-1} – which ensures that the semiconductor returns to thermal equilibrium. For simplicity we assume the entire light power to be absorbed in the detector volume. Then the charge carrier density at constant light intensity is $n_{\text{el,ph}} = \eta P_{\text{L}} \tau_{\text{rec}} / h\nu V_{\text{D}}$.

However, the measured quantity is the conductivity σ and the current $I = A\sigma U / \ell$, respectively, flowing through a photoconductor of length ℓ with effective diameter A when there is a voltage U across over it. It depends not only on the charge carrier densities n_{el} and p_{h} but also on the mobilities μ_{el} and μ_{h} of the electrons and holes, respectively. Owing to their low mobility, the holes contribute only negligibly to the conductivity, and hence

$$\sigma \simeq en\mu_{\text{el}}. \tag{10.26}$$

By means of the photo-effect, conductivity is generated within the photoconductor. It lasts until the electron–hole pair has recombined either still in the photoconductor itself, or at the interfaces to the metallic connections. On the other hand, during the recombination time, a current flows that is determined by the mobility of the electrons. In the semiclassical Drude model the drift velocity of the electrons can be connected with the applied voltage $v_{\text{el}} = \mu_{\text{el}} U / \ell$, and also with the time $\tau_{\text{d}} = \ell / v_{\text{el}}$ that it takes an electron to drift out of the photoconductor via the metallic leads. From $I = Aen_{\text{el}} v_{\text{el}}$ the sensitivity can be calculated as

$$\mathcal{R} = \frac{\eta e}{h\nu} \frac{\tau_{\text{rec}}}{\tau_{\text{d}}}.$$

Thus a photoconductor has an intrinsic gain $G = \tau_{\text{rec}}/\tau_{\text{d}}$, which can sometimes be smaller than unity. Moreover, the gain is obtained at the expense of a reduced detector bandwidth since the recombination rate τ_{rec} determines the temporal behaviour of the photo-cell as well.

Noise properties

The conductivity that is generated by thermal motion can be suppressed by routine cooling of the detector. So strictly speaking Eq. (10.26) has photoelectric and thermal parts,

$$\sigma = e(n_{\text{ph}} + n_{\text{th}})\mu_{\text{el}}.$$

The steady state of the conductivity in a photoconductor is determined by charge carrier generation and balanced by the recombination rate, which itself is a random mechanism. The shot noise of a photoconductor is called *generation–recombination noise* and is larger by a factor of 2 compared to the photomultiplier tube or the photodiode,

$$i_{\text{GR}}^2 = 4e\bar{I} \frac{\tau_{\text{rec}}}{\tau_{\text{d}}}.$$

At wavelengths of $10\,\mu\text{m}$ and beyond, the detectivity is generally limited by the thermal radiation background. Real detectors to a large extent operate at this limit.

10.6.2   Photodiodes or photovoltaic detectors

Semiconductor photodiodes are among the most common optical detectors altogether because they are compact components and have many desirable physical properties, e.g. high sensitivity, a fast rise time and a large dynamical range. In addition, they come in numerous layouts and are straightforwardly interfaced with electronic semiconductor technology.

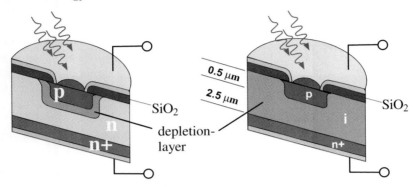

Fig. 10.8: *Layouts of Si photodiodes. Left: conventional design. Right: in the pin layout, the separation of the charge carriers is reached particularly fast.*

Their effect is based on the pn junction layer, which forms the so-called depletion layer where free charges are eliminated (see below). New electron–hole pairs are generated by absorption of light and accelerated by the internal electric field and thus cause a current flow in the test circuit. The *depletion region* acts as a nearly perfect current source, i.e. with high internal impedance.

pn and pin diodes

A depletion layer is formed close to the pn junction (Fig. 10.8). Holes in the p-doped material and electrons in the n-doped material, respectively, diffuse to the opposite side and recombine there. The holes cause a positive space-charge zone at the n side; since the electrons are in general more mobile than the holes, the corresponding negative zone is more extended on the p side. This process is finished when the electric field caused by the space charge prevents further diffusion of electrons and holes, respectively. A Si diode generates a known voltage drop of 0.7 V across the depletion layer.

The construction of an efficient photodiode has the goal to absorb as much light as possible in the barrier layer so that the electric field, which can be increased still further by an external bias voltage, rapidly separates the electron–hole pairs. In contrast to a photoconductor, then recombination can no longer occur. This process can constructively be supported by inserting an insulating layer between the n and p layers, making the detector a pin photodiode. With this, the absorbing volume is increased and additionally the capacitance of the barrier layer limiting the rise time is decreased.

Operating modes

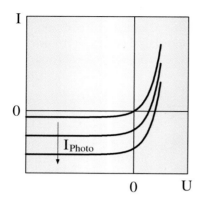

Fig. 10.9: *Family of characteristics of a photodiode.*

In Fig. 10.9 a family of electrical characteristic curves of a photodiode is presented. The diagram results by adding the negative photocurrent $-I_{\mathrm{ph}}$ to the characteristic curve with $I = I_{\mathrm{s}}(e^{eV/kT} - 1)$ for a common diode. There are three operating modes:

• **Photovoltaic mode.** When the photodiode is connected to an open circuit, then it is operated in the photovoltaic mode. Current flow is negligible ($I = 0$), and the responsivity is given in $\mathrm{V\,W^{-1}}$. This operating mode is also used in solar cells.

• **Short-circuit mode.** In the short-circuit mode, the current generated by the photo-electrons is measured and given in $\mathrm{A\,W^{-1}}$.

• **Voltage bias operating mode.** In this most common operation mode, the barrier layer is further extended by a bias voltage so that a higher quantum efficiency and shorter rise times are achieved.

10.6.3 Avalanche photodiodes

The principle of the *avalanche photodiode* (APD) has been known for a long time. However it was not possible to manufacture technically stable products until recently. In a way the APD realizes a photomultiplier based on semiconductor devices. If a very large bias voltage of several 100 V (in the reverse direction) is applied across the depletion region, then photo-electrons can be accelerated so strongly that they generate another electron–hole pair. Exactly as in the photomultiplier a large amplification of the photo-electron can be achieved by a cascade of such ionization events. Therefore also the term 'solid-state photomultiplier' is used occasionally.

The gain of APDs reaches 250 or more. Like in the usual pin Si photodiode, the photo-electrons are released in the depletion region with a correspondingly high quantum efficiency. Therefore the responsivity of APDs can exceed $100\,\mathrm{A\,W^{-1}}$.

For high light intensities, avalanche photodiodes are operated like photomultiplier tubes in current mode. The gain, however, is sufficient to operate them in the Geiger mode for photon counting, also. With the ionization not only electrons but also holes are generated. If both charge carriers are generated with the same efficiency, then the detector is 'ignited' by a first charge carrier pair and does not lose its conductivity since new electron–hole pairs are generated continuously.

In silicon the ionization coefficient for electrons is very much larger than that for holes. The current flow, however, cannot be stopped until all holes have left the depletion layer and only then can a new charge pulse be generated. In order to keep the resulting dead time as short as possible, the discharge can be passively quenched

by a current-limiting resistor. Better conditions can be provided by interrupting the
discharge current actively through suitable servo-loops.

10.7   Position and image sensors

The application of the highly integrated concepts of
semiconductor technology to photodetectors, not only
to Si, but also to other materials, is quite obvious.
Typically four photodiodes are combined on a Si substrate with relatively large area, forming a 'quadrant
detector'. This serves, for example, to determine the
position of a light beam. By means of difference amplifiers, the detection of slight motions is possible with
remarkable sensitivity. In another layout, photodiodes
are used line-wise or column-wise with 'diode arrays'
in order, for example, to measure simultaneously the
spectrum of a monochromator (see Fig. 5.7) without
mechanically actuating a grating. In a line camera a
movable mirror provides the line feed and thus allows
recording of a full two-dimensional picture.

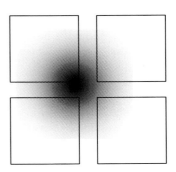

Fig. 10.10: *Quadrant detectors
for the localization of a laser
beam.*

A two-dimensional array of photo-capacitors without movable parts can be applied
for image formation. In such an array, the intensity distribution of a real image
is stored as a two-dimensional charge distribution. The technical challenge is to
'read out' the information saved in the capacitor charges on demand using electronic
devices and at the same time to convert it into a time sequence of electrical impulses
that are compatible with conventional video standards. For this purpose the concept
of CCD (charge-coupled device) sensors developed in the 1970s based on MOS
(metal–oxide–semiconductor) capacitors has gained wide acceptance, since such a
sensor exhibits particularly low noise. Only in the infrared spectral range, when the
sensors have to be cooled and the MOS capacitance decreases, do conventional pn
capacitances equipped with MOS switches have advantages.

10.7.1   Photo-capacitors

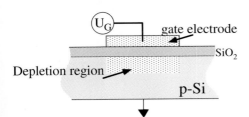

Fig. 10.11: *MOS photo-capacitor. Electrons optically generated are stored in the
depletion region.*

The charge generated by illumination in a common pn photodiode in the photovoltaic
operating mode and with an open circuit does not drain but is stored in the capacitance

of the space charge region. It operates as a potential well for the electrons released nearby, and we can call it a 'photo-capacitor'. Such devices are of particular interest for image sensors since the image information can be first saved in the photo-capacitances and then be read out serially. Of course the charge will drain eventually by thermal motion, but the storage time is from several seconds up to minutes or hours depending on the system and temperature.

The MOS capacitors have proven themselves as photo-capacitors. At the metal–oxide–semiconductor interface, which is also known as a Schottky contact, a potential is generated that serves to store photo-electrons. With MOS capacitors, large capacitance values are achieved. They prevent the stored charges from reducing the potential well, and thus the capacitor does not saturate with just a few photo-electrons or holes. A model of a MOS capacitor consisting of a metallic or polycrystalline Si gate, an SiO_2 oxide layer and p-Si is presented in Fig. 10.11. For positive gate voltage U_G a potential well for electrons is formed. Electrons ejected in the space charge region and stored in the potential well can later be released by decreasing the gate voltage. The storage time of photo-capacitors is limited by thermal relaxation and varies at room temperature from seconds up to several minutes.

10.7.2   CCD sensors

The heart of modern digital cameras is the CCD chip, which in its detector array generates a charge proportional to the intensity of the incident radiation and stores it in photo-capacitors until it gets read out by control electronics [15]. In comparison to the photographic plate, the CCD camera has the advantages of a large linear range, high quantum efficiency of 50–80% and direct generation of a voltage signal that can be digitized and processed by computer.

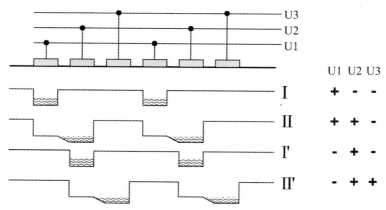

Fig. 10.12: *Three-phase operation of a CCD cell.*

The key to the success of CCD sensors is the read-out method, presented in Fig. 10.12 using the example of a three-phase system. It is organized such that, by voltage control sequences of the gate electrodes, the charge stored in a sensor or *pixel* is transferred to the adjacent capacitor. The clock frequency of this periodic sequence can be more than 20 MHz. The average of the charge loss during the transmission is below 10^{-6}. Therefore, even for many hundreds of transmission steps generally more than 99.99% of the charge content of a pixel arrives at the read-out amplifier.

An image sensor has to be read out line-wise. In order to prevent a long dead time being caused by this, and in addition to be able to accumulate more charges, the CCD sensors consist of an illuminated 'image region' and a dark 'storage region'. The formation of an image is finished by transferring all columns from the illuminated part in parallel and within 1 ms to the adjacent storage zone. While being transferred line-wise through a read-out register step by step to the read-out amplifier, already the next picture can be taken in the illuminated part.

The sensitivity of a CCD sensor is determined by the noise properties of each pixel, which on the one hand depend on the fluctuation of the thermally generated electrons, and on the other hand are mostly dominated by the so-called 'read-out noise'. This is added to the charge content of a pixel by the read-out amplifier. Since this noise contribution occurs only once per read-out process, it is often favourable to accumulate charges generated photo-electronically on the sensor for as long as possible. For this, though, only slow image sequences can be achieved. The noise properties of a CCD sensor are often given in the unit 'electrons/pixel' indicating the r.m.s. width of the dark current amplitude distribution.

The spatial resolution of a CCD sensor is determined by the size of the pixels, whose edge length today is typically 1–$25\,\mu$m. The resolution of course cannot be better than the optical image system, i.e. the camera lens. The determination of the positions of small objects is, however, sometimes possible with subpixel resolution. If the point spread function of the optical imaging system is known, it can be fitted to the distribution extending over several pixels. The centre value can then be evaluated with subpixel resolution.

10.7.3   Image intensifiers

For image amplifiers, the extremely sensitive properties of a photomultiplier based on the conversion of light into electrons are used in detectors with spatial resolution. The potential for applications of image amplifier tubes and their variants is quite high since they allow more than just the taking of pictures of extremely faint objects. The concept can be transferred to many kinds of radiation, e.g. infrared radiation or X-rays, which are not visible to the human eye and common cameras at all, but can cause localized ejection of electrons. Such devices are also called *image converters*.

In Fig. 10.13 we present two widely used concepts for optical image intensifiers. On the left is the first-generation concept, in which a picture is guided through fibre optics to a photo-cathode. The electrons emitted there are accelerated by electro-optics and projected onto a luminescent screen. Its luminescence can be observed by eye or by camera. There can be up to $150\,\text{lm}\,\text{lm}^{-1}$ of image intensification.[1]

---

[1] Here the SI unit lumen (lm) is used, which measures the light current emitted by a point source with one candela (cd) light intensity into a solid angle of one steradian (sr): $1\,\text{lm} = 1\,\text{cd}\,\text{sr}^{-1}$. Light intensity is measured in the SI unit candela (cd). At $555\,\text{nm}$ wavelength its value is $1\,\text{cd} = (1/683)\,\text{W}\,\text{sr}^{-1}$; at other wavelengths, it is referred to the spectrum of a blackbody radiator operated

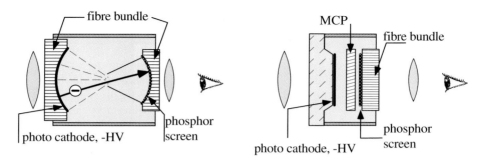

Fig. 10.13: *Concepts for image amplifiers of first and second generation.*

On the right, a model of the so-called second generation is shown. Here by means of a channel plate (MCP, see p. 308) a gain of 10^4 and more can be achieved. The spatial resolution of the incident optical image is slightly decreased by the spread of the electron bunches emitted from the MCP.

Image intensifiers not only allow the observation of very faint signals. The high voltage necessary at the channel plate for amplification can be switched on and off on a nanosecond scale and this makes it possible to realize cameras with extremely high shutter speed.

---

at the melting point of platinum.

11 Laser spectroscopy

In Chapter 6 on light and matter, we theoretically investigated the occupation number and the polarizability of an ensemble of atomic or other microscopic particles. In experiments, these quantities though are not observed directly but through their effect on certain physical properties of a sample. Here, we restrict ourselves to all optical methods, such as the fluorescence of an excited sample or the absorption and dispersion of a probe beam. There are also numerous alternative methods of detection, e.g. the effect on acoustic or electrical properties of the sample. For a wider overview over the extended field of laser spectroscopy, we refer the reader to [24], for example.

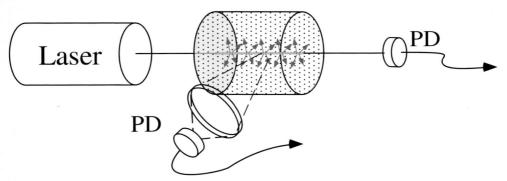

Fig. 11.1: *Laser spectroscopy. The spectral properties of a sample can be detected by laser-induced fluorescence (LIF) or by absorption. For the detection of dispersive properties, interferometric experiments are necessary.*

11.1 Laser-induced fluorescence (LIF)

Fluorescence is caused by spontaneous emission. We observe it, for example, when a laser beam passes through a gas cell. It corresponds to radiation damping and can only occur when an atom is in an excited state. In the Bloch equations (6.28) we have taken the fluorescence into account phenomenologically with the decay rates γ, γ'. A single particle in the excited state emits energy $\hbar\omega$ during its lifetime γ, and an intense resonant laser beam keeps half of all particles excited on average. We can also express the saturated fluorescence power P_{sat} by the saturation intensity I_0 according

Optics, Light and Laser. Dieter Meschede
Copyright © 2004 Wiley-VCH Verlag GmbH & Co. KGaA
ISBN: 3-527-40364-7

to Eq. (6.34):

$$P_{\text{sat}} = \hbar\omega\frac{\gamma}{2} = \frac{\gamma/2}{\gamma'}\sigma_Q I_0.$$

More generally, the intensity of the fluorescence to be observed is proportional to the excitation probability $(w + 1)/2$, where w is the z-component of the Bloch vector (see Section 6.2.3), and to the particle density N/V. In addition, we have to take the experimental setup (losses, solid angle of observation, ...) into account with a geometry factor G. The observed fluorescence intensity may then be defined by $I_{\text{fl}} = GP_{\text{sat}}$,

$$I_{\text{fl}} = G\frac{N}{V}\hbar\omega\gamma\frac{1}{2}(1 + w) = G\frac{N}{V}P_{\text{sat}}\frac{s}{1 + s}.$$

The saturation parameter s is proportional to the intensity of the exciting laser field I according to Eq. (6.31). For large excitation field intensity ($s \gg 1$) it can be found immediately that

$$I_{\text{fl}} \simeq G\frac{N}{V}P_{\text{sat}}.$$

In the limiting case of low excitation ($s \ll 1$) laser-induced fluorescence (LIF) allows linear mapping of selected properties of a sample such as particle density, damping rates γ, γ', and so on. The spectral dependence of the low intensity resonance line at ω_0 is Lorentz-shaped in the stationary case,

$$I_{\text{fl}}(\omega) \simeq G\frac{N}{V}P_{\text{sat}}s = G\frac{N}{V}\frac{\gamma\gamma'}{2}\frac{I}{(\omega - \omega_0)^2 + \gamma'^2},$$

and a fluorescence profile like in Fig. 6.2 is obtained. With laser-induced fluorescence, for example, spatially resolved density measurements of known atomic or molecular gases can be carried out.

11.2   Absorption and dispersion

Like fluorescence, linear absorption and dispersion of the driving laser field occurs at low saturation intensities only. Therefore we determine the more general absorption coefficient and the real part of the refractive index according to Eq. (6.22),

$$\begin{aligned}
\alpha(\omega) &= -\frac{\omega}{2I(z)}\text{Im}\{\mathcal{E}(z)\mathcal{P}^*(z)\} = \frac{N}{V}\frac{\omega}{2I(z)}d_{\text{eg}}\mathcal{E}_0 v_{\text{st}}, \\
n'(\omega) - 1 &= \frac{c}{2I(z)}\text{Re}\{\mathcal{E}(z)\mathcal{P}^*(z)\} = \frac{N}{V}\frac{c}{2I(z)}d_{\text{eg}}\mathcal{E}_0 u_{\text{st}}.
\end{aligned} \tag{11.1}$$

When we insert the dipole components ($d = d_{\text{eg}}(u_{\text{st}} + iv_{\text{st}})$) from eqs. (6.37), we again obtain the relations

$$\begin{aligned}
\alpha(\omega) &= -\frac{N}{V}\frac{\gamma}{2\gamma'}\frac{w_0\sigma_Q}{1 + I/I_0 + [(\omega - \omega_0)/\gamma']^2}, \\
n'(\omega) - 1 &= -\frac{N}{V}\frac{\gamma}{2\gamma'}\frac{\lambda}{2\pi}\frac{w_0\sigma_Q(\omega - \omega_0)/\gamma'}{1 + I/I_0 + [(\omega - \omega_0)/\gamma']^2},
\end{aligned} \tag{11.2}$$

by also accounting for Eqs. (6.37). From these relations the limiting case of small intensities can be reduced again to the classical case (6.19) without any further difficulties ($I/I_0 \ll 1$ and $w_0 = -1$). There, the absorption coefficient and refraction coefficient depend only on atomic properties (decay rates γ, γ', detuning $\delta = \omega - \omega_0$, particle density N/V) and not on the incident intensity. Conversely, these physical quantities can be determined using absorption spectroscopy. Since the determination of the refractive index generally requires an interferometric method, and thus considerable instrumental effort, the absorption measurement is the preferred method.

11.2.1   Saturated absorption

For increasing intensity ($I/I_0 \simeq 1$) the 'saturation' of a resonance plays a more and more important role since the absorption coefficient becomes nonlinear: it itself depends on the intensity. For the sake of clarity we introduce the resonant unsaturated absorption coefficient $\alpha_0 = -\sigma_Q w_0 (N/V) \cdot (\gamma/2\gamma')$ ($= \sigma_Q (N/V) \cdot (\gamma/2\gamma')$ for optical frequencies), and with the new linewidth $\Delta\omega = 2\gamma_{\text{sat}}$ according to Eq. (6.35) we write

$$\alpha(\omega) = \alpha_0 \frac{\gamma'^2}{(\omega - \omega_0)^2 + \gamma'^2(1 + I/I_0)} = \alpha_0 \frac{\gamma'^2}{(\omega - \omega_0)^2 + \gamma_{\text{sat}}^2}. \tag{11.3}$$

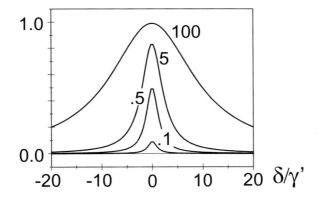

Fig. 11.2: *Saturation of resonance lines. Normalized fluorescence intensity as a function of normalized detuning δ/γ'. The parameter gives the incident laser power normalized to the saturation intensity, I/I_0. The maximum fluorescence intensity occurs at even occupation of the atomic levels.*

So, in spite of the saturation at large intensity $I \geq I_0$, the Lorentz shape of the resonance line is preserved, though it becomes wider. It is also straightforward to show that on resonance ($\omega = \omega_0$) the intensity no longer decreases exponentially following Beer's law but decreases linearly for large I/I_0 according to

$$\frac{dI}{dz} = -\alpha(I)I \simeq -\alpha_0 I_0.$$

11.3   The width of spectral lines

The observation of fluorescence and absorption spectra are among the simplest and thus most common methods of spectroscopy. Physical information is contained in the centre frequency value of a line as well as in its shape and width. As a measure for the width (Fig. 11.3), usually the full width at half-maximum is used, i.e. the full frequency width between the values for which the resonance line reaches the half-maximum value.[1] For intensities below the saturation value, $I/I_0 \ll 1$, the transverse relaxation rate γ' can be inferred from Eqs. (6.31) and (11.2),

FWHM $\Delta\omega = 2\pi\Delta\nu = 2\gamma'.$

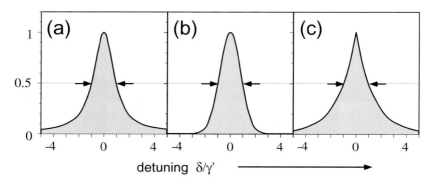

Fig. 11.3: *Important spectral line shapes: (a) Lorentz line, (b) Gaussian profile, and (c) profile of time-of-flight broadening are presented for identical half-widths.*

For a free atom, which can release its energy only by radiative decay, we have $\gamma' = \gamma/2$ and thus

$$\Delta\omega = 2\pi\Delta\nu = \gamma. \tag{11.4}$$

In dilute atomic or molecular samples, the Q-value of the resonance, i.e. the ratio between resonance frequency and FWHM, can easily assume very large values of 10^6 and more for optical frequencies of 10^{14}–10^{15} Hz,

$$Q = \nu/\Delta\nu.$$

It is obvious that, for decreasing linewidth $\Delta\nu$ of a spectral line, the Q-value and thus the 'definition' of the centre wavelength or frequency of a resonance line increases. The precise experimental preparation and measurement of such 'sharp' resonances is a goal for spectroscopists. This goal requires a deep understanding of the physical mechanisms determining the position of a line, its width and shape. Usually the natural linewidth caused by the spontaneous decay of excited states is considered to be the lower limit. It has been known, though, for a long time that this decay rate can be modified by properties of the environment. For instance, the results of

---

[1] Often the short forms FWHM and HWHM for full width and half-width at half-maximum, respectively, are used.

measurements are systematically influenced in the vicinity of conductive or reflecting walls (see excursion on p. 187, [71]).

Let us now present the most important limiting cases only; an extensive microscopic theory would go far beyond the scope of this chapter. Also the interaction of different broadening mechanisms is often complex, has to be described by mathematically elaborate convolutions and so is neglected here.

11.3.1 Natural width, homogeneous linewidth

The dream of the precision spectroscopist is a motionless particle in free space [23] whose resonance linewidth is limited only by the finite lifetime τ of an excited state according to Eq. (11.4). It is called the 'natural linewidth' $\Delta\nu = \Delta\omega/2\pi = \gamma_{\text{nat}}/2\pi$ and is identical with the Einstein A coefficient of the spontaneous decay rate,

$$\gamma_{\text{nat}} = A_{\text{Einstein}} = \frac{1}{\tau}.$$

For an estimate of the natural width of typical atomic resonance lines, a characteristic dipole can be estimated with the Bohr radius, $d_{\text{eg}} = er_{\text{eg}} = ea_0$. For a red atomic resonance line ($\lambda = 600\,\text{nm}$) we find from Eq. (6.41):

$$A_{\text{Einstein}} \simeq 10^8\,\text{s}^{-1}.$$

The resonance frequency of a free undisturbed particle is still shifted by the Doppler effect ($\Delta\omega = kv = 2\pi v/\lambda$; see below), which we discuss in the next section. However, for a long time it has been possible to prepare almost motionless atoms and ions routinely in atom and ion traps using the method of *laser cooling*, see section 11.6. Since the motion-induced frequency shift is caused only by the component of motion in the direction of the exciting or emitting light, the natural linewidth of an atomic or molecular resonance can be observed also with atomic beams.

The natural linewidth is identical for all particles of an ensemble. Such line broadening is called 'homogeneous'.

11.3.2 Doppler broadening, inhomogeneous linewidth

During the emission of a photon, not only the energy difference between the internal excitation states of the atom is carried away but additionally the momentum $\hbar\mathbf{k}$. For low velocities ($v/c \ll 1$) we can take the difference between the resonance frequency in the laboratory frame (ω_{lab}) and in the rest frame ($\omega_{\text{rest}} = (E-E')/\hbar$) from momentum and energy conservation,

$$m\mathbf{v}' + \hbar\mathbf{k} = m\mathbf{v},$$
$$E' + \tfrac{1}{2}mv'^2 + \hbar\omega_{\text{lab}} = E + \tfrac{1}{2}mv^2.$$

In nearly all cases the atomic momentum is much larger than the recoil experienced by the emission process, $mv/\hbar k \gg 1$. Thus we can neglect the term $\hbar^2 k^2/mc^2$ in

$$\frac{m}{2}v^2 = \frac{m}{2}v'^2 + \frac{\hbar^2 k^2}{2m} + \hbar\mathbf{k}\mathbf{v} \simeq \frac{m}{2}v'^2 + \hbar\mathbf{k}\mathbf{v},$$

and arrive at the linear Doppler shift:

$$\omega_{\text{lab}} = \omega_{\text{rest}} + \mathbf{k}\mathbf{v}. \tag{11.5}$$

The direction within the laboratory frame ($\mathbf{k}$) is determined either by the observer (in emission) or by the exciting laser beam (in absorption). The radiation frequency of a source appears to be blue-shifted towards shorter wavelengths if it travels towards the observer, and red-shifted towards longer wavelengths if it moves away.

In a gas the molecular velocities are distributed according to the Maxwell–Boltzmann law. The probability $f(v_z)$ of finding a particle at temperature T with velocity component v in an interval dv_z is

$$f_{\text{D}}(v_z)dv_z = \frac{1}{\sqrt{\pi}\,v_{\text{mp}}}\, e^{-(v_z/v_{\text{mp}})^2}\, dv_z, \tag{11.6}$$

where $\int_{-\infty}^{\infty} dv_z\, f_{\text{D}}(v_z) = 1$. The most probable velocity is (k_{B} is Boltzmann constant, T is absolute temperature)

$$v_{\text{mp}} = \sqrt{2k_{\text{B}}T/m}.$$

For common temperatures the velocities of the molecular parts of a gas generally are between 100 and 1000 m s^{-1} so that typical shifts of $kv/\omega = v/c \simeq 10^{-6}$–$10^{-5}$, or some 100 to 1000 MHz are expected. The natural linewidth of atomic or molecular resonance transitions is in general much smaller and therefore masked by the Doppler shift. For this reason the methods of *Doppler-free spectroscopy* (Section 11.4) have been an important topic of research for many years.

If the emission of the particles is otherwise undisturbed, the spectral line shape and width of the absorption line of the gas can be obtained from the superposition of all contributing undisturbed absorption profiles according to Eq. (11.3),

$$\alpha_{\text{D}}(\omega) = \int_{-\infty}^{\infty} dv_z\, f_{\text{D}}(v_z)\alpha(\omega + kv_z).$$

If $\alpha(\omega)$ has Lorentz shape, the line profile α_{D} described by this mathematical convolution is called a *Gauss–Voigt profile*. At room temperature in many gases the decay rate γ of an optical transition is much smaller than the typical Doppler shift kv_{mp}. Then the distribution function $f_{\text{D}}(v_z)$ virtually does not change in that range where $\alpha(\omega + kv_z)$ differs significantly from zero. It can be replaced by its value at $v_z = (\omega - \omega_0)/k$ and pulled out of the integral. The integration over the remaining Lorentz profile results in a constant factor,

$$\alpha_{\text{D}}(\omega) = \alpha_0 f_{\text{D}}\left(\frac{\omega - \omega_0}{k}\right)\frac{\pi\gamma'}{k\sqrt{1 + I/I_0}},$$

and with $\sqrt{\pi\ln 2} = 2.18$ we arrive eventually at the Gaussian profile

$$\alpha_{\text{D}}(\omega) = \frac{2.18\alpha_0}{\sqrt{1 + I/I_0}}\frac{\gamma'}{\Delta\omega_{\text{D}}}\exp\left[-\ln 2\left(\frac{\omega - \omega_0}{\Delta\omega_{\text{D}}/2}\right)\right]. \tag{11.7}$$

Here we have already introduced the Doppler FWHM or Doppler width

$$\Delta\omega_{\text{D}} = \omega_0\sqrt{\frac{8k_{\text{B}}T\ln 2}{mc^2}}.$$

The absorption coefficient is reduced by approximately the factor $\gamma'/\Delta\omega_D$ since the line intensity is now spread over a very much bigger spectral range. It is useful to express the Doppler width in units of the dimensionless atomic mass number M and the absolute temperature T in *Kelvin*,

$$\Delta\nu_D = \Delta\omega_D/2\pi = 7.16 \times 10^{-7}\sqrt{T/M}\,\nu_{\rm rest}, \tag{11.8}$$

where $\nu_{\rm rest}$ is the resonance frequency of the particle at rest.

The Doppler broadening is an example of an 'inhomogeneous' linewidth. In contrast to the homogeneous line, each particle contributes to the absorption of fluorescence line with a different spectrum depending on its velocity.

11.3.3   Pressure broadening

In a gas mixture atoms and molecules continuously experience collisions with neighbouring particles that disturb the motion of the orbital electrons for a short time. During the collision the frequency of the emission is slightly changed compared to the undisturbed case. For neutral atoms or molecules the interaction can be described, for example, by a van der Waals interaction causing a mutual polarization of the collision partners. In a plasma the interaction of the charged particles is much stronger.

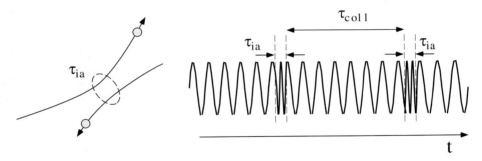

Fig. 11.4: *Disturbance of radiative processes by collisions in a neutral gas. The duration of the collisions is very short compared to the collision rate ($\tau_{\rm coll}^{-1}$) and to the lifetime of the excited state. The influence of collisions can be modelled through random phase jumps of an otherwise undisturbed wave.*

It is useful to consider first the relevant time scales that determine collisional processes and are compiled in Tab. 11.1.

The interaction between neutral particles is generally *short-range*, i.e. it is significant only over a short distance of the order of the diameter of the atom or the molecule. The duration of the *interaction time* $\tau_{\rm ia}$ can therefore be estimated from the typical transit time across an atomic diameter. For thermal velocities according to this some 10–1000 oscillation cycles occur during the collision. The mean time interval between collisions (or the inverse collision rate) $\tau_{\rm coll}$ can be determined from the collision cross-section σ_A and the mean velocity v following the known formula $\tau_{\rm coll} = n\sigma_A v$. It is much larger than the interaction time $\tau_{\rm ia}$ even under atmospheric conditions, and thus

Tab. 11.1: *Relevant times for collisional broadening.*

Process	Formula	Conditions	Duration
Optical cycle	$\tau_{\mathrm{opt}} = 1/\nu_{\mathrm{opt}}$		10^{-14}–10^{-15} s
Interaction time	$\tau_{\mathrm{ia}} = d_{\mathrm{atom}}/v_{\mathrm{therm}}$	$T = 300\,\mathrm{K}$	10^{-12}–10^{-13} s
Time between collisions	$\tau_{\mathrm{coll}} = n\sigma_A v_{\mathrm{therm}}$	$T = 300\,\mathrm{K}$, $n = 10^{19}\,\mathrm{cm}^{-3}$	10^{-7}–10^{-9} s
Natural lifetime	$\tau = A_{\mathrm{Einstein}}^{-1}$		10^{-8} s

$d_{\mathrm{atom}} = 2\,\text{Å}, \ \sigma_A = \pi d_{\mathrm{atom}}^2/4.$

electronic motion is rarely disturbed by the collisions. In a simple model all details of the molecular interaction are therefore negligible and the effect of the collision can be reduced to an effective random phase shift of the otherwise undisturbed optical oscillation.

Let us first consider the intensity spectrum $\delta I(\omega)$ of a damped harmonic wavetrain that starts at t_0 and is simply aborted after a randomly chosen time τ:

$$\delta I = I_0 \left| \int_t^{t_0+\tau} e^{[-i(\omega_0-\omega)-\gamma']t}\, dt \right|^2 = I_0\, e^{-2\gamma' t_0} \left| \frac{e^{[i(\omega_0-\omega)-\gamma']\tau}-1}{i(\omega_0-\omega)-\gamma'} \right|^2.$$

The dependence on the start time t_0 can be eliminated immediately by integration, $I(\omega) = 2\gamma' \int \delta I(\omega, t_0)\, dt_0$. The phase jumps (and thus the periods of the undisturbed radiation times) are distributed randomly and occur with a mean rate $\gamma_{\mathrm{coll}} = \tau_{\mathrm{coll}}^{-1}$. Then we can calculate the shape of the collision-broadened spectral line with the probability distribution $p(\tau) = e^{-\tau/\tau_{\mathrm{coll}}}/\tau_{\mathrm{coll}}$,

$$I(\omega) = I_0 \int_0^\infty \left| \frac{e^{[i(\omega_0-\omega)-\gamma']\tau}-1}{i(\omega_0-\omega)-\gamma'} \right|^2 \frac{e^{-\tau/\tau_{\mathrm{coll}}}}{\tau_{\mathrm{coll}}}\, d\tau.$$

The result is

$$I(\omega) = \frac{I_0}{\pi} \frac{\gamma' + \gamma_{\mathrm{coll}}}{(\omega_0-\omega)^2 + (\gamma'+\gamma_{\mathrm{coll}})^2}.$$

The Lorentzian line shape is maintained; the effective collisional broadening rate γ_{coll} though has to be added to the transverse relaxation rate γ'. Since all particles of an ensemble are subject to the same distribution of collisions, this line broadening is homogeneous like the natural line shape.

Spectral lines are affected not only by pressure broadening but also by a pressure shift of the centre of mass of a line. With increasing pressure, the number of collisions between the particles of a gas increases. Naively we can imagine that the volume available for the binding orbital electrons is reduced, and in quantum mechanical systems volume reduction is always associated with an increase of the binding energy. The pressure shift therefore generally causes a shift to blue frequencies.

11.3.4   Time-of-flight (TOF) broadening

Light–matter interaction of atoms and molecules in a gas or in an atomic beam is mostly restricted to a finite period. For example, for $v = 500\,\mathrm{m\,s^{-1}}$, an atom needs $\tau_{\mathrm{tr}} = 2\,\mu\mathrm{s}$ to pass a beam of diameter $d = 1\,\mathrm{mm}$. However, the relaxation of many optical transitions occurs on the nanosecond scale, during which time an atom travels a few micrometres at most. The stationary solutions for (6.28) are a good approximation in these cases. However, in focused laser beams or for slowly decaying transitions equilibrium is never reached, and the line shape is dominated by transient interaction corresponding to the finite time of flight τ_{tr}. Slow or long-lived transitions are of particular interest since the corresponding very sharp resonance lines are excellent objects for precision measurements at low intensities. The two-photon spectroscopy of the hydrogen atom (see the example on p. 331) is an exceptionally beautiful example of this.

Slow transitions which are only briefly subjected to a weak light field have $\Omega_R < \gamma'$, τ_{tr}^{-1}, and we can assume that the population of the ground state is virtually unchanged $(w(t) \simeq w(t = 0) = -1)$. Let us consider atoms or molecules crossing a laser beam which is assumed to have Gaussian envelope with $1/e^2$ radius w_0. The Rabi frequency $\Omega_R(z) = (d_{\mathrm{eg}}\mathcal{E}_0/\hbar)\exp[-(z/w_0)^2]/\sqrt{\pi}$ is now a function of position, and we have to solve the first optical Bloch equation of (6.29),

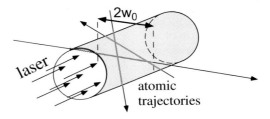

Fig. 11.5: *Atomic trajectories crossing a laser beam with Gaussian profile.*

$$\frac{d}{dt}\rho_{\mathrm{eg}} = v\frac{d}{dz}\rho_{\mathrm{eg}} = -(\gamma' + i\delta)\rho_{\mathrm{eg}} + i\Omega_R(z).$$

We calculate the mean absorption coefficient of a single dipole with velocity v from eq. (6.22),

$$\langle \alpha(v) \rangle = \frac{\omega}{2I}\frac{1}{w_0}\int_{-\infty}^{\infty} dz\,\mathfrak{Im}\{d_{\mathrm{eg}}\mathcal{E}(z)\rho_{\mathrm{eg}}(z,v)\}. \tag{11.9}$$

Before the particle enters the light field there is no dipole moment, thus $\rho_{\mathrm{eg}}(z=-\infty) = 0$, and the general solution can be given by

$$\rho_{\mathrm{eg}}(z,v) = i\,\frac{d_{\mathrm{eg}}\mathcal{E}_0}{\hbar}\,e^{-(\gamma'+i\delta)z/v}\int_{-\infty}^{z}\frac{dz'}{v}\,e^{(\gamma'+i\delta)z'/v}\,e^{-(z'/w_0)^2}.$$

If the typical time of flight is small compared to the typical decay time $\gamma' \ll \tau_{\mathrm{tr}}^{-1}$, we can neglect γ'. By inserting $\rho_{\mathrm{eg}}(z,v)$ then the integral in Eq. (11.9) can be evaluated analytically,

$$\langle \alpha(v,\delta) \rangle = \frac{\omega}{2I}\frac{|d_{\mathrm{eg}}\mathcal{E}_0|^2}{\hbar}\frac{w_0}{v}\,e^{-(\delta w_0/2v)^2}.$$

In order to determine the absorption coefficient of a gaseous sample with a cylindrical laser beam, we would still have to average over all possible trajectories, but this results only in a modified or effective beam cross-section whose details we skip over here.

The summation of the velocity distribution in a two-dimensional gas ($f(v)\,dv = (v/\overline{v}^2)\exp[-(v/\overline{v})^2]\,dv$, as the velocity component along the direction of the laser beam does not play any role here) results in

$$\alpha(\delta) = \int_0^\infty dv\, f(v)\langle\alpha(v)\rangle = \alpha_0\, e^{-|\delta w_0/\overline{v}|} = \alpha_0\, e^{-|\delta\tau_{\rm tr}|},$$

whose form has already been presented in Fig. 11.3(c). The effective width of this line is determined by $\tau_{\rm tr} = w_0/\overline{v}$.

11.4 Doppler-free spectroscopy

The linewidth of atomic and molecular resonances at room temperature is usually dominated by the Doppler effect. The intrinsic and physically attractive properties of an isolated particle are revealed only at velocity $v = 0$. Laser spectroscopy offers several nonlinear methods where light–matter interaction is effective for selected velocity classes only. The result is called 'Doppler-free' spectroscopy.

11.4.1 Spectroscopy with molecular beams

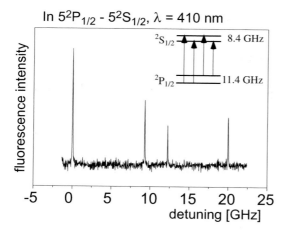

As soon as tunable lasers became available in the 1970s, high-resolution optical spectra were obtained with molecular or atomic beams. In such an apparatus, where the transverse velocities of molecules are reduced to near zero by geometric collimation, resolutions of $\Delta\nu/\nu \simeq 10^8$ and better are routinely achieved.

The example in Fig. 11.6 was recorded with an indium atomic beam. The transverse velocities were limited to $v \le 5\,{\rm m\,s^{-1}}$ by appropriate apertures so that the residual Doppler effect $kv \le 10\,{\rm MHz}$ was significantly smaller than the natural linewidth of 25 MHz. Blue

Fig. 11.6: *Fluorescence spectrum of an an indium atomic beam obtained with a blue diode laser.*

diode lasers (see excursion on p. 263) have been used for this purpose for a short time only – before the year 2000 they were still hardly imaginable equipment for such experiments.

11.4.2   Saturation spectroscopy

By a resonant laser light field, atoms are promoted to the excited state, and as a result the occupation number difference is modified. In an inhomogeneously broadened spectral line profile such as the Doppler profile, then for not too large intensities a spectral hole is 'burnt' into the velocity distribution, which is qualitatively presented in Fig. 11.7.

Generally the distribution modified by a laser beam can now be 'queried' spectroscopically by means of another light field. It is even simpler to excite the sample with two opposite laser beams.

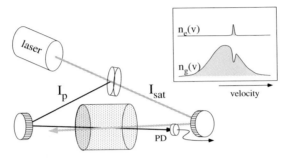

Fig. 11.7: *Principle of saturation spectroscopy. Inset: A laser beam with frequency ω is used to 'burn' a spectral hole into the ground-state velocity distribution at $kv = \omega_0 - \omega$; simultaneously, an excited state population with a narrow velocity distribution is generated.*

Fig. 11.7 presents one of the simplest possible arrangements for the so-called 'saturation spectroscopy'. In order to simplify the theoretical description, we assume that the intensities of saturation (I_{sat}) and probe beam (I_{p}) are small in comparison with the saturation intensity (Eq. (6.34)), $I_{\mathrm{sat,p}}/I_0 \ll 1$, and do not directly influence each other. Let us calculate the absorption coefficient according to Eq. (11.2) by again using the Maxwell–Gauss velocity distribution $f_{\mathrm{D}}(v)$ from Eq. (11.6) and carrying out the Doppler integration:

$$\alpha_{\mathrm{p}}(\delta) = \frac{\omega}{2I} \int_{-\infty}^{\infty} dv\, f_{\mathrm{D}}(v) d_{\mathrm{eg}} \mathcal{E} v_{\mathrm{st}}^{+}(\delta, v).$$

We now distinguish the forward ('+') and the backward ('−') travelling laser beams and from Eq. (6.36) we use

$$v_{\mathrm{st}}^{+}(\delta, v) = -\gamma' d_{\mathrm{eg}} \mathcal{E} w_{\mathrm{st}}^{-} / \{1 + [(\delta - kv)/\gamma']^2\},$$

but following (6.30) we insert $w_{\mathrm{st}}^{-} = -1/(1 + s^{-}) \simeq -(1 - s^{-})$ in order to account for the modification of the occupation number by the second counter-propagating laser beam which has the saturation parameter

$$s^{-} = (I_{\mathrm{sat}}/I_0)/\{1 + [(\delta + kv)/\gamma']^2\}.$$

Since the Doppler profile varies only slowly compared to the narrow Lorentzian contributions of each velocity class, at the detuning $\delta = \omega_0 - kv$ we can again pull $f_{\mathrm{D}}(v)$

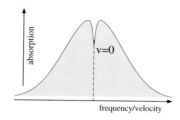

Fig. 11.8: *Doppler profile with Lorentz-shaped saturation resonance. The Doppler-free line leads to an increased transparency.*

out of the integral:

$$\alpha_{\mathrm{p}}(\delta) = \alpha_0 f_{\mathrm{D}}(\delta/k) \left(1 - \frac{1}{\pi} \frac{I_{\mathrm{sat}}}{I_0} \int_{-\infty}^{\infty} dv \, \frac{\gamma'^2}{\gamma'^2 + (kv - \delta)^2} \frac{\gamma'^2}{\gamma'^2 + (kv + \delta)^2} \right).$$

The evaluation of the integral [63] again results in a Lorentzian curve which, due to our assumption of a very low saturation ($s^{\pm} \ll 1$), has the natural linewidth $2\gamma'$:

$$\alpha_{\mathrm{p}}(\delta) = \alpha_0 f_{\mathrm{D}}(\delta/k) \left(1 - \frac{I_{\mathrm{sat}}}{I_0} \frac{\gamma'^2}{\gamma'^2 + \delta^2} \right).$$

The saturation resonance occurs exactly at the velocity class with $v = 0$. A more complete calculation shows the width corresponding to the width saturated by both laser fields according to Eq. (6.35) [63],

$$\gamma_{\mathrm{sat}} = \gamma'[1 + (I_{\mathrm{sat}} + I_{\mathrm{p}})/I_0].$$

The concept of saturation spectroscopy explains the occurrence of spectral holes in the Doppler profile (or in other inhomogeneously broadened spectral lines). In realistic experiments, though, it is influenced by further phenomena such as, for example, optical pumping or magneto-optical effects, all of which are collected a bit less precisely under the term 'saturation spectroscopy'.

A simple experiment, though complex in its interpretation, can be carried out with diode lasers and a caesium- or rubidium-vapour cell.. Their vapour pressure at room temperature already leads to absorption lengths of only a few centimetres. In Fig. 11.9 characteristic absorption lines are presented together with an energy diagram of the caesium D2 line at 852.1 nm.

From the $^2S_{1/2}$, F=3 hyperfine state three transitions with different frequencies to $^2P_{3/2}$, F'=2,3,4 are available. From our simple analysis we thus expect three line-shaped incursions in the absorption, but we observe six instead! And not only this, if the magnetic field is manipulated – in the upper spectra the geomagnetic field of 0.5 G is reduced to below 0.01 G by means of compensating coils – then even a reversal of selected lines can be observed. The reasons for this complex behaviour are explained in detail in [93] and can only be indicated here:

1. *Number of lines.* For velocities $v \neq 0$ two different excited states can be coupled at the same time if the frequency differences are compensated by the Doppler

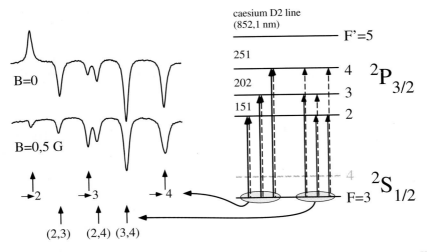

Fig. 11.9: *Saturation spectrum of a caesium-vapour cell at the 852.1 nm D2 line. Here the $F = 3 \rightarrow F = 2, 3, 4$ lines of the D2 line are presented. The second hyperfine transition from the ground state $(F = 3 \rightarrow F = 3, 4, 5)$ is at a distance of 9.2 GHz and cannot be seen here. The transition $F = 3 \rightarrow F = 5$ is forbidden according to dipole selection rules $(\Delta F = 0, \pm 1)$. The separations of the hyperfine structure levels in the excited state are given in mega-Hertz.*

effect. They also cause velocity-dependent population redistribution and lead to additional resonances called *cross-over* lines. In Fig. 11.9 three of those cases are presented for example at $\omega = (\omega_{F=3 \rightarrow F=4} + \omega_{F=3 \rightarrow F=3})/2$. Here they are particularly prominent since one of the laser fields can empty one of the two lower hyperfine levels $(F = 3)$ in favour of the other one $(F = 4)$ by optical 'depopulation pumping', which effectively removes these absorbers from the other light beam.

2. *Line reversal.* In simple laboratory setups there is no care taken about compensating the geomagnetic field of 0.5 G. Then the lower form of the spectrum in Fig. 11.9 is observed. The geomagnetic field, which does not have any well-defined direction relative to the laser polarization, is too small to split the lines visibly. But atoms are microscopic gyromagnets, and they can rapidly change their orientation by precession, and therefore all of them without exception can be excited by the light field. The m quantum number is in fact not a 'good' quantum number in the geomagnetic field.

If this precession is suppressed, atoms can be trapped in 'dark states' due to optical pumping, and thus no longer participate in the absorption process and increase the transparency. But also the opposite effect occurs if they are pumped back to absorbing sub-states with the right choice of frequencies or polarizations (in Fig. 11.9: orthogonal linear polarizations of pump and probe beam induce repopulation pumping) and by this increase the absorption. A detailed understanding here requires detailed knowledge of the level structure.

11.4.3   Two-photon spectroscopy

In the interaction of light and matter usually electric dipole transitions are of interest because their relative strength dominates all other types. We understand these processes as absorption or emission of a photon, without actually having defined the term 'photon' [84][2] more precisely at all.

Besides the dipole interaction, also higher multipole transitions or multiphoton processes occur. The latter are nonlinear in the intensities of the participating light fields. A simple and illustrative example is two-photon spectroscopy. For this in an atom or a molecule a polarization $P_{2ph} \propto E_1(\omega_1)E_2(\omega_2)$ is induced causing absorption of radiation. Two-photon transitions follow different selection rules regarding the participating initial and final states – for example $\Delta\ell = 0, \pm 2$ has to be fulfilled for the angular-momentum quantum number. Furthermore the calculation of the transition probabilities may raise problems where we expect ad hoc from second-order perturbation theory of quantum physics that matrix elements have to have the form [94]

$$M_{if} = \sum_s \left(\frac{\langle i|dE_1|s\rangle\langle s|dE_2|f\rangle}{E_i - E_s - \hbar\omega_1} + \frac{\langle i|dE_2|s\rangle\langle s|dE_1|f\rangle}{E_i - E_s - \hbar\omega_2} \right).$$

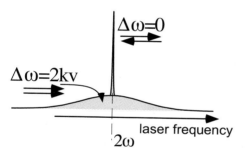

Fig. 11.10:   *Two-photon spectroscopy: Doppler background and Doppler-free resonance line.*

Transition rates are proportional to $|M_{if}|^2$, and the square value will also be proportional to the product $I_1 I_2$ of both participating fields. A more detailed calculation shows that, as in the one-photon process, a Lorentz line with width $2\gamma' = 2/T_2$ is obtained which in the case of free atoms is identical with the natural linewidth. A simplified model for anharmonic oscillators as well conveys an impression about the origin of the two-photon absorption (Section 12.1).

Like in saturation spectroscopy, two-photon spectroscopy allows the nonlinear generation of signals at velocity $v = 0$. Here the absorption has to occur from two exactly counter-propagating laser beams with identical frequency since that way the linear Doppler shift is just compensated:

$$\begin{aligned} (E_1 - E_2)/\hbar &= \omega_1 + \mathbf{kv} + \omega_2 - \mathbf{kv} \\ &= \omega_1 + \omega_2. \end{aligned}$$

As a result Doppler-free spectra are obtained whose linewidths are limited by the natural lifetime or the time of flight for very long-lived states (see Section 11.3.4). In contrast to saturation spectroscopy, though, not only one selected velocity class at $v = 0$ contributes to the signal with width $\Delta v = \gamma/k$ but all velocity classes! The

---

[2]The term 'photon' was introduced by Gilbert N. Lewis, 1926, *Nature,* **118**, 874: '... I therefore take the liberty of proposing for this hypothetical new atom, which is not light, but plays an essential part in every process of radiation, the name *photon.*'

total strength of the Doppler-free resonance therefore is as large as that of the Doppler-broadened one and can be very easily separated from it (Fig. 11.8).

Example: The mother of all atoms – two-photon spectroscopy of the hydrogen atom

The hydrogen atom is an atom of outstanding interest for spectroscopists. In contrast to all other systems, it is a two-body system and allows direct comparison with theoretical predictions, especially of quantum electrodynamics.[3] Its energy levels are principally determined just by the Rydberg constant, which as a result of the two-photon spectroscopy is today the most exactly measured physical constant of all.

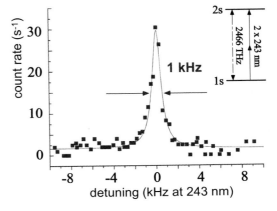

Fig. 11.11: *Two-photon resonance of the 1s–2s transition of atomic hydrogen. See text. By permission of T. W. Hänsch [49].*

The most interesting transition wavelength for precision measurements is the 1s–2s transition driven by ×243 nm. This wavelength can be generated experimentally in a much more convenient way than the 121.7 nm of the directly adjacent 1s–2p Lyman α line. Furthermore, and in contrast to the adjacent 2p state (lifetime 0.1 ns), the decay rate of this metastable level is only about $7\,\text{s}^{-1}$ and promises a very unusually narrow linewidth of just 1 Hz!

For many years T. W. Hänsch and his coworkers have studied more and more exactly the 1s–2s transition of atomic hydrogen and are steadily approaching this ultimate goal of spectroscopy. At present their best published value is about $\Delta\nu \simeq 1\,\text{kHz}$ at 243 nm [49], i.e. for a transition frequency of $\nu_{1s2s} = 2466\,\text{THz}$ already a Q-value of more than 10^{12}! By a phase coherent comparison of the optical transition frequency with the time standard of the caesium atomic clock, the 1s–2s transition frequency has meanwhile become the best-known optical frequency of all (and thus wavelength as well, see p. 32, [73]):

$$f_{1s2s} = 2\,466\,061\,413\,187.103(46)\ \text{kHz}.$$

During detection of the 1s–2s spectrum, another interesting spectroscopic effect occurs: the observed lines are asymmetric and slightly shifted to red frequencies with increasing velocity of the atoms. The reason for this is the Doppler effect of second order, which is not suppressed in two-photon spectroscopy. For the hydrogen atom it plays an

---

[3]This assertion though is challenged since at present the physical significance of the extremely precise measurement is limited by the relatively insufficient knowledge of the structure of the proton, which consists of several particles and is in fact not point-like as assumed by Dirac theory.

important role due to its low mass and therefore high velocity. Only for very low velocities the relatively symmetric signal of Fig. 11.11 is observed.

The line shift caused by the second-order Doppler effect is proportional to $\Delta\nu_{2o} = \omega(v/c)^2/2$ and can be explained by the time dilation known from the special theory of relativity. In a moving atomic inertial reference frame time seems to run more slowly than for an observer at rest in the laboratory frame.

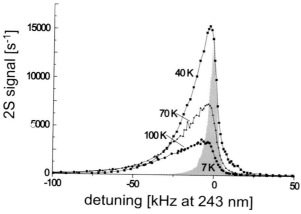

Fig. 11.12: *Second-order Doppler effect of the two-photon spectrum of atomic hydrogen.*

In the experiment the observation has been made that the different line shapes of Fig. 11.12 are a function of the temperature of the nozzle from which the hydrogen atoms are ejected into the evacuated spectrometer. They have a velocity distribution corresponding to the nozzle temperature, and they travel through the exciting UV laser beam for a length of about 30 cm. The linewidth is determined by the time of flight.

11.5   Transient phenomena

So far we have considered the interaction between a light field and matter particles by means of the optical Bloch equations (Eqs. (6.28)) and mostly concentrating on stationary solutions. In the last section, though, we had to investigate the dynamic behaviour in order to describe time-of-flight broadening of long-lived states. It is in general always necessary to take the dynamic properties into account whenever the interaction time scale is short compared to the relevant damping times $T_{1,2}$.

Let us study important special cases as examples of dynamic light–matter interaction: π pulses, rapid turn-on processes, and the effect of a sequence of short light pulses.

11.5.1 π pulses

First, we consider once again the undamped case of the optical Bloch equations of Eq. (6.27). For the frequent case of an atom initially in the ground state ($w(t{=}0) = -1$), for $\delta = 0$ the resonant solution,

$$(u, v, w)(t) = (0, \, \sin\theta(t), \, \cos\theta(t)),$$

can be easily found. The light field simply causes rotation of the Bloch vector in the vw plane. The rotation angle $\theta(t)$ is determined by the pulse area,

$$\theta(t) = \int_{-\infty}^{t} \Omega_R(t') \, dt' = -\frac{d_{eg}}{\hbar} \int_{-\infty}^{t} \mathcal{E}_0(t') \, dt', \tag{11.10}$$

where $\Omega_R(t) \propto \sqrt{I(t)}$. If the pulse area assumes the value $\theta = \pi$, then the atom is promoted exactly from the ground state to the excited state. If the value is 2π, then the atom finishes the interaction again in the ground state.

Let us estimate what kind of light pulse is required to drive a π pulse for an atomic resonance line that has dipole moment $d_{eg} \simeq ea_0 = 0.85 \times 10^{-29}\,\mathrm{C\,m}$. For a light pulse with constant intensity and period T, the necessary intensity and pulse duration can be determined according to (6.26), $\pi = (ea_0/\hbar)\mathcal{E}_0 T$. The numerical value for the corresponding intensity seems to be enormously high at first,

$$I_0 \simeq 120\,\mathrm{kW\,mm^{-2}}\,(T/\mathrm{ps})^{-2}.$$

But it has to be taken into consideration that the pulses are very short, so that the average power of a picosecond laser does not need to be very high. Standard pulsed lasers of the mode-locked type (see Section 8.5.2) operate at a pulse rate of 80 MHz, and for an area of $1\,\mathrm{mm}^2$ an average total power of $\langle P \rangle = 80\,\mathrm{MHz} \times T \times P_0 \simeq 10\,\mathrm{W} \times (T/\mathrm{ps})^{-1}$ is necessary. Commercial laser systems offer average output powers exceeding $1\,\mathrm{W}$, which is quite sufficient if the pulse lengths are slightly increased to $10\,\mathrm{ps}$. Even then the excitation time is only about $1/1000$ of the lifetime of an excited atomic state.

11.5.2 Free induction decay

At both the beginning and the end of an interaction period in light–matter coupling, transient oscillations can occur like for the classical damped oscillator of section 6.1.1. While stationary behaviour is characterized by an oscillation at the driving frequency ω, immediately after turn-on (or turn-off) we also expect dynamic evolution at the eigenfrequency ω_0 of the system, which though is damped out very rapidly (with time constant γ^{-1}). General time-dependent solutions of the (optical) Bloch equations have already been given by Torrey [2] in 1950. However, they are transparent and easily understood only for special cases, such as for example at exact resonance ($\delta = 0$). The dynamic phenomena are also known as 'optical nutation'.

An interesting case occurs for the so-called *free induction decay* (FID). It describes especially the decay of the macroscopic polarization of a sample in the absence of laser light, for example after the application of a very short laser pulse with large intensity. The polarization of an individual particle may live for a much longer time than the macroscopic polarization of an ensemble, which is affected by 'dephasing' of the individual particles.

The evolution of the Bloch vector components depends of course on the detuning, $\mathbf{u} = \mathbf{u}(t, \delta)$. We conveniently use Eq. (6.29) for analysis. For very large intensity ($\Omega_R \gg \delta$) and very short time, we can neglect the detuning at first, since, during the coupling period, the Bloch vector does not have any time to precess by a significant angle. Thus a short and strong initial pulse rotates the dipole ρ_{eg} to the angle given by Eq. (11.10) and we arrive at

$$\rho_{eg}(0,\delta) = u(0,\delta) + iv(0,\delta) = i \sin \theta, \tag{11.11}$$

where we fully neglect the time elapsed. Once the light pulse is turned off, free precession occurs according to

$$\rho_{eg}(t,\delta) = i \sin \theta \, e^{-(\gamma'+i\delta)t}. \tag{11.12}$$

In a large sample, there often exists an inhomogeneous distribution $f(\omega_0)$ of eigenfrequencies of the individual particles and thus of the detunings. In a gas cell this distribution is determined, for example, by the Doppler shift, with $\delta_D = \Delta\omega_D/2\sqrt{\ln 2}$,

$$f(\delta) = \frac{1}{\sqrt{\pi}} e^{-(\delta/\delta_D)^2}.$$

Following an excitation with a $\pi/2$ pulse with $\sin \theta = 1$, the free evolution of the macroscopic polarization is calculated from

$$P(t) = \frac{N_{At}}{V} d_{eg} \, e^{-i\omega_0 t} \int_{-\infty}^{\infty} f(\delta) \, e^{-(\delta/\delta_D)^2} \, e^{-(\gamma'+i\delta)t} \, d\delta. \tag{11.13}$$

If $\gamma' \ll \delta_D$ we can neglect the slow decay, and the integral straightforwardly yields

$$P(t) = \frac{N_{At}}{V} d_{eg} \, e^{-i\omega_0 t} \, e^{-(\delta_D t/2)^2}.$$

The macroscopic polarization thus decays with the lifetime T_2^*,

$$T_2^* = \frac{4 \ln 2}{\Delta\omega_D} \ll \gamma'^{-1} = T_2,$$

much faster than the microscopic polarization whose relaxation determines the fastest time scale for individual particles. This rapid decay is a consequence of the dephasing of the precession angles of the microscopic dipoles. Experimental observation for typical atomic resonance transitions in a gas cell must have resolution better than 1 ns and requires considerable effort; thus slower and weaker transitions are more appropriate to observe this phenomenon.

In the middle row in Fig. 11.13 the time evolution of the radiation field is presented, which is caused by the macroscopic polarization and contains the cooperative radiation field of all excited microscopic dipoles of the sample. At the beginning, constructive interference of the microscopic dipole fields generates a radiation field propagating exactly in the direction of the exciting laser beam. For a perfectly synchronized phase evolution, a well-directed, accelerated and exhaustive emission of the excitation energy would be observed due to the so-called 'super-radiance'. In an inhomogeneous sample, however, this emission ceases very rapidly as a result of the destruction of the phase synchronization ('dephasing'); the stored excitation energy is then released only by common spontaneous emission with a lower rate and isotropically.

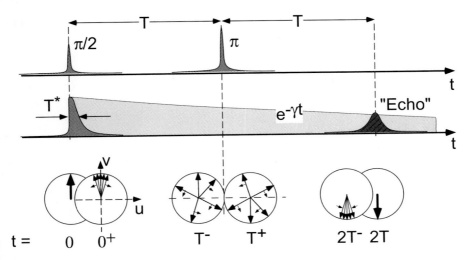

Fig. 11.13: *Free induction decay (FID) and photon echo. In this example a sample is excited by a π/2 pulse and after a time $T \ll \gamma^{-1}$ is subject to another π pulse that causes inversion of precession (top row). After the excitation light pulse, free induction decay is observed at first, which emerges from the initially cooperative emission of all excited atoms and is emitted into the direction of the excitation laser (middle row). Afterwards the polarization of microscopic particles continues to decay by spontaneous emission. After time 2T an echo pulse is observed in the direction of the excitation pulses. The precession of the Bloch vector components in the u–v plane is marked in the bottom row.*

11.5.3   Photon echo

The method of 'photon echoes' for inhomogeneously broadened lines – like many other optical phenomena – has been stimulated by the 'spin echo' method at radio frequencies, which was discovered by I. Hahn for nuclear magnetic resonance. If a sample is excited by two or more short light pulses ($T \ll \gamma'^{-1}$), under certain conditions it emits an 'echo pulse' that follows the excitation pulses in their direction and seems to appear from nowhere. This contradiction is again due to the different evolutions of the microscopic and macroscopic polarizations in a large sample of atoms, molecules, or other microscopic objects, which we have already just met in free induction decay.

The photon echoes can of course be observed only within the natural lifetime of the microscopic polarization. Let us consider the evolution of an individual single dipole with detuning δ under the effect of two resonant light pulses. After time T the dipole has reached the value

$$\rho_{\mathrm{eg}}(t, \delta) = i \sin\theta \, e^{-(\gamma'+i\delta)T}$$

according to Eqs. (11.11) and (11.12). The application of a π pulse now generates an inversion of the (v, w) components ('phase reversal'). Formally this situation is identical with an inversion of the detuning, i.e. after the π pulse we have

$$\rho_{\mathrm{eg}}(t, \delta) = i \sin\theta \, e^{-(\gamma'-i\delta)T} e^{-(\gamma'+i\delta)(t-T)}.$$

The development of the macroscopic polarization can now be given again by eq. (11.13),

$$P(t) = \frac{N_{\text{At}}}{V} d_{\text{eg}} \, e^{-i\omega_0 t} \, e^{-[\delta_0(t-2T)/2]^2} \, e^{-\gamma' t}.$$

After time $t = 2T$ the precession phase angle of each of the microscopic dipoles coincides again; macroscopic polarization is thus restored and once more causes cooperative emission of a macroscopic and coherent radiation field in the direction of the exciting light beam. This pulse is called a 'photon echo'.

11.5.4   Quantum beats

Simultaneous excitation of two or more electronic states by a short light pulse causes observation of a damped oscillation in the fluorescence. These oscillations are usually called *quantum beats*.

In order to realize coherent superposition of several adjacent quantum states, the inverse period of the light pulse T^{-1} (or in other words its 'bandwidth' $\Delta\nu = 1/T$) has to be larger than the frequency separation of the states from each other. Thus the spectral structure of the system is in fact not resolved by the exciting light pulse!

A simple quantum mechanical description assumes that the coherent superposition of two excited states decays freely and spontaneously after the excitation. For a single decaying channel it can be shown that one can describe the time evolution of the excited state with the wavefunction $|\Psi(t)\rangle = e^{-\gamma' t} e^{-i\omega t} |e\rangle$. Furthermore the observed fluorescence intensity is proportional to the square of the induced dipole moment $|\langle g|\hat{d}_{\text{eg}}|e(t)\rangle|^2$ and the following can be easily calculated:[4]

$$I_{\text{fl}} = I(0) \, e^{-2\gamma' t}.$$

If two states $|e_{1,2}\rangle$ with excitation frequencies $\omega_{1,2}$ are prepared in a coherent superposition $|\Psi(t{=}0)\rangle = |e_1\rangle + |e_2\rangle$, then one has

$$|\Psi(t)\rangle = |e_1\rangle \, e^{-i(\omega_1 - \gamma_1')t} + |e_2\rangle \, e^{-i(\omega_2 - \gamma_2')t},$$

and the emitted field contains also the beat frequency $\Delta\omega = \omega_1 - \omega_2$. For the special case $\gamma_1' = \gamma_2'$ one can calculate

$$I_{\text{fl}} = I(0) \, e^{-\gamma' t} (A + B \cos \Delta\omega t).$$

The quantum-beat method has proven to be very useful, e.g. to investigate the fine structures of excited atomic or molecular states with broadband pulsed laser light, which themselves do not provide the necessary spectral resolution. For a systematic experiment it is, though, necessary to use laser pulses of good quality, i.e. precisely known shape (so-called 'transform-limited pulses') to guarantee coherence conditions.

---

[4] For a rigorous theoretical treatment, one has to consider quantum states $|e\rangle|0\rangle$, the product states for the excited atom and the electromagnetic vacuum field, and all states $|g\rangle|1_{\mathbf{k}}\rangle$ for the ground state and field modes with wavevector $\mathbf{k}$. Here we restrict ourselves to an ad hoc treatment of the time evolution of the excited state only.

11.5.5   Wavepackets

A natural extension of quantum beats is offered by wavepackets in microscopic systems which are generated by coherent superposition of many quantum states. With extremely short laser pulses (10 fs corresponds to a bandwidth of 17 THz!), for example, in a molecule, numerous vibrational states can be superimposed coherently [8]. A large density of electronic states is also offered from Rydberg states in atomic systems. Rydberg atomic states have very large principal quantum numbers $n > 10$ [34]. Neither the atomic Rydberg states nor the molecular vibrational states are usually very strongly radiating states, and therefore it is quite difficult to detect them with common fluorescence detectors. In vacuum, though, the weakly bound Rydberg states can be detected by field ionization, and the molecular states by multiphoton ionization. These charged products can be detected with such high sensitivity and selectivity that only a few excited particles are required for such experiments.

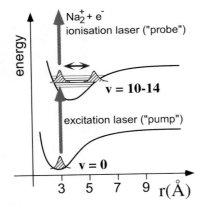 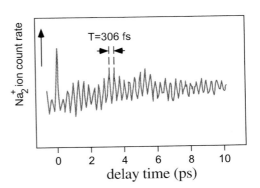

Fig. 11.14: *Two-photon stepwise ionization of Na₂ molecules as a function of the delay time between the two pulses. The ion current is plotted as a function of the delay of the ionization or 'probe' pulse from the excitation or 'pump' pulse. The duration of the laser pulses was 70 ps. The oscillation shows a beat signal which originates from the superposition of two contributions with periods 306 and 363 fs [8].*

With this evolution of the old quantum-beat method, it is conceivable that a wavepacket is prepared from excited quantum states by a light pulse and subsequently propagates freely, i.e. undisturbed by further light interaction. As long as we use a perfect harmonic oscillator the wavepacket will even propagate dispersion-free and return periodically to the origin.

Real molecules though have a strong anharmonicity, which leads to the loss of phase coherence of the atomic wavefunction like for the free induction decay. The total wavefunction is then more or less spread over the energetically allowed space. For many systems however – in this case without application of an external pulse – the wavepacket re-occurs. The phenomenon of collapse and revival of an oscillation was already predicted by Poincaré for classical oscillators. It occurs always when a finite number of oscillations are superimposed; the larger the number, the more time this return takes.

The dynamic evolution of a wavepacket in molecules or Rydberg atoms can be investigated experimentally by so-called 'pump–probe' experiments. With the first pulse a physical excitation is generated; with the second one the dynamic evolution is probed after a variable time delay. We introduce a transparent example, multiphoton ionization of the model system of Na_2 molecules, in a qualitative way.

A molecular beam with Na_2 molecules is excited by a sequence of laser pulses (pulse period 70 fs, $\lambda = 627$ nm). The first laser pulse transfers molecules from the ground state ($v = 0$) to an excited state in which several oscillation states ($v \simeq 10$–14) are superimposed. A further laser pulse generated by the same laser in this experiment generates Na_2^+ molecules by two-photon ionization. These ions can be detected by a secondary electron multiplier, e.g. a channeltron with a probability approaching 100%. In the experiment more filters such as mass spectrometers are used to separate the Na_2^+ signal from the background. If the ionization pulse is delayed, an oscillation of the ion current can be observed as a function of the delay time. Since a beat is observed, the spectrum has to consist of two oscillation frequencies. The first one at 306 fs is caused by the oscillation of the wavepacket in the molecular potential, the second one at 363 fs by the interaction of the detection laser with yet another, higher-lying molecular potential.

Using laser pulses of an extremely short period, it has become possible to resolve the dynamics of molecular wavepackets directly on the femtosecond time scale. These and other methods are used more and more in the so-called 'femto-chemistry'.

11.6   Light forces

When light–matter interaction is analysed, usually the influence on the internal dynamics, e.g. of atoms and molecules, is in the foreground. Absorption and emission of light, though, also changes the external mechanical state of motion of a particle. Photons have momentum $\hbar\mathbf{k}$ and during absorption and emission this momentum has to be transferred to the absorber as a result of momentum conservation. For these processes we expect recoil effects, and the corresponding forces are called *light forces*. Though the photon picture derived from quantum mechanics is very useful, light forces are known from classical light–matter interaction in an analogous way – for example, the Poynting vector describes the momentum density of the propagating electromagnetic field. So let us begin with a study of the mechanical effect of a planar electromagnetic wave on a classical Lorentz oscillator.

An inhomogeneous electric field exerts a force on a particle carrying a dipole, whether induced or permanent, which we may describe component-wise, $\mathbf{d} = (d_x, d_y, d_z)$. For an oscillating dipole, we furthermore have to average over an oscillation period $T = 2\pi/\omega$ of the field, $\langle F \rangle = T^{-1} \int_0^T F(t)\, dt$:

$$F_i^{\mathrm{el}} = \left\langle \sum_j d_j(t) \frac{\partial}{\partial X_j} E_i(t) \right\rangle \qquad \text{or} \qquad \mathbf{F}^{\mathrm{el}} = \langle (\mathbf{d}(t) \cdot \boldsymbol{\nabla})\mathbf{E}(t) \rangle . \tag{11.14}$$

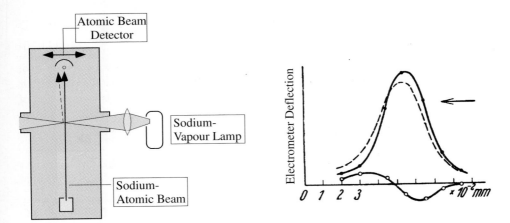

Fig. 11.15: *First observation of the deflection of an atomic beam by light forces. The data are taken from the original publication: R. Frisch, 1933, Z. Phys.,* **86**, *42. The solid line shows the atomic beam profile without the light beam, and the dashed line that with the light beam. The difference is given in the lower part.*

While this analysis seems straightforward, there *is* a problem. In a planar wave travelling in free space, the electromagnetic field is transverse, and thus in the linear Lorentz model the induced dipole has to be transverse. The electric field of a planar wave, on the other hand, can change only in the propagation direction $\mathbf{k}$; hence $\mathbf{d} \perp \nabla$ and one should not expect any electrical force at all from Eq. (11.14). In a realistic light beam, though, with, for example, a Gaussian-shaped envelope, of course transverse electric dipole forces do occur, which we illustrate for the case of a standing-wave field in Section 11.6.4.

We must not forget, however, that there are also magnetic forces in general acting on neutral polarizable atoms. They are caused by the Lorentz force on the electric current in the atom, which is given by the time derivative of the dipole moment,

$$\mathbf{F}^{\mathrm{mag}} = \langle \dot{\mathbf{d}} \times \mathbf{B} \rangle = \frac{1}{c} \langle \dot{\mathbf{d}} \times (\mathbf{e_k} \times \mathbf{E}) \rangle, \tag{11.15}$$

and these magnetic forces exert a net force on the entire atom.

11.6.1   Radiation pressure in a propagating wave

Let us calculate this latter force for a linear electronic Lorentz oscillator with eigen-frequency ω_0 subject to a planar, transverse wave $\mathbf{E} \perp \mathbf{k}$. By using the complex polarizability $\alpha = \alpha' + i\alpha''$ (see Section 6.1), we find

$$\dot{\mathbf{d}}(t) = -i\omega\alpha(\delta)\mathbf{E}(t) \qquad \text{with} \qquad \alpha(\delta) = \frac{q^2/2m\omega_0}{\delta - i\gamma/2},$$

and the average over an electromagnetic cycle is evaluated by means of the Poynting theorem (Appendix A.2), which picks out the imaginary or absorptive part of the polarizability. We finally arrive at

$$\mathbf{F}^{\mathrm{mag}} = \mathbf{k}\alpha''(\delta)|\mathbf{E}|^2 = \mathbf{k}\alpha''(\delta)I/c\epsilon_0. \tag{11.16}$$

The force derived here predicts a light force that is parallel to the wavevector $\mathbf{k}$ of propagation and the intensity I of the light field. It is called 'radiation pressure' or 'spontaneous force' since it depends on the absorption and spontaneous re-emission of photons.

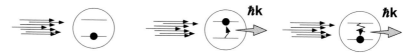

Fig. 11.16: *Absorption–emission cycle and momentum transfer of the spontaneous force. During absorption, momentum is always transferred in the direction of the laser beam. The recoil of spontaneous emission is exerted in random directions. Thus on average for many cycles there is no momentum transferred on emission.*

We expect that the classical treatment is a good approximation for low intensities ($I/I_0 \ll 1$, where I_0 is the saturation intensity from Eq. (6.33)). For larger intensities, we have to treat the internal atomic dynamics according to the Bloch equations. We may use a shortened transition to the results of the semi-classical treatment by replacing the classical Lorentz oscillator in an ad hoc way by the Bloch oscillator through $\mathbf{d} \cdot \mathbf{E} = \alpha\mathbf{E} \cdot \mathbf{E} \to (u + iv)\hbar\Omega_{\mathrm{R}}$. Using the normalized intensity s_0 from Eq. (6.33), we now obtain

$$\mathbf{F}^{\mathrm{mag}} = M\mathbf{a} = \hbar\mathbf{k}\frac{\gamma}{2}\frac{s_0}{1 + s_0 + (2\delta/\gamma)^2} \qquad \text{with} \qquad s_0 = I/I_0. \tag{11.17}$$

This result lends intuitive support for the interpretation of radiation pressure as a spontaneous force. The force grows linearly with 'photon momentum' $\mathbf{k}$ and (for small s_0) with the intensity I of the light field. It is proportional to the absorptive component with the characteristic Lorentz line shape. The force is also proportional to the rate of spontaneous emission γ, and for large intensities ($s \gg 1$) it saturates at the value $\mathbf{F}^{\mathrm{sp}} \to \hbar\gamma\mathbf{k}/2$ while exerting maximum acceleration

$$a_{\mathrm{max}} = \hbar k\gamma/2M. \tag{11.18}$$

Tab. 11.2: *Overview of mechanical parameters for important atoms subject to light forces (see text for details).*

Atom	λ (mm)	γ ($10^6\,\mathrm{s}^{-1}$)	v_{th} ($\mathrm{m\,s}^{-1}$)	a/g	τ (ms)	ℓ (cm)	N
^{1}H	121	600	3000	1.0×10^8	0.003	4.5	1 800
^{7}Li	671	37	1800	1.6×10^5	1.2	112	22 000
^{23}Na	589	60	900	0.9×10^5	0.97	42	30 000
^{133}Cs	852	31	320	0.6×10^4	5.9	94	91 000
^{40}Ca	423	220	800	2.6×10^5	0.31	13	34 000

On average, a strongly driven atom is excited with a probability of 50% and can take on the momentum $\hbar\mathbf{k}$ with each emission cycle.

In Tab. 11.2 we have collected together the essential parameters for the mechanical effect of light on some important atoms. For their 'cooling transitions' with wavelength λ and decay rate of spontaneous emission γ, we list the following: v_{th} = initial thermal velocity of the atomic beam; a/g = maximum acceleration a caused by radiation pressure (light force), normalized to the gravitational acceleration $g = 9.81\,\mathrm{m\,s^{-2}}$; τ = stopping or deceleration time for thermal atoms; ℓ = stopping or deceleration distance for thermal atoms; and N = number of photons scattered during this stopping time.

Excursion: Zeeman slowing

The spontaneous force is perfectly suited to slow down atoms from large thermal velocities (several $100\,\mathrm{m\,s^{-1}}$) to extremely low ones (some $\mathrm{mm\,s^{-1}}$ or $\mathrm{cm\,s^{-1}}$), provided we can exploit the maximum acceleration. In the laboratory system, however, the atomic resonance frequency ω_0 is shifted by the Doppler effect, $\omega_{\mathrm{lab}} = \omega_0 + \mathbf{k}v$, and an atom would lose resonance with the slowing laser – which is effective only within a natural linewidth – after only a few cycles.

Fig. 11.17: *A Zeeman slower for the deceleration of atomic beams. The magnetic field compensates the variation of the Doppler shift caused by deceleration.*

This problem can be overcome by either tuning the laser synchronously with the slowing-down process ('chirp slowing'), or by compensating the Doppler shift by means of a spatially variable magnetic field in which the Zeeman effect[5] ($\delta_{\mathrm{Zee}} = \mu B/\hbar$, with μ = effective magnetic moment, and typically $\mu/\hbar = 2\pi \times 14\,\mathrm{MHz\,mT^{-1}}$) compensates the change by the Doppler shift ('Zeeman slowing'):

$$\delta = \omega_{\mathrm{L}} - (\omega_0 + kv - \mu B/\hbar).$$

For Zeeman slowing along the atomic trajectory, a constant acceleration $v = -a_{\mathrm{sp}}t$ as large as possible is desirable. The compensation field is formed according to

$$B(z) = B_0\sqrt{1 - z/z_0}.$$

The construction length z_0 is generally given and after a short calculation it is found that only velocities with $v \le v_0 = (2a_{\mathrm{sp}}z_0)^{1/2}$ can be slowed down. Moreover, also the magnetic field strength is limited, $B_0 \le \hbar k v_0/\mu$.

---

[5] More precisely the Zeeman shift depends on the magnetic quantum number m and the Landé g-factors of the excited and ground states: $\mu = \mu_{\mathrm{B}}(m_e g_e - m_g g_g)$ (here μ_{B} = Bohr magneton). Optical pumping with circularly polarized light leaves only the highest m value with $mg \simeq 1$ of significance.

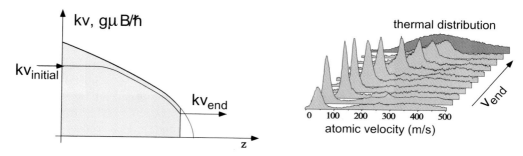

Fig. 11.18: *Left: Evolution of the velocity in the Zeeman slower. Velocity and magnetic field are given in terms of their corresponding Doppler- and Zeeman-shifts. Right: Experimental velocity profiles of an atomic beam at the exit of the Zeeman slower.*

In Fig. 11.18 the effect of laser slowing on the initial thermal distribution is shown. At the end of the Zeeman slower a narrow distribution is generated whose mean velocity is tunable by the laser frequency and the magnetic field. Its width is limited by the so-called Doppler temperature (see Eq. (11.21)). The Zeeman slower is well suited to prepare 'cold' atomic beams with large intensities [65]. A 'cold' atomic beam not only exhibits a low mean velocity but also has much smaller velocity spread than the initial thermal beam.

11.6.2   Damping forces

Let us consider the effect of the spontaneous light force exerted by two counter-propagating laser beams with identical frequency. For this effectively one-dimensional situation, we assume that the forces according to Eq. (11.17) simply add up, i.e. we neglect interference effects:

$$F = F_+ + F_- = \frac{\hbar k \gamma}{2} \left(\frac{s_0}{1 + s_0 + (2\delta_+/\gamma)^2} - \frac{s_0}{1 + s_0 + (2\delta_-/\gamma)^2} \right). \qquad (11.19)$$

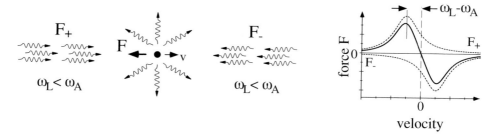

Fig. 11.19: *Light forces for counter-propagating laser fields depending on the velocity or detuning.*

The Doppler detuning $\delta_\pm = \delta_0 \pm kv$ now depends on the direction of the light wave (Fig. 11.19). The atomic resonance frequency Doppler-shifted towards red frequencies always lies closer to the laser frequency of the counter-propagating beam. Therefore the atom is always slowed down – its motion is damped like by a damping force. Very low velocities with $kv/\delta_0 \ll 1$ are of special interest in order to estimate ultimate velocities. We use an expansion of the force equation (Eq. (11.19)) in terms of the velocity v and find

$$\frac{dp}{dt} = F \simeq -\frac{8\hbar k^2 \delta_0}{\gamma} \frac{s_0}{[1 + s_0 + (2\delta_0/\gamma)^2]^2} v = -\alpha m v. \tag{11.20}$$

For $\delta_0 < 0$ we find a damping force with the coefficient α. While the radiation pressure causes only retardation or acceleration, true laser cooling relies on such damping forces.

The one-dimensional concept of laser cooling can be extended to three dimensions by exposing an atom to counter-propagating laser beams in all directions of space. For this at least four tetrahedrally arranged laser beams have to be used. This situation corresponds to the strongly damped motion in a highly viscous liquid and is called 'optical honey' or 'optical molasses'.

11.6.3 Heating forces, Doppler limit

The spontaneous light force not only causes an acceleration in the direction of the beam (which can be combined with cooling) but also a fluctuating force leading to heating of an atomic ensemble.[6] In a simple model we can consider the heating effect by the stochastic effect of the photon recoil $\hbar k$ caused by spontaneous emission in analogy with the Brownian motion or diffusion of molecules. If N photons are randomly scattered, then for the average $\bar{p}_N$ and the variance $(\Delta^2 p)_N = \overline{p_N^2} - \bar{p}_N^2$ of the atomic momentum change by the isotropic emission, we have

$$\bar{p}_N = 0 \qquad \text{and} \qquad (\Delta^2 p)_N = \overline{p_N^2} = N\hbar^2 k^2.$$

The heating force or power can now be estimated from the scattering rate for photons, $dN/dt = (\gamma/2)s_0/[1 + s_0 + (2\delta/\gamma)^2]$, so

$$\left(\frac{d}{dt}\overline{p_N^2}\right)_{\text{heat}} = \frac{\hbar^2 k^2 \gamma}{2} \frac{s_0}{1 + s_0 + (2\delta/\gamma)^2} = 2D,$$

where the relation with the diffusion constant D is taken from the theory of Brownian motion. In equilibrium, we expect the heating and the cooling or damping power to exactly compensate each other,

$$(d\,\overline{p_N^2}/dt)_{\text{heat}} + (d\,\overline{p^2}/dt)_{\text{cool}} = 0.$$

For the cooling power we use relation (11.20),

$$\frac{d}{dt}p = -\alpha p \qquad \text{and} \qquad \left(\frac{d}{dt}p^2\right)_{\text{cool}} = -2\alpha p^2,$$

and for the stationary state we thus obtain

$$p^2 = D/\alpha = M k_{\text{B}} T.$$

---

[6]This fact reflects the very fundamental law that dissipative processes (here, damping forces) are always associated with fluctuations and thus heating processes.

It is associated with a characteristic temperature that is obtained by explicit insertion of D and α,

$$k_B T_{\text{Dopp}} = -\frac{\hbar\gamma}{2}\frac{1 + (2\delta/\gamma)^2}{4\delta/\gamma}. \tag{11.21}$$

When this temperature reaches its lowest value at $2\delta/\gamma = -1$ with $k_B T_{\text{Dopp}} = \hbar\gamma/2$, it is called the *Doppler temperature*.

The Doppler limit has played an important role for many years since it was considered the fundamental limit of laser cooling. It was therefore a big surprise when in experiments considerably lower, so-called sub-Doppler, temperatures were observed. Optical pumping processes are the origin of sub-Doppler laser cooling, and thus they occur only in atoms with a complex magnetic fine structure.

Excursion: Magneto-optical trap (MOT)
In optical molasses, atomic gases are cooled down to the millikelvin (mK) range and lower by very efficient laser cooling. However, atoms cannot be stored in the intersection region of four or more laser beams only by radiation pressure, because they diffuse out of the overlap volume – the dissipative forces do not define a binding centre. This problem has been solved by the invention of the *magneto-optical trap* (MOT), in which the radiation pressure is spatially modified by a quadrupole field.

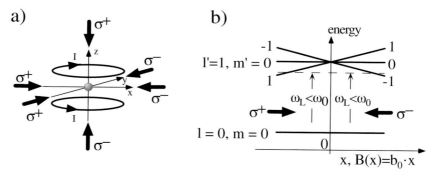

Fig. 11.20: *Magneto-optical trap. (a) Schematic setup with 3 pairs of counterpropagating laser beams. A set of two coils carries opposing currents which generate a quadrupolar magnetic field with a zero at the center. (b) Spatial dependence of energy levels for a model J=0 to J=1 atomic transition. In given direction, the magnetic field increases linearly from the center.*

In one dimension, the MOT can be explained using the simplified example of an atom with a $J = 0 \rightarrow J = 1$ transition (Fig. 11.20(b)) which is exposed to a pair of counter-propagating light beams with circular polarization of opposite handedness in a linearly increasing magnetic field ($\sigma^+\sigma^-$ configuration). Away from the centre and for red detuning ($\omega_L < \omega_0$), a sufficiently slow atom is always much more in resonance with the laser beam whose radiation pressure is directed to the centre of the quadrupole field. Thus the atom will experience a force directed towards this centre.

In three dimensions, a spherical quadrupole field has to be used. It is generated by two coils with currents flowing in opposite directions ('anti-Helmholtz coils'). The handedness of the circular polarizations has to be chosen in accordance with the magnetic field (Fig. 11.20(a)). The simple one-dimensional concept has led to success in three dimensions as well. The realization of the MOT with simple vapour cells has significantly contributed to its widespread use. The MOT is used in numerous laboratories in experiments for laser cooling. In the MOT, an equilibrium between loading rate (by capture of atoms from the slow part of the thermal distribution) and loss rate (by collisions with 'hot' atoms) is built up, which typically contains some 10^8 atoms and has a volume of $0.1\,\mathrm{mm}$ diameter. The residual pressure of the cell must not be too high, because the loading of atoms into the magneto-optical trap must not be interrupted by collisions with fast atoms during the capture process (which takes some milliseconds).

11.6.4   Dipole forces in a standing wave

Let us now evaluate the magnetic force, Eq. (11.15), for the case of a standing wave field generated from counter-propagating plane waves. In this standing wave the B field is shifted by $90°$ with respect to the electric field,

$$\mathbf{E}(z) = 2\mathbf{E}(t)\cos(kz) \qquad \text{and} \qquad \mathbf{B}(z) = (2i/c)\mathbf{e_k} \times \mathbf{E}(t)\sin(kz),$$

and the time average picks out the real part in this case (compare Eq. (11.16)):

$$\mathbf{F}^{\mathrm{mag}} = k\alpha'(\delta)\sin(2kz)\,|\mathbf{E}|^2.$$

This force is called the *dipole force* and can be derived from a potential

$$U_{\mathrm{dip}} = \alpha'(\delta)I(z)/2c\epsilon_0.$$

The interpretation is obvious as well: the force shows a dispersive frequency characteristic, i.e. it changes sign with detuning from the resonance frequency. An interesting application of dipole forces in a standing wave is realized with 'atom lithography' and is described on p. 346.

   In order to proceed to the semi-classical description, we may again use the trick from the previous section (see Eq. (11.17)), which yields

$$U_{\mathrm{dip}} = \tfrac{1}{2}\hbar\delta\ln(1+s).$$

Dipole forces derive from a conservative potential and thus should be disturbed by spontaneous events as little as possible. Therefore, in applications a large detuning $\delta \gg \gamma'$ is chosen and correspondingly small saturation parameters $s \simeq (I/I_0)/(\delta/\gamma')^2$ (6.31) are obtained, so that to a good approximation the dipole potential results in

$$U_{\mathrm{dip}}(\mathbf{r}) \simeq \frac{I}{I_0}\frac{\hbar\gamma'^2}{2\delta}.$$

   Dipole forces, though, only exist if the intensity of the electromagnetic field depends on position, for example in the standing wave mentioned above. Also a Gaussian beam profile makes an inhomogeneous light field and indeed provides an optical dipole trap for atoms and molecules [37], closely resembling the macroscopic optical tweezers of section 11.6.6. Dipole forces always occur when coherent fields are superimposed. The details can be complicated because of the three-dimensional vector nature of the fields, and can cause the appearance of 'optical lattices' [51]. These are standing-wave fields with periodicity in one to three dimensions in which laser-cooled atoms move like in a crystal lattice.

Excursion: Atom lithography

By means of standing-wave fields, apparently strong forces can be exerted on the motion of atoms. Direct experimental proof is not very simple since the motion occurs even on a microscopically small scale. A transparent example for the application, though, is the so-called 'atom lithography', which is an example of atom nanofabrication [72]. With this method the dipole forces of a standing wave serve to modulate the intensity of an atomic beam when it is deposited onto a surface. The surface is physically or chemically modified only where the atoms hit. The experimental concept is presented in Fig. 11.21 and is as outlined below:

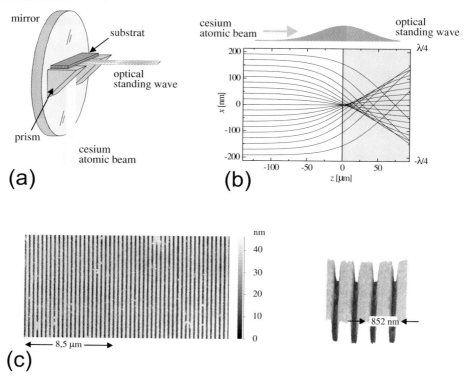

Fig. 11.21: *Atomic lithography – see text for details.*

(a) A substrate is exposed to an atomic beam that has just passed a standing wave generated by a mirror arranged behind the substrate.

(b) The simulation of atomic trajectories in a half-wave shows that the atoms are focused onto the surface in close analogy to an optical lens. Even spherical aberrations are visible. The periodic field of micro-lenses generates changes on the substrate by either growing layers ('direct deposition') or causing chemical reactions ('neutral atom lithography') with dimensions considerably below optical wavelengths. Therefore atom lithography is included in the class of methods allowing structuring at nanometre scales.

(c) An example of nanoscale grooves manufactured with a Cs atomic beam and a standing-wave light field at $\lambda = 852$ nm.

11.6.5 Generalization

Let us briefly discuss the relation of electric and magnetic forces again that we started at the beginning of this section. We can also express the magnetic force (Eq. (11.15)) according to

$$\mathbf{F}^{\text{mag}} = \left\langle \frac{d}{dt}(\mathbf{d}\times\mathbf{B}) \right\rangle - \left\langle \mathbf{d}\times\dot{\mathbf{B}} \right\rangle .$$

The first term vanishes when averaged over a period. If the particle velocity is small, $\dot{\mathbf{R}} \ll c$, furthermore $d\mathbf{B}/dt \simeq \partial\mathbf{B}/\partial t = \boldsymbol{\nabla}\times\mathbf{E}$ can be replaced, yielding

$$\mathbf{F}^{\text{mag}} = \langle -\mathbf{d}\times\boldsymbol{\nabla}\times\mathbf{E} \rangle ,$$

or component-wise

$$F_i^{\text{mag}} = \left\langle \sum_j d_j \left[\frac{\partial}{\partial X_i}E_j - \frac{\partial}{\partial X_j}E_i \right] \right\rangle .$$

Comparison with Eq. (11.14) shows that the total force can generally be determined from

$$\mathbf{F} = \mathbf{F}^{\text{el}} + \mathbf{F}^{\text{mag}} = \left\langle \sum_j d_j \boldsymbol{\nabla} E_j \right\rangle .$$

11.6.6 Optical tweezers

In the last section we investigated the mechanical effect of light beams on microscopic particles such as atoms. Especially for dipole forces we can give a macroscopic analogue that is used more and more widely, the so-called 'optical tweezers' [82].

The dispersive properties of an atom are in fact similar in many ways to those of a transparent dielectric glass sphere for which we can describe the effect of macroscopic light forces qualitatively and in terms of ray optics.

In Fig. 11.22 the position of a glass sphere is either transversely shifted away from the axis of a Gaussian laser beam (left) or longitudinally shifted away from the focus of a focused beam. Taking into account that the refraction of light beams causes transfer of momentum just as for atoms, we can infer from the directional changes of the beams that there is a mechanical force exerted on the glass sphere.

Optical tweezers are useful as non-material micro-manipulators in microscopy, for instance bacteria in liquids can be caught, trapped and moved.

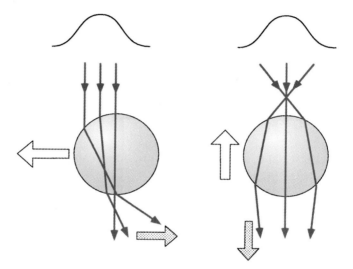

Fig. 11.22: *The effect of optical tweezers in terms of ray optics. White arrows: forces on the glass sphere. Dotted arrows: momentum change of the light rays. The transversal profile of a laser beam is indicated above the spheres.*

12 Nonlinear optics I: Optical mixing processes

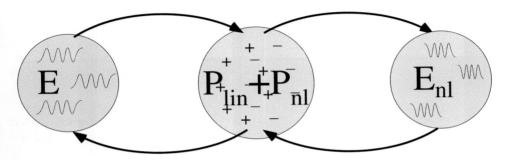

Fig. 12.1: *A nonlinear polarization P_{NL} is generated in a dielectric material at large intensities, in addition to linear interactions. It acts as the source of a new electromagnetic field E_{NL}, which acts back onto the polarization.*

Up to now we have mostly considered polarizations connected linearly with the driving field. The theory of the *linear response* was completely sufficient as long as only classical light sources were available. Since the invention of the laser, we have been able to drive matter so strongly that, besides linear contributions to the polarization (like in eq. (6.12)), nonlinear ones also become noticeable.

12.1 Charged anharmonic oscillators

We can modify the classical model of Section 6.1.1 to obtain a simplified microscopic model of the properties of nonlinear interactions of light and matter. For this purpose, we add a weak anharmonic force $m\alpha x^2$ to the equation of motion of the linear oscillator. This model reflects, for example, the situation of the potential of a charge in a crystal with a lack of inversion symmetry. At the same time we neglect the linear damping by absorption and scattering, which are undesired for the application and the study of nonlinear processes and, as we will see, make the formal treatment even more complex. So we consider the undamped equation

$$\ddot{x} + \omega_0^2 x + \alpha x^2 = \frac{q}{m}\mathcal{E}\cos(\omega t).$$

Optics, Light and Laser. Dieter Meschede
Copyright © 2004 Wiley-VCH Verlag GmbH & Co. KGaA
ISBN: 3-527-40364-7

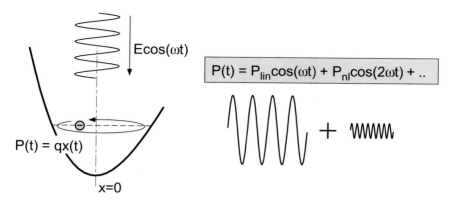

Fig. 12.2: *Charged oscillator in an anharmonic potential. By the anharmonic motion, the harmonics of the driving frequency ω are excited. In a real crystal, $x(t)$ has to be replaced by an appropriate normal coordinate.*

We now seek a solution $x(t) = x^{(1)}(t) + x^{(2)}(t)$ with $x^{(1)}$ as the already known linear part, $x^{(1)}(t) = x_\mathrm{L}\cos(\omega t)$.[1] The amplitude is $x_\mathrm{L} = q\mathcal{E}/m(\omega_0^2 - \omega^2)$ and the small nonlinear perturbation ($|x^{(2)}| \ll |x^{(1)}|$) approximately fulfils the equation

$$\ddot{x}^{(2)} + \omega_0^2 x^{(2)} = -\alpha[(x^{(1)})^2 + 2x^{(1)}x^{(2)} + \cdots] \simeq -\alpha x_\mathrm{L}^2 \cos^2(\omega t).$$

We now split the nonlinear polarization into a constant and a term oscillating at twice the frequency of the driving field 2ω,

$$x^{(2)} = x^{(2)}_\mathrm{DC} + x^{(2)}_{2\omega}.$$

We find the solution

$$x^{(2)}_\mathrm{DC} = -\frac{\alpha x_\mathrm{L}^2}{2\omega_0^2},$$

$$x^{(2)}_{2\omega} = -\frac{\alpha x_\mathrm{L}^2}{2(\omega_0^2 - 4\omega^2)}\cos(2\omega t).$$

The first term describes the shift of the mean position of the charge caused by the asymmetry of the potential. So the optical wave causes a constant, macroscopic polarization of the sample, which we can as well interpret as 'optical rectification' or as 'inverse Kerr effect' (see Section 3.6.1).

The second term describes the first harmonic of the charge at frequency 2ω. For suitable conditions, which will be discussed in more detail in Section 12.4 on frequency doubling, the sample emits a coherent electric field at this frequency!

In analogy with the linear case, we can introduce a nonlinear susceptibility describing nonlinear light–matter interaction. It causes a harmonic wave at frequency 2ω and is connected with a new polarization at this frequency,

$$P_{2\omega}(t) = -\frac{\alpha(q/m)^2}{2(\omega_0^2 - \omega^2)^2(\omega_0^2 - 4\omega^2)}\mathcal{E}^2 \cos(2\omega t).$$

---

[1] In transparent materials the electronic resonances are far away and we can neglect the absorptive contribution ($\propto \sin\omega t$) to a good approximation.

We thus obtain a nonlinear susceptibility

$$\chi(2\omega) = -\frac{1}{\epsilon_0}\frac{\alpha(q/m)^2}{2(\omega_0^2 - \omega^2)^2(\omega_0^2 - 4\omega^2)}. \tag{12.1}$$

It is interesting to note that it shows a resonance at $\omega_0 = 2\omega$ which may be interpreted through two-photon absorption.

12.2   Second-order nonlinear susceptibility

We can generally describe the response of a sample to one or more optical waves by means of nonlinear susceptibilities. In the following we only consider monochromatic electric fields, which we split into positive and negative frequency parts using the complex notation,

$$
\begin{aligned}
\mathbf{E}(\mathbf{r}, t) &= (\mathbf{E}^{(+)} + \mathbf{E}^{(-)})/2, \\
\mathbf{E}^{(+)}(\mathbf{r}, t) &= \mathcal{E} e^{-i(\omega t - \mathbf{k}\mathbf{r})}, \\
\mathbf{E}^{(-)}(\mathbf{r}, t) &= (\mathbf{E}^{(+)}(\mathbf{r}, t))^*,
\end{aligned}
$$

and correspondingly for the dielectric polarization $\mathbf{P}^{(\pm)}$. If the field is linearly polarized, the amplitude is calculated from

$$|\mathcal{E}| = \sqrt{\frac{I}{nc\epsilon_0}},$$

because, in this definition, $|\mathbf{E}|^2 = \mathbf{E}^{(+)}\mathbf{E}^{(-)}$. The linear relation of field strength and polarization is already known from Eq. (6.14). In order to avoid the elaborate presentation using the convolution integral, we symbolize it here by the $\odot$ sign,

$$\mathbf{P}(\mathbf{r}, t) = \epsilon_0\chi^{(1)} \odot \mathbf{E}(\mathbf{r}, t).$$

Furthermore we use an additional superscript index '(1)' to identify the linear or first-order contribution. In the most important case of monochromatic fields, a simple product is recovered from the temporal convolution (Eq. (6.14)).

At high field intensities the nonlinear contributions of the polarization also lead to perceptible effects,

$$
\begin{aligned}
\mathbf{P}(\mathbf{r}, t) &= \mathbf{P}^{\mathrm{lin}}(\mathbf{r}, t) + \mathbf{P}^{\mathrm{NL}}(\mathbf{r}, t) \\
&= \epsilon_0[\chi^{(1)} \odot \mathbf{E}(\mathbf{r}, t) + \chi^{(2)} \odot \mathbf{E}(\mathbf{r}, t) \odot \mathbf{E}(\mathbf{r}, t) + \cdots].
\end{aligned}
$$

Terms of second order and higher are the topics of nonlinear optics; they are also called 'nonlinear products'. In general, the interaction is anisotropic ($\chi^{(1)} = \chi^{(1)}_{ij}$, $\chi^{(2)} = \chi^{(2)}_{ijk}$, etc.) and depends on the individual vector components; thus nonlinear products for all the relevant field components can occur ($(\mathbf{E} \odot \mathbf{E})_{ij} = E_i \odot E_j$, etc.).

12.2.1   Mixing optical fields

For each order of $\chi^{(n)}$, a series of new frequencies is generated through the 'mixing' products, the $e^{-i\omega_i t} e^{-i\omega_j t} \ldots e^{-i\omega_n t}$ terms, . It is thus much simpler for nonlinear optics to sort the contributions to the polarization by their frequency components $\omega = \omega_i \pm \omega_j \pm \ldots \pm \omega_n$. A general term for the polarization of second order can, for instance, be given componentwise by

$$P_i(\omega) = \sum_{jk} \sum_{mn} \chi^{(2)}_{ijk}(\omega; \omega_m \omega_n) E_j(\omega_m) E_k(\omega_n). \tag{12.2}$$

For a simplified one-dimensional and isotropic case $(j = k)$, we can extract all the frequency components of nonlinear polarization from

$$E(\mathbf{r}, t)^2 = \left[\sum_m (E_m^{(+)} + E_m^{(-)}) \right]^2$$

$$= \sum_i \left[(E_m^{(+)})^2 + E_m^{(+)} E_m^{(-)} + 2 \sum_{n \neq m} (E_m^{(+)} E_n^{(+)} + E_m^{(+)} E_n^{(-)}) \right] + \text{c.c.}$$

Already by irradiation with just two optical waves $(m = 1, 2)$ of different frequency $(\omega_{1,2}$ in Eq. (12.2)) nonlinear polarizations at five different sum and difference frequencies are produced, which act as the driving force for generation of a new wave at the mixing frequency:

$$\begin{array}{llll}
\mathbf{P}(2\omega_1) & \text{second harmonic frequency (1)} & \text{(SHG)}, \\
\mathbf{P}(2\omega_2) & \text{second harmonic frequency (2)} & \text{(SHG)}, \\
\mathbf{P}(\omega_1 + \omega_2) & \text{sum frequency} & \text{(SUM)}, & (12.3) \\
\mathbf{P}(\omega_1 - \omega_2) & \text{difference frequency} & \text{(DIF)}, \\
\mathbf{P}(\omega = 0) & \text{optical rectification} & \text{(OR)}.
\end{array}$$

Two field components each with frequencies ω_1, ω_2 generate a polarization at frequency ω. The corresponding susceptibility is characterized by the notation

$$\chi^{(2)}_{ijk}(\omega; \omega_1, \omega_2), \qquad \omega = \omega_1 + \omega_2.$$

The indices 'ijk' can represent every Cartesian coordinate (x, y, z) and take the tensorial character of the susceptibility into account. Therefore for each frequency combination in principle there are 27 tensor elements in second order.

Neglecting the Cartesian dependence for now, the following relations can be found by splitting the polarization into Fourier components $P_i(\mathbf{r}, t) = (P_i^{(+)} + P_i^{(-)})/2$ and by comparison with Eq. (12.3):

$$\begin{aligned}
P^{(+)}(\omega = 2\omega_1) &= \epsilon_0 \chi^{(2)}(\omega, \omega_1, \omega_1)(E_1^{(+)})^2, \\
P^{(+)}(\omega = 2\omega_2) &= \epsilon_0 \chi^{(2)}(\omega, \omega_2, \omega_2)(E_2^{(+)})^2, \\
P^{(+)}(\omega = \omega_1 + \omega_2) &= 2\epsilon_0 \chi^{(2)}(\omega, \omega_1, \omega_2) E_1^{(+)} E_2^{(+)}, \\
P^{(+)}(\omega = \omega_1 - \omega_2) &= 2\epsilon_0 \chi^{(2)}(\omega, \omega_1, -\omega_2) E_1^{(+)} E_2^{(-)}, \\
P^{(+)}(\omega = 0) &= 2\epsilon_0 [\chi^{(2)}(0; \omega_1, -\omega_1) E_1^{(+)} E_1^{(-)} \\
&\quad + \chi^{(2)}(0; \omega_2, -\omega_2) E_2^{(+)} E_2^{(-)}].
\end{aligned}$$

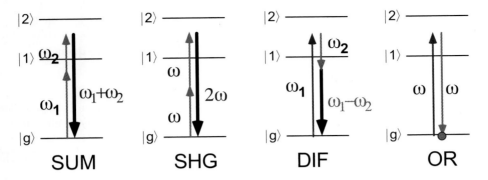

Fig. 12.3: *Passive* $\chi^{(2)}$ *processes. SUM = sum frequency generation; SHG = second harmonic generation; DIF = difference frequency generation; OR = optical rectification.*

12.2.2 Symmetry properties of susceptibility

The search for crystals with large nonlinear coefficients is a matter of continued scientific research. The symmetry properties of real crystals play an important role [14] and will be here subject to a short consideration with regard to nonlinear optics. For the sake of simplicity, we restrict this discussion to the second-order effects.

Intrinsic permutation symmetry

Using two fundamental waves and one polarization wave, six different mixing products can be generated if we additionally require $\omega = \omega_1 + \omega_2$,

$$\chi^{(2)}_{ijk}(\omega;\omega_1,\omega_2);\ \chi^{(2)}_{ijk}(\omega_1;-\omega_2,\omega);\ \chi^{(2)}_{ijk}(\omega_2;\omega,-\omega_1);$$

$$\chi^{(2)}_{ijk}(\omega;\omega_2,\omega_1);\ \chi^{(2)}_{ijk}(\omega_1;\omega,-\omega_2);\ \chi^{(2)}_{ijk}(\omega_2;-\omega_1,\omega).$$

The upper row is identical to the lower one when coordinates (i,j) are permuted along with the corresponding frequencies,

$$\chi^{(2)}_{ijk}(\omega;\omega_1,\omega_2) = \chi^{(2)}_{ikj}(\omega;\omega_2,\omega_1).$$

Real electromagnetic fields

Since the harmonic time dependence of $P^{(-)}$ is connected with $P^{(+)}$ by replacement of $\omega_i \rightarrow -\omega_i$, the following has to be valid:

$$\chi^{(2)}_{ijk}(\omega_i;\omega_j,\omega_k) = \chi^{(2)}_{ijk}(-\omega_i;-\omega_k,-\omega_j)^*.$$

Loss-free media

In loss-free media the susceptibility is real. Then we have

$$\chi^{(2)}_{ijk}(\omega_i;\omega_j,\omega_k) = \chi^{(2)}_{ijk}(-\omega_i;-\omega_k,-\omega_j). \tag{12.4}$$

In addition, 'complete permutation symmetry' holds, i.e. all frequencies can be permuted if the corresponding Cartesian indices are permuted at the same time. For this it has to be taken into account that the sign of the commuted frequencies has to change in order to meet the condition $\omega = \omega_1 + \omega_2$,

$$\chi_{ijk}^{(2)}(\omega; \omega_1, \omega_2) = \chi_{jik}^{(2)}(-\omega_1; -\omega, \omega_2) = \chi_{jik}^{(2)}(\omega_1; \omega, -\omega_2).$$

In the last step we used Eq. (12.4). A proof of this symmetry can be based on the quantum mechanical calculation of χ or the energy density in a nonlinear medium.

12.2.3   Two-wave polarization

In the previous section we have seen that one or more new polarization waves result as a mixing product of two input fields,

$$\begin{aligned}
P^{(+)} &= \epsilon_0 \chi^{(2)}(\omega; \omega_1, \omega_2) E_1^{(+)} E_2^{(+)}, \\
P_1^{(+)} &= \epsilon_0 \chi^{(2)}(\omega_1; -\omega_2, \omega) E_2^{(-)} E^{(+)}, \\
P_2^{(+)} &= \epsilon_0 \chi^{(2)}(\omega_2; \omega, -\omega_1) E^{(+)} E_1^{(-)}.
\end{aligned} \tag{12.5}$$

At the same time a new field at the frequency of the polarization wave has to emerge, which by nonlinear interaction now itself contributes to the polarization at the already existing frequencies. This nonlinear coupling describes the back-action of the nonlinear polarization onto the fundamental waves, e.g. the exchange of energy. With the symmetry rules of section 12.2.2, we can confirm that for the approximation of loss-free media the $\chi^{(2)}$ coefficients in Eq. (12.5) are identical! In Section 12.3.1 we shall go further to investigate the coupling of *three* waves.

Contracted notation

In nonlinear optics the 'contracted notation' is used very often, which at first is defined by the tensor

$$d_{ijk} = \tfrac{1}{2}\chi_{ijk}^{(2)}.$$

The notation is now simplified and the number of possible elements is reduced from 27 for $\chi_{ijk}^{(2)}$ to 18 by contracting the last two indices (j, k) to a single index l, i.e. $d_{ijk} \to d_{il}$. So because of the intrinsic permutation symmetry we have

$$\begin{aligned}
jk &: 11\ 22\ 33\ 23, 32\ 31, 13\ 12, 21 \\
l &:\ 1\ \ \ 2\ \ \ 3\ \ \ \ 4\ \ \ \ \ \ 5\ \ \ \ \ \ 6.
\end{aligned}$$

For example, the matrix equation describing frequency doubling reads with the d_{ij} tensor

$$\begin{pmatrix} P_x(2\omega) \\ P_y(2\omega) \\ P_z(2\omega) \end{pmatrix} = 2 \begin{pmatrix} d_{11} & d_{12} & d_{13} & d_{14} & d_{15} & d_{16} \\ d_{21} & d_{22} & d_{23} & d_{24} & d_{25} & d_{26} \\ d_{31} & d_{32} & d_{33} & d_{34} & d_{35} & d_{36} \end{pmatrix} \begin{pmatrix} E_x(\omega)^2 \\ E_y(\omega)^2 \\ E_z(\omega)^2 \\ 2E_y(\omega)E_z(\omega) \\ 2E_x(\omega)E_z(\omega) \\ 2E_x(\omega)E_y(\omega) \end{pmatrix}. \tag{12.6}$$

Kleinman symmetry

Often the resonance frequencies of a nonlinear material are much higher than those of the driving fields. Then the susceptibilities – which typically have forms similar to our classical model of Eq. (12.1) – depend only weakly on the frequency and are subject to the approximate *Kleinman symmetry*. If furthermore the susceptibility does not even depend on the frequency, the Cartesian indices can be permuted without permuting the corresponding frequencies at the same time. The Kleinman symmetry reduces the maximum number of independent matrix elements from 18 to 10.

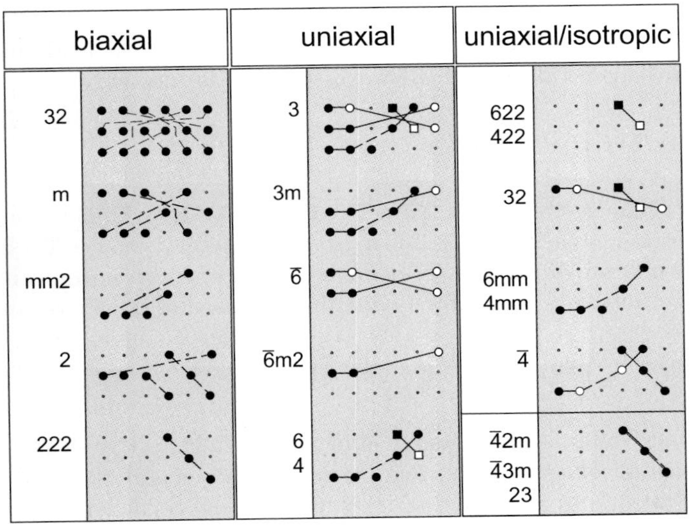

Fig. 12.4: *Non-vanishing second-order coefficients* d_{eff} *(according to Zernike and Midwinter 1973, [118], see also [14]). Identical coefficients are connected by lines (dashed: only for Kleinman symmetry). Full and open symbols indicate opposite signs; square symbols vanish at Kleinman symmetry.*

12.2.4 Crystal symmetry

A crystal with inversion symmetry cannot show any susceptibility of second order at all. With the inversion of all coordinates, the sign of the field amplitude changes as well as that of the polarization,

$$P_i(\mathbf{r}) = d_{ijk} E_j(\mathbf{r}) E_k(\mathbf{r}) \quad \overset{\mathbf{r} \to -\mathbf{r}}{\longrightarrow} \quad -P_i(\mathbf{r}) = d_{ijk} E_j(-\mathbf{r}) E_k(-\mathbf{r}).$$

Thus the inversion symmetry leads to $d_{ijk} = \chi^{(2)} = 0$, and from 32 crystal classes those 11 exhibiting inversion symmetry are eliminated. The symmetry properties of the remaining crystal classes significantly reduce the number of non-vanishing non-linear d coefficients that are independent of each other. In Fig. 12.4 the non-zero coefficients for the different crystal classes are given in the standard notation.

12.2.5   Effective value of the nonlinear d coefficient

In general nonlinear crystals are anisotropic and birefringent; indeed we are going to detail that the asymmetry of birefringence really makes their efficient application possible. Frequently one finds quotations of effective values d_{eff} depending on the so-called phase matching angles θ and ϕ given for the d_{il} coefficients, which are also tabulated [27].

12.3   Wave propagation in nonlinear media

In order to understand the propagation of waves in a nonlinear medium [17], we first consider again the general form of the wave equation in matter,

$$\nabla \times \nabla \times \mathbf{E}(\mathbf{r}, t) + \frac{1}{c^2} \frac{\partial^2}{\partial t^2} \mathbf{E}(\mathbf{r}, t) = -\frac{1}{\epsilon_0 c^2} \frac{\partial^2}{\partial t^2} \mathbf{P}(\mathbf{r}, t).$$

The first term of the vector identity $\nabla \times \nabla \times \mathbf{E} = \nabla(\nabla \cdot \mathbf{E}) - \nabla^2 \mathbf{E}$ cannot be removed in nonlinear optics as easily as for linear isotropic media because $\nabla \cdot \mathbf{E} = 0$ can no longer be inferred from $\nabla \cdot \mathbf{D} = 0$. Fortunately the first contribution can be neglected in many cases of interest, especially for the limiting case of planar waves:

$$\left(\nabla^2 - \frac{1}{c^2} \frac{\partial^2}{\partial t^2} \right) \mathbf{E}(\mathbf{r}, t) = \frac{1}{\epsilon_0 c^2} \frac{\partial^2}{\partial t^2} \mathbf{P}(\mathbf{r}, t).$$

The polarization contains linear and nonlinear parts, $\mathbf{P} = \mathbf{P}^{(1)} + \mathbf{P}^{\text{NL}}$. The linear contribution has an effect only on one or more fundamental waves $\mathbf{E}^{\text{F}}$ driving the process and is taken into account through the refraction coefficient $n^2 = 1 + \chi^{(1)}$, i.e. $\mathbf{P}^{(1)} = \epsilon_0(n^2 - 1)\mathbf{E}^{\text{F}}$. Then a new wave equation is obtained driven by the nonlinear polarization, $\mathbf{P}^{\text{NL}}$,

$$\left(\nabla^2 - \frac{n^2}{c^2} \frac{\partial^2}{\partial t^2} \right) \mathbf{E}(\mathbf{r}, t) = \frac{1}{\epsilon_0 c^2} \frac{\partial^2}{\partial t^2} \mathbf{P}^{\text{NL}}(\mathbf{r}, t).$$

If this vanishes, the already known equation for the propagation of a wave in a dielectric medium is found. In a dispersive medium, the refraction coefficient depends on the frequency, $n = n(\omega)$. We now consider again each frequency component ω_i separately and also split the oscillating part from the positive and negative polarization components,

$$\mathbf{P}^{\text{NL}}(\mathbf{r}, t) = \sum_i [\widetilde{\boldsymbol{\mathcal{P}}}_i(\mathbf{r})\, e^{-i\omega_i t} + \widetilde{\boldsymbol{\mathcal{P}}}_i^{*}(\mathbf{r})\, e^{i\omega_i t}]/2.$$

With this notation the wave equation separates into single-frequency Helmholtz equations and can be written as

$$\left(\nabla^2 + \frac{n(\omega)^2 \omega_i^2}{c^2} \right) \boldsymbol{\mathcal{E}}_i(\mathbf{r})\, e^{i\mathbf{k}\mathbf{r}} = -\frac{\omega_i^2}{\epsilon_0 c^2} \widetilde{\boldsymbol{\mathcal{P}}}_i(\mathbf{r}). \tag{12.7}$$

12.3.1 Coupled amplitude equations

To simplify equations (12.7) we first consider only planar waves propagating in the z direction. Additionally, it is generally realistic to assume again that the amplitudes of the waves change only slowly compared with the wavelength or that the curvature of the amplitude is much smaller than the curvature of the wave:

$$\left| \frac{\partial^2 \mathcal{E}(z)}{\partial z^2} \right| \ll k \left| \frac{\partial \mathcal{E}(z)}{\partial z} \right|.$$

Then with

$$\frac{\partial^2}{\partial z^2} [\mathcal{E}(z) \, e^{ikz}] \simeq e^{ikz} \left[2ik \frac{\partial}{\partial z} - k^2 \right] \mathcal{E}(z),$$

the wave equation is reduced to an approximate form

$$\left[2ik \frac{\partial}{\partial z} - k^2 + \frac{n^2(\omega)\omega^2}{c^2} \right] \mathcal{E}(z) = -\frac{\omega^2}{\epsilon_0 c^2} \widetilde{\mathcal{P}}(\omega) \, e^{-ikz}.$$

With $k^2 = n^2(\omega)\omega^2/c^2$ we can furthermore identify the wavevector of propagation in a dielectric medium and arrive at

$$\frac{d}{dz} \mathcal{E}(z) = \frac{\omega^2}{\epsilon_0 c^2} \frac{i}{2k} \widetilde{\mathcal{P}}(\omega) \, e^{-ikz}. \tag{12.8}$$

Incidentally, a more exact consideration shows that not only a forward-running but also a backward-running wave is generated, but only the forward-running wave couples significantly to the fundamental wave ([94], chapter 33).

For each of the complicated wave equations from (12.7), we can therefore draw up a simpler equation according to (12.8) replacing the polarization by its explicit form, e.g. according to (12.5). The most important problems of nonlinear optics can be solved with this standard method.

12.3.2 Coupled amplitudes for three-wave mixing

The nonlinear polarization describes the coupling between the fundamental waves $\mathbf{E}_1(\omega_1)$ and $\mathbf{E}_2(\omega_2)$ and their mixing product $\mathbf{E}_3(\omega)$. For symmetry reasons according to Eq. (12.5) we use the same $\chi^{(2)}$ coefficients for the nonlinear susceptibility in all three cases.

For this purpose we write down the three amplitude equations according to Eq. (12.8) by inserting the polarizations from Eq. (12.5). Introducing the abbreviation

$$\Delta k = k - k_1 - k_2,$$

we get

$$\widetilde{\mathcal{P}}_3(z) \, e^{-ikz} = 4\epsilon_0 d_{\text{eff}} \mathcal{E}_1 \, e^{ik_1 z} \mathcal{E}_2 \, e^{ik_2 z} \, e^{-ikz} = 4\epsilon_0 d_{\text{eff}} \mathcal{E}_1 \mathcal{E}_2 \, e^{-i\Delta k z}.$$

The factor 4 occurs here because we have to sum over all contributions to the polarization according to Eq. (12.2). We insert in Eq. (12.8), and for the sake of improved transparency we use the complex conjugate of the equations for $\mathcal{E}_{1,2}$:

$$\frac{d}{dz}\mathcal{E}_3(\omega) = \frac{2i\omega d_{\text{eff}}}{cn(\omega)}\mathcal{E}_1\mathcal{E}_2\,e^{-i\Delta kz},$$

$$\frac{d}{dz}\mathcal{E}_1^*(\omega_1) = \frac{-2i\omega_1 d_{\text{eff}}}{cn(\omega_1)}\mathcal{E}_3^*\mathcal{E}_2\,e^{-i\Delta kz}, \tag{12.9}$$

$$\frac{d}{dz}\mathcal{E}_2^*(\omega_2) = \frac{-2i\omega_2 d_{\text{eff}}}{cn(\omega_2)}\mathcal{E}_1\mathcal{E}_3^*\,e^{-i\Delta kz}.$$

In principle with these equations all important $\chi^{(2)}$ processes can be addressed. For passive processes, including frequency doubling, sum and difference frequency mixing, and optical rectification, initial conditions $\mathcal{E}_1, \mathcal{E}_2 \neq 0$ and $\mathcal{E}_3 = 0$ have to be considered. It is also possible to understand the parametric oscillator using this set of equations. It resembles an active medium in a way similar to the laser, where the initial conditions now have the form $\mathcal{E}_1, \mathcal{E}_2 = 0$ and $\mathcal{E}_3 \neq 0$.

12.3.3   Energy conservation

The intensity I of a linearly polarized wave in a dielectric medium with refraction coefficient $n(\omega)$ is

$$I = \tfrac{1}{2}n(\omega)c\epsilon_0|E|^2.$$

By multiplying equations (12.9) with the respective conjugate amplitude $n(\omega_i)c\epsilon_0\mathcal{E}_i^*/2$, the *Manley–Rowe relation* is obtained,

$$\frac{1}{\omega}\frac{d}{dz}I_3(\omega) = -\frac{1}{\omega_1}\frac{d}{dz}I_1(\omega_1) = -\frac{1}{\omega_2}\frac{d}{dz}I_2(\omega_2).$$

This describes the conservation of energy, since the expression is equivalent to

$$I_3(\omega) + I_1(\omega_1) + I_2(\omega_2) = 0$$

because $\omega = \omega_1 + \omega_2$. This fact is also called 'photon conservation', for in this interpretation two photons with frequencies ω_1 and ω_2 are combined to make one photon with frequency ω. This term, though, is just another expression for energy conservation. Nonlinear optics does not at all have to invoke quantum physics for theoretical explanations.

Nevertheless the fact of the conservation of photon number is convenient and we can transform equations (12.9) to normalized amplitudes

$$\mathcal{A}_i = \sqrt{\frac{n(\omega_i)}{\omega_i}}\mathcal{E}_i.$$

The amplitude of the electromagnetic wave is now

$$I = c\epsilon_0\omega|\mathbf{A}(\mathbf{r}, t)|^2,$$

and we find

$$\frac{d}{dz}\mathcal{A}_3(\omega) = i\kappa\mathcal{A}_1\mathcal{A}_2\,e^{-i\Delta kz},$$

$$\frac{d}{dz}\mathcal{A}_1^*(\omega_1) = -i\kappa\mathcal{A}_3^*\mathcal{A}_2\,e^{-i\Delta kz}, \qquad (12.10)$$

$$\frac{d}{dz}\mathcal{A}_2^*(\omega_2) = -i\kappa\mathcal{A}_1\mathcal{A}_3^*\,e^{-i\Delta kz},$$

with the material dependent coupling coefficient

$$\kappa = \frac{2d_{\mathrm{eff}}}{c}\sqrt{\frac{\omega\omega_1\omega_2}{n(\omega)n(\omega_1)n(\omega_2)}}. \qquad (12.11)$$

12.4   Frequency doubling

The first important special case of the coupled amplitude equations (12.9) is frequency doubling. It has a particularly great significance because with this method coherent harmonics of a fundamental wave can be generated. By this means, for example, coherent ultraviolet radiation becomes available at wavelength where no tunable laser system exists. The equations (12.9) are in this case reduced to two equations due to the degeneracy of ω_1 and ω_1. We once again recapitulate the form for the field intensity of the fundamental wave $\mathcal{E}_{\mathrm{FUN}}$ and for the second harmonic $\mathcal{E}_{\mathrm{SHG}}$,

$$\frac{d}{dz}\mathcal{E}_{\mathrm{SHG}}(2\omega) = \frac{i2\omega}{cn(2\omega)}d_{\mathrm{eff}}\mathcal{E}_{\mathrm{FUN}}^2(\omega)\,e^{-i\Delta kz},$$

$$\frac{d}{dz}\mathcal{E}_{\mathrm{FUN}}(\omega) = \frac{i\omega}{cn(\omega)}2d_{\mathrm{eff}}\mathcal{E}_{\mathrm{SHG}}(2\omega)\mathcal{E}_{\mathrm{FUN}}^*(\omega)\,e^{i\Delta kz}.$$

Because of the degeneracy, the term for frequency doubling appears only once in eq. (12.2), and therefore the first equation is smaller by a factor of 2 than in eq. (12.9). The so-called phase mismatch,

$$\Delta k = k_{2\omega} - 2k_\omega = \frac{2\omega}{c}(n_{2\omega} - n_\omega), \qquad (12.12)$$

apparently depends on the difference of the refraction coefficients of the fundamental and harmonic waves. Owing to the fact that dispersion of common materials is always present, we have $n_{2\omega} \neq n_\omega$. For simplification we again use normalized equations (12.10) with $\mathcal{A}_{\mathrm{FUN}}(\omega) = (n_\omega/\omega)^{1/2}\mathcal{E}_{\mathrm{FUN}}(\omega)$ and $\mathcal{A}_{\mathrm{SHG}}(\omega) = (n_{2\omega}/\omega)^{1/2}\mathcal{E}_{\mathrm{SHG}}$,

$$\frac{d}{dz}\mathcal{A}_{\mathrm{SHG}} = i\kappa\mathcal{A}_{\mathrm{FUN}}^2\,e^{-i\Delta kz},$$

$$\frac{d}{dz}\mathcal{A}_{\mathrm{FUN}} = i\kappa\mathcal{A}_{\mathrm{SHG}}\mathcal{A}_{\mathrm{FUN}}^*\,e^{i\Delta kz}. \qquad (12.13)$$

The coupling coefficient, $\kappa = (2d_{\mathrm{eff}}/c)/(\omega^3/n_{2\omega}n_\omega^2)^{1/2}$, here is also slightly modified compared to (12.11),

12.4.1   Weak conversion

Usually only the fundamental wave enters a crystal of length ℓ, i.e. we have $\mathcal{A}_{\mathrm{SHG}}(z{=}0) = 0$. In the weak conversion approximation, we assume that the fundamental wave is only slightly weakened, i.e. $\mathcal{A}_{\mathrm{FUN}} \simeq$ const. Then we need to solve only the first equation of the system (12.13) and finally achieve the harmonic amplitude at $z = \ell$,

$$\mathcal{A}_{\mathrm{SHG}} = \kappa\ell\mathcal{A}_{\mathrm{FUN}}^2(\omega)\, e^{i\Delta k\ell/2}\frac{\sin(\Delta k\ell/2)}{\Delta k\ell/2}.$$

The quantities that depend on the material are combined into the conversion coefficient Γ,

$$\Gamma^2 = \frac{\kappa^2}{c\epsilon_0\omega} = \frac{4d_{\mathrm{eff}}^2\omega^2}{c^3\epsilon_0 n_\omega^2 n_{2\omega}}. \tag{12.14}$$

Further, the intensity of the harmonic wave depends only on the crystal length ℓ, the incident intensity and the phase mismatch Δk:

$$I_{\mathrm{SHG}} = \Gamma^2\ell^2 I_{\mathrm{FUN}}^2\frac{\sin^2(\Delta k\ell/2)}{(\Delta k\ell/2)^2}.$$

Depending on the magnitude of the phase mismatch Δk, it obviously oscillates between the fundamental radiation field and the harmonic wave when propagating through the crystal.

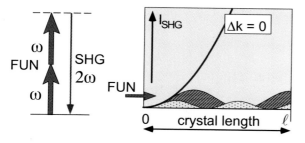 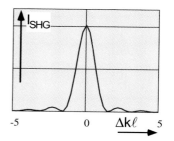

Fig. 12.5: *Evolution of the intensity I_{SHG} of the second harmonic for the limiting case of weak conversion. Only for perfect phase matching can continuous growth of the nonlinear product be achieved. Otherwise the radiation power oscillates between the fundamental and the harmonic like the wave in the middle picture.*

The phase mismatch according to Eq. (12.12) is $|n_\omega - n_{2\omega}| \simeq 10^{-2}$ in typical crystals with normal dispersion. That is why the intensity of the harmonic oscillates with a period of a few $10\,\mu\mathrm{m}$, which is called the 'coherence length',

$$\ell_{\mathrm{coh}} = \frac{\pi}{\Delta k} = \frac{\lambda}{4(n_{2\omega} - n_\omega)}. \tag{12.15}$$

Only in the case of perfect 'phase matching' at $(n_\omega - n_{2\omega}) = 0$ does the intensity grow continuously with crystal length,

$$I_{\mathrm{SHG}} = \Gamma^2 I_{\mathrm{FUN}}^2\ell^2. \tag{12.16}$$

This relation suggests that for frequency doubling it is worth while to increase the intensity by focusing and to make the crystal longer. Though by focusing, as we know from the description of Gaussian beams (Section 2.3), a constant intensity, a quasi-planar wave, is generated only in a narrow range around the focus, so that a compromise between the demand for strong focusing and long crystals has to be found.

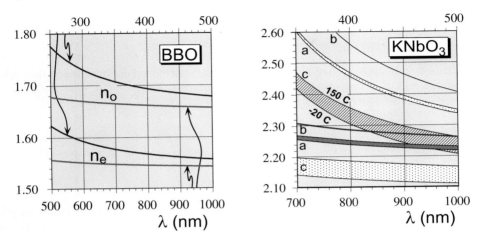

Fig. 12.6: *Refractive index for BBO and KNbO₃ as a function of the wavelength. In the uniaxial BBO crystal the ordinary refraction coefficient (n_o) of a fundamental wave is found in between the ordinary and the extraordinary coefficient (n_e) of half the wavelength, which corresponds to the frequency-doubled wave and makes the angle phase matching possible. In the triaxial KNbO₃ (refractive indices n_a, n_b, n_c) phase matching can be achieved by temperature tuning.*

12.4.2 Strong conversion

In the extreme case of strong conversion, the decrease of the intensity of the pumping wave cannot be neglected any more. We consider the case of perfect phase matching $\Delta k = 0$. In order to get real equations, let us introduce the quantities $\tilde{A}_{SHG} = A_{SHG}\, e^{i\phi_{2\omega}}$ and $\tilde{A}_{FUN} = A_{FUN}\, e^{i\phi_\omega}$ with real amplitudes $\tilde{A}$. Then we have

$$\frac{d}{dz}\tilde{A}_{SHG} = i\kappa\tilde{A}_{FUN}^2\, e^{-i(2\phi_\omega - \phi_{2\omega})},$$

and we are now free to choose the relative phase of the amplitudes, e.g. $e^{-i(2\phi_\omega - \phi_{2\omega})} = -i$. As a result of energy conservation we have

$$\frac{d}{dz}(|A_{SHG}|^2 + |A_{FUN}|^2) = 0.$$

Then for the case of a harmonic vanishing at the input facet at $\mathcal{A}_{\mathrm{SHG}}(z{=}0) = 0$ and $\mathcal{A}_{\mathrm{FUN}}(z{=}0) = \mathcal{A}_0$ (we immediately remove the $\sim$ marks) the real equations are:

$$\frac{d}{dz}\mathcal{A}_{\mathrm{SHG}} = \kappa \mathcal{A}_{\mathrm{FUN}}^2 = \kappa|\mathcal{A}_{\mathrm{FUN}}|^2 = \kappa(\mathcal{A}_0^2 - \mathcal{A}_{\mathrm{SHG}}^2),$$

$$\frac{d}{dz}\mathcal{A}_{\mathrm{FUN}} = -\kappa\mathcal{A}_{\mathrm{SHG}}\mathcal{A}_{\mathrm{FUN}}.$$

The first equation can be solved by standard techniques and results in

$$\mathcal{A}_{\mathrm{SHG}}(z) = \mathcal{A}_{10}\tanh(\kappa\mathcal{A}_0).$$

Thus in principle 100% conversion efficiency can be achieved for frequency doubling since at the end of a long crystal the harmonic intensity

$$I_{\mathrm{SHG}}(z) = I_0\tanh^2(\Gamma I_0^{(1/2)}z)$$

should be found. This result is important particularly for frequency doubling with powerful pulsed lasers since it promises very efficient second harmonic generation.

12.4.3 Phase matching in non-linear crystals

We have already seen in Eq. (12.15) that frequency conversion takes place only over a certain length depending on the dispersion $n(\omega)$. Birefringent crystals already introduced in Section 3.5.1 make it possible to realize $\ell_{\mathrm{coh}} \to \infty$ by choosing a direction of propagation in which the refraction coefficients of fundamental and harmonic wave are identical. Also, in Section 12.4.6 we shall discuss the method of 'quasi-phase matching', which has arisen more recently as a successful method of outwitting dispersion.

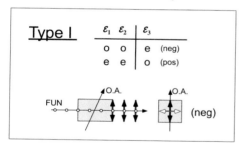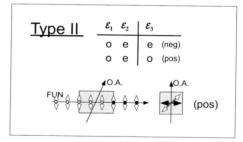

Fig. 12.7: *Polarization directions of fundamental and harmonic wave for phase matching. In a crystal with negative (positive) birefringence, the shortest wavelength has to propagate on the extraordinary (ordinary) beam. For type I matching all polarization directions are orthogonal. For type II matching one polarization direction is used to achieve equally strong projections onto the optical principal axes.*

The simplest situation occurs in uniaxial crystals. For the ordinary beam, polarization and propagation direction are perpendicular to the optical axis, and the phase velocity is characterized by the linear ordinary refraction coefficient $n_{\mathrm{o}}(\omega)$. Since frequency conversion usually takes place in crystals with normal dispersion for the harmonic wave, the smaller refractive index must always be chosen, i.e. in a *negatively* uniaxial crystal ($n_e < n_o$) the harmonic has to be chosen as the extraordinary beam, in a *positively* uniaxial crystal ($n_o < n_e$) as the ordinary one. Then phase matching can be achieved by choosing the polarization of the fundamental wave orthogonal to the harmonic ('type I phase matching'). Alternatively, according to Eq. (12.6) the polarization of the fundamental wave can as well be spread over ordinary and extraordinary beams (i.e. incidence under 45° to the crystal axes) using the 'type II phase matching', so that the four alternatives of Fig. 12.7 are available.

Angle or critical phase matching

As we have already investigated in Section 3.5.1, the refraction coefficient $n_e(\theta)$ of the extraordinary beams depends on the angle between the optical axis and the beam direction according to the 'indicatrix' since the polarization has components parallel as well as perpendicular to the optical axis (Eq. (3.36)),

$$\frac{1}{n_e(\theta)} = \frac{\cos^2\theta}{n_o^2} + \frac{\sin^2\theta}{n_e^2}.$$

Phase matching can now be achieved by choosing appropriately the angle between the fundamental wave and the optical axis. For a negatively (positively) uniaxial crystal, the angles of phase matching for type I/II are determined from the conditions

$$
\begin{array}{lll}
\text{type I} & \text{neg} & n_e(\theta, 2\omega) = n_o(\omega), \\
& \text{pos} & n_e(\theta, \omega) = n_o(2\omega), \\
\text{type II} & \text{neg} & n_e(\theta, 2\omega) = \tfrac{1}{2}[n_o(\omega) + n_e(\theta, \omega)], \\
& \text{pos} & n_o(2\omega) = \tfrac{1}{2}[n_o(\omega) + n_e(\theta, \omega)].
\end{array}
$$

From Eq. (3.36) we have for the case of negative type I phase matching

$$\frac{\cos^2\theta_m}{n_0(2\omega)} + \frac{\sin^2\theta_m}{n_e(2\omega)} = \frac{1}{n_0(\omega)},$$

from which we can deduce the phase matching angle

$$\sin^2\theta_m = \frac{n_o^{-2}(\omega) - n_o^{-2}(2\omega)}{n_e^{-2}(2\omega) - n_o^{-2}(2\omega)}.$$

Similar relations are deduced for the other cases.

For applications nonlinear crystals are cut with respect to the input facet (see Fig. 12.8) in such a way that for normal incidence propagation occurs near to the ideal phase matching angle from the beginning. In order to minimize losses these facets are frequently atireflection coated, too, sometimes even for both the fundamental and the harmonic wave.

Once angle phase matching is achieved, the *walk-off* problem occurs, since the ordinary and extraordinary beams propagate with the same phase velocity but not in the same direction. The *walk-off angle* ρ has already been discussed with Eq. (3.37) for uniaxial crystals,

$$\tan\rho = \frac{n^2(\theta)}{2}\left(\frac{1}{n_o^2} - \frac{1}{n_e^2}\right)\sin 2\theta.$$

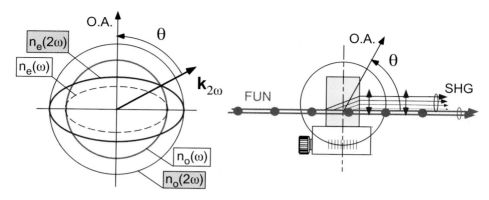

Fig. 12.8: *Phase matching by angle tuning ('critical phase matching'). On the left, the 'indicatrix' for a uniaxial crystal is presented. In order realize angle phase matching, there has to be an intercept between the ellipsoids of the refractive indices for the ordinary (n_o) and extraordinary (n_e) beams. The fundamental must propagate at the phase matching angle with respect to the optical axis. On the right a typical set-up is shown, in which the crystal angle can be adjusted. Fundamental and harmonic wave deviate from each other since they correspond to the ordinary and extraordinary beams, respectively. The angle between fundamental and harmonic wave is called the walk-off angle.*

The harmonic wave therefore leaves the nonlinear crystal with an elliptical beam profile. Additionally, the intensity does not grow quadratically with the crystal length but just linearly according to Eq. (12.16), since the harmonic already generated does not overlap any more with the fundamental wave after a certain propagation length.

Non-critical or 90° phase matching

The disadvantages of angle phase matching can be avoided if one succeeds in matching the ordinary and extraordinary refraction coefficients at the condition $\theta = 90°$. This situation is realized in special crystals where one of the two refractive indices can be tuned over a quite large range by controlling the temperature. Because of the long interaction length this method provides particularly large conversion efficiency. For this reason $KNbO_3$ is a very important nonlinear material: it has a large nonlinear coefficient and it allows 90° phase matching in the important near-infrared range. In Fig. 12.6 the refractive indices for the three axes (a, b, c) were presented. It shows that for the a cut phase matching can be achieved for frequency doubling from 840 to 960 nm, and in the b cut from 950 to 1060 nm. Of course, the methods of angle phase matching can also be employed with these crystals.

Besides '90° phase matching', also the terms *temperature* and *non-critical* phase matching are used for this type of phase matching.

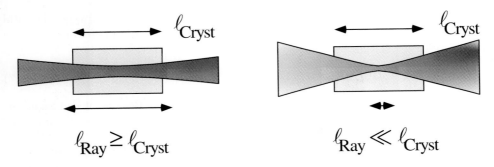

Fig. 12.9: *Focusing of a fundamental wave into a nonlinear crystal. If the Rayleigh zone of the Gaussian beam is larger than the crystal length, an almost planar wave propagates in the crystal volume. If the focusing is too tight, phase matching is again violated in the strongly divergent sections of the beam.*

12.4.4   Frequency doubling with Gaussian beams

Having looked at the principle of phase matching from the point of view of plane waves, we now have to study the influence of realistic laser beams. The conversion efficiency increases with the intensity of the fundamental wave, and so focusing is an obvious choice. On the other hand, too strong focusing leads to large divergence, and reduces the effect again (Fig. 12.9). So intuitively an optimum effect is anticipated if the Rayleigh length roughly corresponds to the crystal length.

A Gaussian beam (details about wave optics can be found in Section 2.3) in the TEM$_{00}$ mode has radial intensity distribution and total power given by

$$\mathcal{E}(r) = \mathcal{E}_0\, e^{-(r/w_0)^2},$$

$$P = \frac{\pi c \epsilon_0}{2} 2\pi \int_0^\infty dr\, r |\mathcal{E}(r)|^2 = I_0 \frac{\pi w_0^2}{2}$$

near the beam waist, where the characteristic parameters are

$$w_0 = \left(\frac{b\lambda}{4\pi n_\omega}\right)^{1/2} \qquad \text{radius of the beam waist,}$$

$$b = 2z_0 \qquad \text{confocal parameter,}$$

$$\theta_{\mathrm{div}} = \frac{\lambda}{\pi w_0 n_\omega} \qquad \text{divergence angle of the Gaussian mode.}$$

Boyd and Kleinman [13] gave a detailed discussion of this problem in the 1960s, and worked out suitable mathematical tools for its treatment. In the limiting case of weak conversion and weak focusing (i.e. $b \ll \ell$), it can be dealt with by simple radial integration. At the end of a crystal of length ℓ and with perfect phase matching $\Delta k = 0$, the following field strength can be found (κ according to eq. (12.11))

$$\mathcal{E}_{\mathrm{SHG}}(r) = i\kappa \mathcal{E}_{\mathrm{FUN}}^2 \ell.$$

With $w_{SHG}^2 = w_{FUN}^2/2$, the beam waists of fundamental and harmonic wave, the total output power depends on the fundamental input power and parameters such as the material constant Γ from Eq. (12.14) and the crystal length ℓ:

$$P_{SHG} = \Gamma^2 \ell^2 I_0^2 \frac{\pi w_{SHG}^2}{2} = \Gamma^2 \ell^2 P_{FUN}^2 \frac{1}{\pi w_{FUN}^2}. \tag{12.17}$$

This corresponds to the already known result of Eq. (12.16). Besides, it can be calculated easily that the fundamental and harmonic wave have the same confocal parameter $b_{SHG} = b_{FUN}$ (see p. 40) under these circumstances.

Boyd et al. have extended this analysis, initially derived for the case of 90° phase matching, to the angle phase matching situation. For this, normalized coordinates for the propagation direction $(z \to t)$ and the walk-off direction (walk-off angle ρ and $x \to u$) are introduced,

$$t = \frac{\sqrt{2\pi}\,z}{\ell_a} \quad \text{with} \quad \ell_a = \sqrt{\pi}w_{FUN}/\rho,$$

$$u = \frac{\sqrt{2}(x - \rho\ell)}{w_{FUN}},$$

and two new functions are defined by

$$\mathcal{F}(u,t) = \frac{1}{t}\int_0^t e^{-(u+\tau)^2}\,d\tau,$$

$$\mathcal{G}(t) = \int_{-\infty}^{\infty} \mathcal{F}^2(u,t)\,du.$$

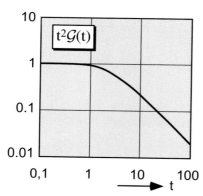 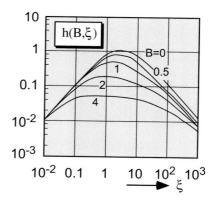

Fig. 12.10: *Graphical representation of the functions $\mathcal{G}(t)$ and $h(B,\xi)$ (according to Boyd and Kleinman [13]).*

The length ℓ_a is called the 'aperture length' and indicates when the harmonic beam has left the volume of the fundamental wave by walk-off. As a result, Eq. (12.17) is modified by the function $\mathcal{G}(t) \leq 1$,

$$P_{SHG} = \frac{\Gamma^2 \ell^2 P_{FUN}^2}{\pi w_{FUN}^2}\mathcal{G}(t).$$

This describes the reduction of the output power caused by the walk-off compared to beams that propagate with perfect overlap. In order also to describe the influence of focusing, it is common to introduce the parameters

$$h(B,\xi) \qquad \text{Boyd–Kleinman reduction factor,}$$
$$B = \tfrac{1}{2}\rho(kl)^{1/2} \qquad \text{birefringence parameter,} \qquad (12.18)$$
$$\xi = \ell/b \qquad \text{normalized crystal length.}$$

The result is

$$P_{\text{SHG}} = \frac{\Gamma^2 \ell^2 P_{\text{FUN}}^2}{\pi w_{\text{SHG}}^2} \frac{1}{\xi} h(B,\xi).$$

For 90° phase matching, we have $B = 0$, and for $\xi = \ell/b < 0.4$ we can approximate $h(0,\xi) = h_0(\xi) \simeq \xi$, so that the previous result of Eq. (12.17) can be reproduced. One generally finds

$$h(0,\xi) = h_0(\xi) \simeq 1 \qquad \text{for} \quad 1 \le \xi \le 6,$$

and the maximum value

$$h_0(\xi) = 1.068 \quad \text{at } \xi = 2.84$$

is realized for a crystal length corresponding to nearly three times the Rayleigh length.
Another useful approximation for $h(B,\xi)$ can be obtained by drawing on the birefringence parameter B (Eq. (12.18)). For $1 \le \xi \le 6$ where $h(B,\xi) \simeq h_{\text{M}}(B)$ and $h_{\text{M}}(0) \simeq 1$ holds, the approximation

$$h_{\text{M}}(B) \simeq \frac{h_{\text{M}}(0)}{1 + (4B^2/\pi)h_{\text{M}}(0)} \simeq \frac{1}{1 + \ell/\ell_{\text{eff}}} \qquad \text{at } \ell/\ell_{\text{eff}} \gg 1$$

is found. Here the effective crystal length ℓ_{eff} has been introduced,

$$\ell_{\text{eff}} = \frac{\pi}{k\rho^2 h_{\text{M}}(0)} \simeq \frac{\pi}{k\rho^2}.$$

12.4.5 Resonant frequency doubling

The low conversion efficiency of nonlinear crystals can be used in a better way if the light is recycled after passing through the crystal. This can be achieved in passive resonators, some of the essential features of which we shall now describe. Alternatively, some nonlinear components can be inserted into active resonators. An important example for *intracavity frequency doubling* is the powerful frequency-doubled neodymium laser from Section 7.8.2.

Passive resonators

To the losses of the resonator by transmission (T) and absorption (A), we now also have to add the conversion of the radiation power of the fundamental into the harmonic

wave. Ashkin et al. [5] have elaborated that the maximum power of the harmonic wave
can be determined from the implicit equation

$$P_{2\omega} = \frac{16T^2 \eta_{SP} P_\omega}{\left[2 - \sqrt{1 - T}\left(2 - A - \sqrt{\eta_{SP} P_{2\omega}}\right)\right]^4}. \tag{12.19}$$

Here the single-pass conversion efficiency (i.e. by single passage of the fundamental
wave) for the crystal is indicated by $\eta_{SP} = P_{2\omega}^0 / P_\omega^2$.

In Fig. 12.11 a ring resonator to enhance the fundamental wave intensity is presented. The nonlinear crystal (NLC) is positioned at the focus of the resonator. For
optimum results, the fundamental wave has to be precisely matched to the Gaussian
mode of the resonator by external optical elements. Furthermore, one of the resonator
mirrors can be adjusted by a piezo-translator (PT). It is controlled by a servo-amplifier
(SA) and ensures that the resonator length is resonantly matched to the fundamental
wave (FUN). The error signal may, for instance, be obtained from the properties of
the light reflected off the input mirror.

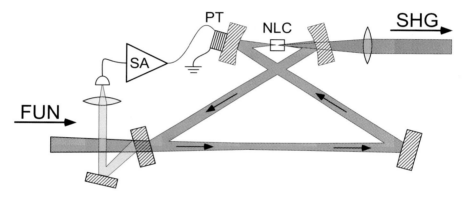

Fig. 12.11: *Frequency doubling in a 'bowtie' ring resonator. FUN = fundamental wave;*
SHG = second harmonic wave; SA = servo-amplifier; NLC = nonlinear crystal; PT = piezo-
translator.

In the best case $P_{2\omega}$ in Eq. (12.19) can be maximized by adjusting the transmission T. This is not possible when mirrors with a fixed reflectivity are used, but the
frustrated total reflection can be used to achieve variable coupling of a resonator with
a driving field (see Fig. 12.12).

A compact layout for frequency conversion is offered by external resonators, which
are directly made from the nonlinear crystal, i.e. 'monolithically' manufactured (Fig.
12.12). They are well suited for temperature-controlled phase matching (p. 364).
The mirrors are integrated through thin layers deposited onto the end facets of the
nonlinear crystal or by means of total internal reflection. The coupling into the ring
can advantageously be achieved by frustrated total internal reflection (FTIR) since
the transmission is set by varying the separation, and therefore optimum conversion
conditions according to Eq. (12.19) can be obtained.

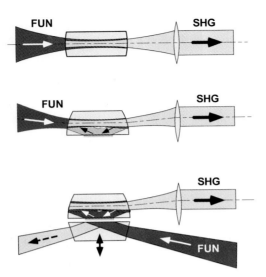

Fig. 12.12: *Frequency doubling in monolithic resonators. In the lower ring resonator the fundamental wave (FUN) is coupled by frustrated total internal reflection (FTIR). The coupling strength is controlled by varying the separation of the monolithic resonator and the coupling prism.*

12.4.6   Quasi-phase matching

For frequency conversion, the proper materials always have to be used under special conditions, such as for example angle phase matching, since the generally small electro-optical coefficients do not allow any large tolerances. The low conversion efficiency of a laser beam at a single pass through a nonlinear material has driven the search for better materials (i.e. especially with higher electro-optical coefficients) or improved methods, like the resonator enhanced frequency doubling from the previous chapter. The search for new materials, though, is laborious, and the effort for servo-controls is quite high with resonant methods.

An alternative route has been opened through successful generation of so-called 'periodically poled' materials where existing and reliable nonlinear materials are tailored in such a way that they allow efficient frequency conversion. The principle of quasi-phase matching is presented in Fig. 12.13. It was already suggested shortly after the invention of the laser [4], but has led to reproducible and robust results only with the manufacturing methods of microelectronics [68].

For production, a periodic pattern of alternating electrodes is deposited onto the crystal. A high-voltage pulse then generates alternating orientation or 'periodic poling' of the ferroelectric domains of certain nonlinear crystals[2] and so leads to a periodic phase jump in the coupling of fundamental and harmonic wave. Successfully used

---

[2]Of the 18 crystal groups allowing phase matching by using birefringence, only 10 are suitable for this method due to symmetry reasons.

crystals include LiNbO$_3$ and KTiOPO$_4$ (KTP). For the periodically manipulated form, also new abbreviations like e.g. PPLN for periodically poled LiNbO$_3$ are used.

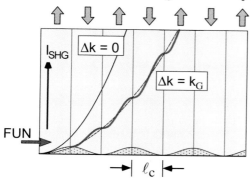

In Section 12.4.1 we have already investigated the coherence length Eq. (12.15) for a material where the phase evolution of fundamental and harmonic wave is not matched with the help of birefringence. The coherence length ℓ_c sets the scale for the period of choice for the artificially induced domain change. At the domain walls phase reversal of the coupling between fundamental and harmonic wave takes place because of the change of the sign of the d coefficient. Thus the retroactive conversion found in a homogeneous material is suppressed by the periodic poling structure and the harmonic wave instead continues to grow.

Fig. 12.13: *Quasi-phase matching in nonlinear crystals. The orientation of the ferroelectric domains is inverted after each coherence length ℓ_c ('periodic poling', PP). The wave in the lower part shows the effect of the crystal without periodic poling. According to [30].*

The theoretical description of frequency doubling in homogeneous materials can straightforwardly be extended to the situation of periodic poling, where the modulation of the sign of the d coefficient is taken into account by a Fourier series,

$$d(z) = d_{\text{eff}} \sum_{m=-\infty}^{\infty} G_m \, e^{-ik_m z} \quad \text{and} \quad G_m = \frac{2}{m\pi} \sin(m\pi\ell/\Lambda). \quad (12.20)$$

Especially $k_m = 2\pi m/\Lambda$ indicates the reciprocal vector of the domain lattice, where Λ indicates the geometrical length of the period. In the end, only one of the Fourier components plays a significant role, all the others contributing only weak conversion similar to the mismatched situation without periodic poling lattice. The important coefficient fulfils the 'quasi-phase matching condition' $\Delta k = k_m$, and it is used in the orders $m = 1, 3, \ldots$. Furthermore one finds [30] that the effective d coefficient is reduced by the Fourier coefficient ($|G_m| < 1$),

$$d_{\text{Q}} = d_{\text{eff}} G_m.$$

We can now adapt the coupled amplitude equations (12.13) to the new situation by the replacements $\Delta k \to \Delta k_{\text{Q}} = \Delta k - k_m$ and $d_{\text{eff}} \to d_{\text{Q}}$ and $\kappa \to \kappa_{\text{Q}}$ respectively in Eq. (12.11). The largest coefficient occurs in first order $m = 1$, thus $d_{\text{Q}}/d_{\text{eff}} = 2/\pi$ from Eq. (12.20); higher orders though allow longer periods and therefore reduce the tolerance requirements in manufacturing.

The quasi-phase matching concept causes a reduction of the nonlinear coefficients but, more importantly, it offers efficient conversion largely independent of the special crystal properties of birefringence. Successful operation of the continuous parametric oscillators, the topic of the following chapter, has been stimulated very much by structured materials with periodic poling [92].

12.5   Sum and difference frequency

12.5.1   Sum frequency

For this case we have to consider the full set of equations (12.10). In the case of sum frequency generation there are already two fields with intensities $I_{1,2}(z{=}0) = I_{10,20}$ present at the entrance of a crystal. For the special case of a very strong pumping field $I_{10} \gg I_{20}$ and perfect phase matching ($\Delta k = 0$), equations (12.10) are greatly simplified:

$$\text{(i)} \qquad \frac{d}{dz}\mathcal{A}_{\mathrm{SUM}} = i\kappa\mathcal{A}_1\mathcal{A}_2,$$

$$\text{(ii)} \qquad \frac{d}{dz}\mathcal{A}_1 \simeq 0, \tag{12.21}$$

$$\text{(iii)} \qquad \frac{d}{dz}\mathcal{A}_2 = i\kappa\mathcal{A}_1^*\mathcal{A}_{\mathrm{SUM}}.$$

The solutions can be easily found by inserting (12.20(iii)) into (12.20(i)),

$$\frac{d^2}{dz^2}\mathcal{A}_{\mathrm{SUM}} = -\kappa^2|\mathcal{A}_1|^2\mathcal{A}_{\mathrm{SUM}},$$

and by applying initial conditions $\mathcal{A}_{1,2}(z{=}0) = \mathcal{A}_{10,20}$. With the inverse scaling length,

$$K = \sqrt{\frac{\kappa^2 I_{10}}{c\varepsilon_0\omega_1}},$$

the normalized amplitudes and intensity evolve along z according to

$$\mathcal{A}_2(z) = \mathcal{A}_{20}\cos(Kz), \qquad I_2(z) = I_{20}\sin^2(Kz),$$

$$\mathcal{A}_{\mathrm{SUM}}(z) = \mathcal{A}_{20}\sin(Kz), \qquad I_{\mathrm{SUM}}(z) = (\omega_{\mathrm{SUM}}/\omega_2)I_{20}\sin^2(Kz).$$

The intensity of the sum frequency wave has naturally to be larger by the factor $\omega_{\mathrm{SUM}}/\omega_2$ than I_2 because energy is drawn from both pump waves. When the weaker input component is entirely converted, difference frequency generation (at the initial frequency ω_2) occurs, until all radiation power is used up again. Thus the intensity oscillates between the sum frequency and the weaker of the two components $I_{1,2}$.

12.5.2   Difference frequency and parametric gain

Let us again consider the case in which a third wave is generated from a strong pumping wave (normalized amplitude $d\mathcal{A}_1/dz \simeq 0$) by difference frequency mixing with a second weaker wave. Then in analogy to Eq. (12.21) and in the case of perfect phase matching ($\Delta k = 0$), the coupled amplitude equations are approximately

$$\text{(i)} \qquad \frac{d}{dz}\mathcal{A}_{\mathrm{DIF}} = i\kappa\mathcal{A}_1\mathcal{A}_2^*,$$

$$\text{(iii)} \qquad \frac{d}{dz}\mathcal{A}_2 = i\kappa\mathcal{A}_{\mathrm{DIF}}^*\mathcal{A}_1. \tag{12.22}$$

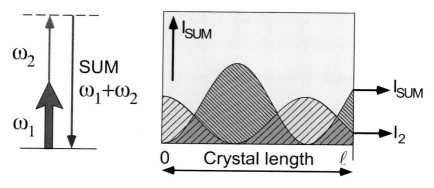

Fig. 12.14: *Weak conversion limit of sum frequency generation as a function of the crystal length. Radiative power oscillates between I_{SUM} and the weaker of the two input components, here I_2.*

The corresponding solutions are

$$\mathcal{A}_2(z) = \mathcal{A}_{20}\cosh(Kz), \qquad\qquad I_2(z) = I_{20}\cosh^2(Kz),$$

$$\mathcal{A}_{\text{DIF}}(z) = -i\mathcal{A}_{20}\sinh(Kz), \qquad I_{\text{DIF}}(z) = (\omega_1/\omega_2)I_{20}\sinh^2(Kz).$$

For $Kz \gg 1$ the intensity dependence shows an interesting behaviour,

$$I_1(z) \simeq I_{20}\,e^{2Kz} \qquad \text{and} \qquad I_2(z) \simeq (\omega_2/\omega_1)I_{10}\,e^{2Kz},$$

where both waves are amplified in this 'parametric process' at the expense of the pumping wave! A more general solution for Eq. (12.22) with coefficients α, β to match initial conditions is given by

$$\mathcal{A}_1(z) = \alpha\sinh(Kz) + \beta\cosh(Kz).$$

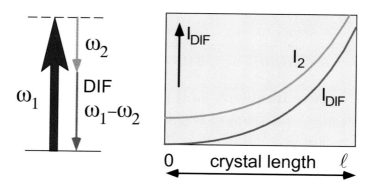

Fig. 12.15: *Parametric gain for difference frequency generation.*

12.5.3 Optical parametric oscillators

Nonlinear generation of coherent radiation is not only interesting for short wavelength production. In principle it promises generation of coherent and fully *tunable* radiation over very wide wavelength ranges. For this purpose the *optical parametric oscillator* (OPO) has been suggested and investigated for a long time. This nonlinear process, more than all others, is affected by loss processes that we have completely neglected up to now but will investigated more deeply here.

First we phenomenologically add the losses γ suffered by the waves passing through the crystal to the coupled amplitude equations from Eq. (12.9), and introduce the specific terms *signal wave* and *idler wave* of the parametric oscillator:

$$\left(\frac{d}{dz} + \gamma\right) \mathcal{A}_P(\omega) = i\kappa \mathcal{A}_S \mathcal{A}_I \, e^{-i\Delta kz} \qquad \text{pumping wave,}$$

$$\left(\frac{d}{dz} + \gamma_S\right) \mathcal{A}_S(\omega_S) = i\kappa \mathcal{A}_P \mathcal{A}_I^* \, e^{i\Delta kz} \qquad \text{signal wave,}$$

$$\left(\frac{d}{dz} + \gamma_I\right) \mathcal{A}_I(\omega_I) = i\kappa \mathcal{A}_S^* \mathcal{A}_P \, e^{i\Delta kz} \qquad \text{idler wave.}$$

Additionally, we again assume the intensity of the pumping wave to be constant ($d\mathcal{A}_P/dz \simeq 0$). From the ansatz $\mathcal{A}_S(z) = \tilde{\mathcal{A}}_S \, e^{(\Gamma + i\Delta k/2)z}$, $\mathcal{A}_I(z) = \tilde{\mathcal{A}}_I \, e^{(\Gamma + i\Delta k/2)z}$, with constant amplitudes $\tilde{\mathcal{A}}_{S,I}$, the condition

$$\left[\left(\Gamma + \gamma_S + i\frac{\Delta k}{2}\right)\left(\Gamma + \gamma_I - i\frac{\Delta k}{2}\right) - \kappa^2 |\mathcal{A}_P|^2\right] \mathcal{A}_S = 0 \tag{12.23}$$

can be obtained. For constant $\mathcal{A}_S \neq 0$ this is exactly fulfilled if the term in square brackets vanishes. This is the case for

$$\Gamma_\pm = -\frac{\gamma_I + \gamma_S}{2} \pm \frac{1}{2}\sqrt{(\gamma_I - \gamma_S - i\Delta k)^2 + 4\kappa^2 |\mathcal{A}_P|^2}.$$

To simplify the interpretation, we consider the special case $\gamma = \gamma_S = \gamma_I$ where the relation for $\Gamma_\pm$ becomes particularly simple,

$$\Gamma_\pm = -\gamma \pm g, \qquad g = \tfrac{1}{2}\sqrt{-\Delta k^2 + 4\kappa^2 |\mathcal{A}_P|^2}. \tag{12.24}$$

The general solution for the coupled waves is

$$\mathcal{A}_S = (\mathcal{A}_{S+}\, e^{gz} + \mathcal{A}_{S-}\, e^{-gz})\, e^{-\gamma z}\, e^{-i\Delta kz/2},$$

$$\mathcal{A}_I^* = (\mathcal{A}_{I+}^*\, e^{gz} + \mathcal{A}_{I-}^*\, e^{-gz})\, e^{-\gamma z}\, e^{-i\Delta kz/2},$$

and obviously for $g > \gamma$ gain is expected. If at the entrance of the crystal there are the amplitudes $\mathcal{A}_{S,I}(z{=}0) = \mathcal{A}_{S0,I0}$, then for the limiting case of weak conversion, i.e. $d\mathcal{A}_P/dz \simeq 0$, we find the following field strengths at the end at $z = \ell$:

$$\mathcal{A}_S(\ell) = [\mathcal{A}_{S0}\cosh(g\ell) - (i/g)(\Delta k \mathcal{A}_{S0} + i\kappa \mathcal{A}_P \mathcal{A}_{I0}^*)\sinh(g\ell)]\, e^{-g\ell}\, e^{i\Delta k\ell/2},$$
$$\mathcal{A}_I(\ell) = [\mathcal{A}_{I0}\cosh(g\ell) - (i/g)(\Delta k \mathcal{A}_{I0} + i\kappa \mathcal{A}_P \mathcal{A}_{S0}^*)\sinh(g\ell)]\, e^{-g\ell}\, e^{i\Delta k\ell/2}. \tag{12.25}$$

For perfect phase matching ($\Delta k = 0$) and for $\mathcal{A}_{S0} = 0$, we reproduce the old result from the difference frequency generation. How the incident fields are really amplified obviously depends on their phase position at the entrance. If there is only one incident field then the second wave 'searches' the right phase position for optimum gain. The solutions of (12.25) depend on the condition that there is at least one field already present at the crystal entrance. In close analogy to the laser, the fulfilment of condition (12.23) can also be understood as a threshold condition. If the parametric gain is generated in a resonator, then the parametric amplifier becomes a parametric oscillator.

Tab. 12.1: *Comparison of laser and optical parametric oscillator (OPO).*

	Laser	OPO
Process	$\chi^{(1)}$, resonant	$\chi^{(2)}, \chi^{(3)}, \ldots,$ non-resonant
Mechanism	occupation number inversion	nonlinear polarization
Pump process	incoherent, energy storable	coherent, not storable

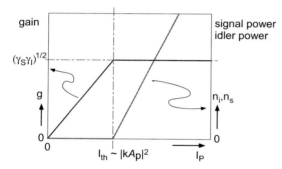

Fig. 12.16: *Gain and power of the parametrically generated fields in a parametric oscillator (see Fig. 8.1).*

Like the laser the OPO starts spontaneously if the gain g overcomes the losses $\sqrt{\gamma_I \gamma_S}$. Parametric oscillators can be operated simply, doubly or even triply resonant, to keep the threshold as low as possible, though again at the expense of large efforts for servo-controlling the optical resonator. It is of course not surprising that according to Eq. (12.24) the gain is proportional to the pumping intensity.

In the operation of tunable lasers (e.g. Ti–sapphire laser, dye laser) inversion is commonly provided by powerful pump lasers. In contrast to the OPO a *coherent* pumping field is, however, not essential for the laser process. In fact incoherent processes, e.g. decay from the pump level, typically take part in the occupation of the upper laser level.

Since the gain depends on the phase mismatch Δk according to Eq. (12.24), the wavelengths of signal and idler wave, λ_S and λ_I, which have to fulfil the equation

$$\lambda_P^{-1} = \lambda_I^{-1} + \lambda_S^{-1}$$

due to energy conservation, can be tuned by varying the angle or the temperature of the birefringent and nonlinear crystal. If the pump wavelength in the *degenerate optical parametric oscillator* (DOPO) is decomposed exactly into two photons at $\omega_S = \omega_I = \omega_P/2$, the reverse process of frequency doubling, then the corresponding phase

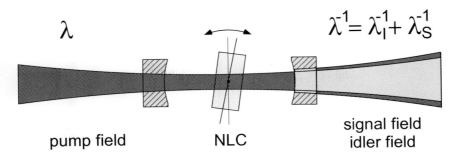

Fig. 12.17: *OPO with linear resonator. The matching of signal and idler wave is achieved by turning the crystal axis if the phase matching is achieved by angular matching. A multiply resonant set-up is in principle difficult to get.*

matching condition has to be valid again, $n_{2\omega}(\omega_P) = n_\omega(\omega_P/2)$. If ordinary dispersion,

$$n_\omega(\omega_{S,I}) \simeq n_\omega(\omega_P/2) + n^{(1)}(\omega_{S,I} - \omega_p/2) + \cdots,$$

is assumed, then a quadratic form for the phase matching condition of the signal and idler frequency is expected near to the degeneracy point:

$$c\Delta k = 0 = n_{2\omega}(\omega_P)\omega_P - [n_\omega(\omega_P/2)\omega_P + n^{(1)}(\omega_S - \omega_I)^2 + \cdots].$$

On the other hand, the difference of the refraction coefficient depends approximately linearly on the angle or the temperature, so that the quadratic behaviour can also be found in the experimental dependence (Fig. 12.18).

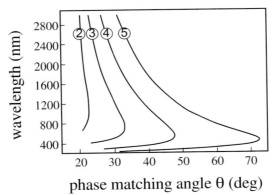

Fig. 12.18: *Tunability of a parametric oscillator driven with a BBO crystal: wavelength of signal and idler waves. The OPO is pumped by the second (532 nm), third (355 nm), fourth (266 nm) or even fifth (213 nm) harmonic of a Nd laser at 1064 nm.*

13 Nonlinear optics II: Four-wave mixing

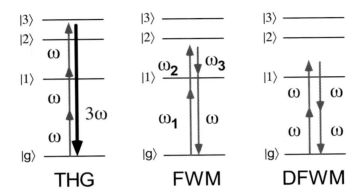

Fig. 13.1: *Selected $\chi^{(3)}$ processes for which the state of the nonlinear material is preserved: third harmonic generation (THG), an example of four-wave mixing (FWM) and degenerate four-wave mixing (DFWM).*

In analogy to the three-wave mixing processes of Section 12.3.2, it is not difficult to compile a typology for four-wave phenomena. Three of the four waves generate a polarization

$$P_i(\omega) = \epsilon_0 \chi^{(3)}_{ijk\ell}(\omega; \omega_1, \omega_2, \omega_3) E_j(\omega_1) E_k(\omega_2) E_\ell(\omega_3), \tag{13.1}$$

which is now characterized by the third-order susceptibility. This fourth-rank tensor describing *four-wave mixing* (FWM) has up to 81 independent components, and therefore is not to be subjected even to general symmetry considerations, which could be described with limited effort for the second-order susceptibility. Instead it is important from the beginning to consider special cases. For the formal consideration, there are basically no new aspects compared to the third harmonic generation (THG) – only the number of coupled amplitude equations is increased by one.

13.1 Frequency tripling in gases

It is obvious in analogy to frequency doubling to ask for frequency tripling by means of the $\chi^{(3)}$ nonlinearity. In Fig. 13.1 it can be seen that *third harmonic generation* (THG) is one of numerous special cases of four-wave mixing.

Optics, Light and Laser. Dieter Meschede
Copyright © 2004 Wiley-VCH Verlag GmbH & Co. KGaA
ISBN: 3-527-40364-7

For practical reasons, this $\chi^{(3)}$ process is really only used when frequencies lying very deep in the ultraviolet spectral range are to be reached. While nonlinear crystals are transparent (i.e. at wavelengths $\lambda > 200\,\text{nm}$), it is an advantage to use frequency doubling and consecutive summation in a two-step $\chi^{(2)}$ process (Fig. 13.2). For example, the 1064 nm line of the Nd laser is transformed, preferably with KTP and LBO materials, to the wavelengths 532 and 355 nm. For this a conversion efficiency of 30% using pulsed light is a matter of routine. The UV radiation generated at 355 nm in this way is very suitable to pump dye lasers in the blue spectral range.

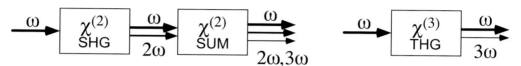

Fig. 13.2: *Frequency tripling with two-step $\chi^{(2)}$ and one-step $\chi^{(3)}$ processes.*

If we neglect geometry effects, the polarization of third order is

$$\mathcal{P}^{3\omega} = \epsilon_0 \chi^{(3)}(3\omega; \omega, \omega, \omega)\mathcal{E}^3.$$

The condition for phase matching, which is

$$\Delta k = k_{3\omega} - 3k_\omega,$$

in this case has to be obtained by adjusting the refraction coefficients of the fundamental wave and harmonic as for the generation of the second harmonic (Section 12.4). As already mentioned above, crystals are of limited use for frequency tripling due to their very small $\chi^{(3)}$ coefficients, poor transparency and the danger of optically induced damage caused by extreme input power and strong absorption of the UV harmonic. Gases, however, have a high threshold of destruction and good transparency below the threshold of photo-ionization, which is at $\lambda \simeq 50\,\text{nm}$ for several noble gases.

The disadvantage of low density in a gas can be compensated by enhancing the nonlinear process using a suitable molecular or atomic resonance in the vicinity of the fundamental wave. Therefore, for the generation of UV light at very short wavelengths, often alkali vapours are used, which allow near-resonant amplification due to their transition frequencies at wavelengths in the visible and near-UV range. They also exhibit a relatively rapidly varying refractive index with normal or anomalous dispersion depending on the position of the fundamental frequency. The resonance lines of the noble gases are in the deep UV ($<100\,\text{nm}$) and mostly in the range of normal dispersion. By adding the 100–10000-fold amount of noble gas atoms to an alkali vapour, the phase velocity of the harmonic can be adjusted. Fig. 13.3 shows a qualitative example of phase matching for frequency tripling of the 1064 nm line of a Nd laser: xenon gas is added to rubidium vapour.

Even if the generation of extreme ultraviolet (XUV) radiation in a gas container were successful, the transport to the planned application still raises special problems since the atmosphere and even the best-known window material, cooled LiF, lose their transparency slightly below 100 nm. That is why very short-wave coherent radiation has in general to be generated very close to the experiment.

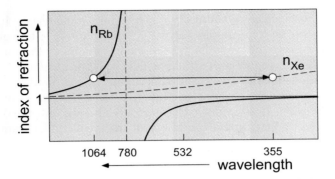

Fig. 13.3: *Matching refractive indices for frequency tripling of the 1064 nm radiation in rubidium vapour (D2 resonance line at 780 nm) by adding xenon gas.*

13.2 Nonlinear refraction coefficient (optical Kerr effect)

In the third order a nonlinear contribution to the polarization at the fundamental wave itself also arises. This is a special case of the *degenerate four-wave mixing* (DFWM), which obviously occurs with well-matched phase propagation because $\Delta \mathbf{k} = \mathbf{k}+\mathbf{k}-\mathbf{k} = \mathbf{k}$ from the beginning. In analogy to the traditional electro-optical Kerr effect, where the refraction coefficient depends on an external electrical field (see Section 3.6.1), nonlinear materials showing this effect are often called *Kerr media*.

The contribution to the polarization of the fundamental wave at the driving frequency ω is[1]

$$\mathcal{P}^{\mathrm{KE}}(\omega) = \epsilon_0 \chi_{\mathrm{eff}}^{(3)}(\omega; \omega, \omega, -\omega)|\mathcal{E}(\omega)|^2 \mathcal{E}(\omega),$$

so that the total polarization is

$$\mathcal{P}(\omega) = \epsilon_0[\chi^{(1)} + \chi_{\mathrm{eff}}^{(3)}|\mathcal{E}(\omega)|^2]\mathcal{E}(\omega) = \epsilon_0 \chi_{\mathrm{eff}} \mathcal{E}(\omega).$$

The total polarization clearly depends on the intensity, and it is convenient to describe this phenomenon transparently anyway by an intensity-dependent refraction coefficient,

$$n = n_0 + n_2 I,$$

with n_0 the common linear refractive index and n_2 a new material constant describing this nonlinearity. By comparing to $n^2 = 1 + \chi_{\mathrm{eff}}$ and with $I = n_0 \epsilon_0 c|\mathcal{E}|^2/2$,

$$n_2 \simeq \frac{1}{n_0^2 c \epsilon_0} \chi_{\mathrm{eff}}^{(3)}.$$

---

[1]There are several definitions of the susceptibility used, which differ from each other mainly by geometry and factors accounting for degeneracy. Here we use an effective susceptibility neglecting such details.

The nonlinear coefficient n_2 naturally depends on the material. Its value varies over a large range and is, for example, just 10^{-16}–$10^{-14}\,\mathrm{cm^2\,W^{-1}}$ for common glasses. However, it can be larger by several orders of magnitude in special materials, e.g. in doped glasses. The propagation of the light field will then strongly depend on the intensity distribution in both space and time. Due to this nonlinearity transverse intensity variations of a light beam cause distortions of optical wavefronts, leading for instance to self-focusing. In section 3.4.2 we have already seen that the self-modulation of the phase caused by longitudinal variations of the intensity, for example in a laser pulse, can lead to the generation of solitons under certain conditions.

13.2.1   Self-focusing

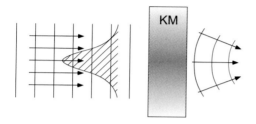

Fig. 13.4: *Self-focusing of a planar wave in a Kerr medium (KM). The intensity profile of a Gaussian beam causes a parabolic transverse variation of the refraction coefficient and therefore works like a lens.*

The transverse Gaussian profile of the $\mathrm{TEM_{00}}$ mode is certainly the best known and most important intensity distribution of all light beams. If the intensity is sufficiently large, e.g. in a short intense laser pulse, then in a Kerr medium it causes an approximately quadratic variation of the refraction coefficient and thus a lens effect, which acts like a converging lens for $n_2 > 0$ and like a diverging lens for $n_2 < 0$ (Fig. 13.4). The focal length depends on the maximum intensity. This effect is actually related to the *thermal lens*.

The change of the refraction coefficient is caused by local temperature variation there, by a nonlinearity here. Temperature modifications can also be generated by a laser beam (e.g. through absorption) but thermal changes are usually very slow (milliseconds) compared to the very fast optical Kerr effect (femto- to nanoseconds) and thus generally not desirable from the practical point of view.

Kerr lens mode-locking

One of the most important applications of self-focusing at present is the so-called *Kerr lens mode-locking* (KLM), which has made the construction of laser sources for extremely short pulses nearly straightforward (see Section 8.5). The self-mode-locking concept was discovered in 1991 [98] with a Ti–sapphire, laser which could be switched from CW to stable pulsed operation by small mechanical disturbances. The laser (Fig. 13.5) consists of just the laser crystal, the mirrors and a pair of prisms for the compensation of the crystal dispersion in the laser crystal and the laser components. Stable pulsed and mode-locked operation can be achieved, for example, by a mode limiter aperture, which increases losses for the CW configuration and thus puts it at a disadvantage against the pulsed mode (see also sect. 8.5.3 and Fig. 8.16).

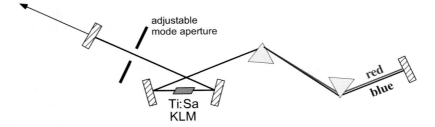

Fig. 13.5: *Ti–sapphire laser with intrinsic Kerr lens mode-locking (KLM) in the laser crystal. The pair of prisms compensates the dispersion. This simple layout generates typical pulse lengths of 50–100 fs. With the mode aperture CW operation is put at an energetic disadvantage against the pulsed operation.*

The trick of self-mode-locking is to align the laser resonator in such a way that during pulsed operation – at which only the induced Kerr lens is active – the resonator field suffers from less losses than during CW operation. There the resonator has to be slightly misaligned. With, for example, an additional aperture at a suitable position in the resonator, these losses can be controlled.

Spatial solitons

Another consequence of self-focusing ought to be mentioned too. As we have investigated in the section about quadratic index media, optical waves are guided in axial media such as a gradient fibre. In a nonlinear medium it is possible that an intense light beam causes 'self-waveguiding' through the nonlinearity of the optical Kerr effect. A propagating beam tends to diverge, as described by Gaussian

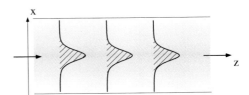

Fig. 13.6: *Propagation of a spatial soliton in a Kerr medium. Note that confinement in the second transverse direction must be achieved by other means.*

beam optics, and as a result of diffraction. In a Kerr medium, however, the intensity distribution may at the same time cause a quadratic transverse index variation prompting a lensing effect. If it exactly compensates diffraction, it allows stable and self-guided beam propagation [58, 103].

We introduce the intensity-dependent transverse variation of the refraction coefficient in the x-direction,

$$n(x) = n_0 + n_2 I(x) = n_0 + \frac{2n_2|\mathcal{A}(x)|^2}{cn_0\epsilon_0}, \qquad (13.2)$$

similar to the paraxial Helmholtz equation (2.30). For the sake of clarity, we introduce $\kappa = 2k^2 n_2/cn_0^2\epsilon_0$, obtaining the *nonlinear Schrödinger equation* (see Section 3.4.2),

$$\left(\frac{\partial^2}{\partial x^2} + 2ik\frac{\partial}{\partial z} + \kappa^2|\mathcal{A}|^2\right)\mathcal{A} = 0,$$

which of course has only its mathematical structure in common with quantum mechanics. It is known that this equation has self-consistent solutions of the form

$$\mathcal{A}(x, z) = \mathcal{A}_0 \operatorname{sech}\left(\frac{x}{w_0}\right) \exp\left(\frac{iz}{4z_0}\right).$$

The properties of this wave are similar to the Gaussian modes with a 'beam waist' $w_0^2 = (\kappa \mathcal{A}_0)^2/2$ and a 'Rayleigh length' $z_0 = kw_0^2/2$. The wave propagates along the z direction and is called a *spatial soliton*.[2] In contrast to the Gaussian beam (see Section 2.3) in a homogeneous medium, the beam parameters (w_0, z_0) now depend on the amplitude $\mathcal{A}_0$! The self-stabilizing mode does not propagate divergently either, but keeps its form undamped over large distances.

Note that from the beginning with Eq. 13.2 we have considered a one-dimensional variation (in x) of the index of refraction only. It turns out that the two-dimensional analogue with variations in both x and y does not yield stable solutions. Two-dimensionally stable modes of propagation can, of course, be obtained if an additional waveguiding effect in the second direction is applied, introduced by, for example, saturation phenomena or other additional nonlinearities.

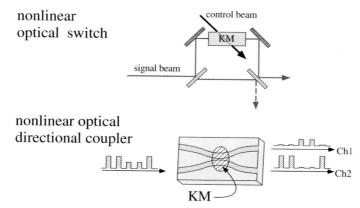

Fig. 13.7: *Applications of the nonlinear optical Kerr effect. Upper part: A Kerr medium (KM) can be used in order to direct a signal beam into one of two exits of a Mach–Zehnder interferometer. In this 'all-optical' switch the refraction coefficient of the Kerr cell causes a phase delay depending on the status of the control beam. Lower part: In a directional coupler (e.g. realized through surface waveguides in LiNbO₃) an incoming signal is distributed into two output channels (Ch1 and Ch2). The coupling efficiency can depend on the input intensity and thus separate pulses of different intensity.*

Nonlinear optical devices

The nonlinear optical Kerr effect is quite interesting for certain applications, e.g. in optical communications. Two examples are presented in Fig. 13.7. A nonlinear switch

---

[2]The 'optical' solitons varying in time discussed in Section 3.4.2 are more widely known though.

is realized by changing the optical length in one branch of a Mach–Zehnder interferometer through a control beam using the Kerr effect. In this way the signal beam can be switched between the two exits. In a nonlinear directional coupler, the coupling efficiency depends on the intensity of the input signal so that pulse sequences with two different intensities can be multiplexed into two channels.

13.2.2   Phase conjugation

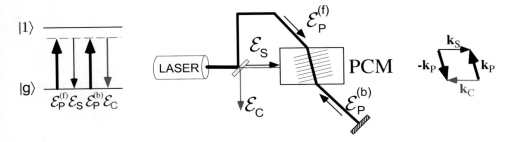

Fig. 13.8: *Left: Phase conjugation as a special case of degenerate four-wave mixing. Middle: Simple set-up for phase conjugation (PCM = phase conjugating medium, e.g. $BaTiO_3$, CS_2). Right: The phase matching condition is always fulfilled in a trivial way.*

Phase conjugation (or 'wavefront reversal') occurs as a special case of degenerate four-wave mixing (DFWM; Figs. 13.1 and 13.8). The phase adjustment is fulfilled intrinsically and ideally since only one optical frequency is involved. The polarization is again calculated according to Eq. (13.1),

$$\mathcal{P}^{PC}(\omega_S) = \epsilon_0 \chi^{(3)}_{\text{eff}}(\omega_S; \omega_P, \omega_P, -\omega_S) \mathcal{E}_P^{(f)} \mathcal{E}_P^{(b)} \mathcal{E}_S^*.$$

Because $\sum_i \mathbf{k}_i = 0$ the phase matching condition is always fulfilled in a trivial way if two waves (in Fig. 13.8 the forward- ($\mathcal{E}_P^{(f)}$) and backward-running ($\mathcal{E}_P^{(b)}$) pump waves) counterpropagate each other. The phase conjugating process can be strongly enhanced by choosing a wavelength in the vicinity of a one-photon resonance.

We now study a simplified theoretical description of phase conjugation, the result of which differs only slightly from the more exact method. In this the nonlinear change of the refraction coefficient for the pumping waves is also taken into account. We especially assume that the intensity of the pump waves does not change, $d\mathcal{E}_P/dz \simeq 0$. Then only two waves instead of four need to be considered,

$$\mathcal{P}_C = \epsilon_0 \chi^{(3)}_{\text{eff}} \mathcal{E}_P^2 \mathcal{E}_S^*,$$

$$\mathcal{P}_S = \epsilon_0 \chi^{(3)}_{\text{eff}} \mathcal{E}_P^2 \mathcal{E}_C^*.$$

We set $\kappa = \omega \chi^{(3)}_{\text{eff}}/2nc\mathcal{E}_P^2$ and consider the signal and conjugate waves propagating in the positive and the negative z directions,

$$\mathcal{A}_C = \mathcal{A}_{C0}\, e^{ikz} \qquad \text{and} \qquad \mathcal{A}_S = \mathcal{A}_{S0}\, e^{-ikz}.$$

They have to fulfil the differential equations

$$\frac{d}{dz}\mathcal{A}_{S0} = i\kappa\mathcal{A}_{C0}^*, \qquad \mathcal{A}_{S0}(z{=}0) = \mathcal{A}(0),$$

$$\frac{d}{dz}\mathcal{A}_{C0} = -i\kappa\mathcal{A}_{S0}^*, \qquad \mathcal{A}_{C0}(z{=}\ell) = 0.$$

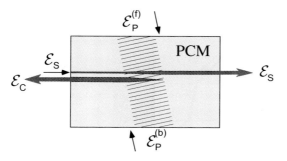

Fig. 13.9: *Signal and conjugated wave within a phase conjugating medium (PCM). Both waves are amplified.*

The boundary conditions at the end of the crystal assume that a signal wave exists at the front end (at $z = 0$) of the crystal but no conjugated wave yet at the rear ($z = \ell$). Here the origin of the phase conjugation is clearly identified as the newly generated conjugated wave $\mathcal{A}_{C0}$, which is driven by the conjugate amplitude $\mathcal{A}_{S0}^*$.

The solutions are found straightforwardly. For the signal wave as well as for the conjugated one there is amplification:

$$\mathcal{A}_{S0} = \frac{\mathcal{A}(0)}{\cos\left(|\kappa|\ell\right)} \qquad \text{and} \qquad \mathcal{A}_{C0} = \frac{i\kappa}{|\kappa|}\tan(|\kappa|\ell)\mathcal{A}^*(0).$$

The phase conjugation has a fascinating application for *wavefront reconstruction* or *wavefront reversal*. Before we study this phenomenon in more detail, we introduce an alternative interpretation derived from conventional holography, which we have discussed already in Section 5.8. In holography a conjugated wave is known to occur too!

The interference of a pump wave with the signal wave causes a periodic modulation of the intensity and thus of the refraction coefficient in the phase conjugating medium (PCM in Fig. 13.10) with reciprocal lattice vector $\mathbf{K}$,

$$\mathbf{K} = \mathbf{k}_P - \mathbf{k}_S \qquad \text{and} \qquad \Lambda = \tfrac{1}{2}\lambda\sin(\theta/2).$$

The counterpropagating pump wave exactly fulfils the Bragg condition,

$$\sin(\theta/2) = \frac{\lambda}{2\Lambda},$$

and is diffracted by this phase grating into the direction precisely opposing the signal wave.

Wavefront reversal is shown in Fig. 13.11 by comparison to a conventional mirror. While a conventional mirror turns a wavefront around on reflection, a phase conjugating mirror (PCM) reverses the wavevector of propagation while maintaining the shape of the wavefront. Any distortion caused by inhomogeneous but linear media will be reversed and hence a laser beam with initially smooth wavefronts will emerge from a

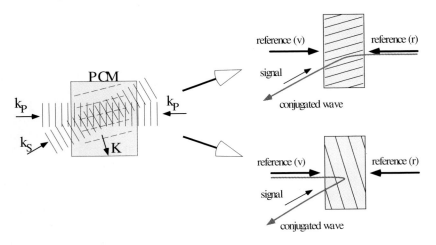

Fig. 13.10: *Real-time holography and phase conjugation. Left: Geometry of relevant waves. Upper right: The forward-propagating pump wave forms a grating by superposition with the signal wave. The backward-running pump wave fulfils the Bragg condition for this grating and is scattered in the direction of the signal wave. Lower right: A similar argument can be used for interference of the backward-propagating pump wave and the signal wave.*

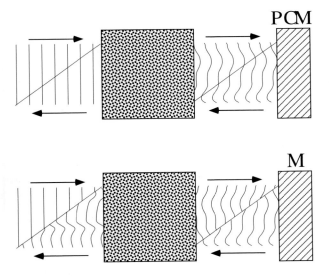

Fig. 13.11: *Wavefront reversal or reconstruction using a phase conjugating mirror (PCM) and a conventional one (M).*

PCM with identical shape. A possible application is efficient focusing of intense laser radiation onto an object with a surface inappropriately matched to conventional, i.e. Gaussian-shaped, laser beams.

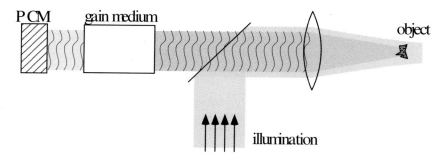

Fig. 13.12: *Application of a phase conjugating mirror for focusing of intense laser radiation onto an optically inappropriately adjusted object.*

13.3   Self-phase-modulation

The nonlinear modification of the refraction coefficient takes effect not only on the spatial wavefronts of laser light but also on the time-variant structure. These nonlinear phenomena are not only important for short-pulse lasers because of their extreme peak intensities but also used for relevant applications. Consider a light pulse with Gaussian amplitude distribution and characteristic pulse length τ,

$$E(t) = E_0\, e^{-(t/\tau)^2/2}\, e^{-i\omega t} \qquad \text{and} \qquad I(t) = I_0\, e^{-(t/\tau)^2},$$

during passage through a nonlinear medium. The phase of the light pulse at the end of a sample of length ℓ develops according to

$$\begin{aligned}
\Phi(t) = nkz\big|_\ell &= n(t)k\ell \\
&= (n_0 + n_2 I_0\, e^{-(t/\tau)^2})ckt.
\end{aligned}$$

The instantaneous frequency is then

$$\omega(t) = \frac{d}{dt}\Phi(t) = [n_0 - n_2 I_0\, 2(t/\tau)\, e^{-(t/\tau)^2}]ck.$$

During the pulse this represents a shift from blue to red frequencies or vice versa, depending of the sign of n_2. This phenomenon is generally called a *frequency chirp*. In the centre at $\exp[-(t/\tau)^2] \simeq 1$ a linear variation can be found

$$\omega(t) \simeq \omega_0 - 2\beta t \qquad \text{with} \qquad \beta = \omega_0 \frac{n_2 I_0}{n_0} \frac{\ell}{\tau}.$$

Frequency chirp is not unusual; in fact, the laser pulses emitted by the simplest versions of Kerr lens mode-locked lasers (Section 13.2.1) always tend to exhibit such frequency variations. In previous chapters we have furthermore encountered other situations where self-phase-modulation is important. It is the origin of soliton propagation in optical fibres described in Section 3.4.2. And the stretchers and compressors introduced in Section 8.5.5 can be used to control, remove or enhance the chirp.

A Mathematics for optics

A.1 Spectral analysis of fluctuating measurable quantities

The Fourier transformation is the 'natural' method to describe the evolution of an optical wave since in the end all optical phenomena can be considered the summation of the action of elementary waves according to Huygens' principle. Exactly this action is calculated with the help of the Fourier transformation.

By *fluctuations* of a physical quantity we understand its irregular variations in time. Deterministic physical predictions can be made not about the actual behaviour of a time-variant quantity but about the probability distribution of its possible values, e.g. the amplitude distribution of a signal voltage. From the theory of probability, it is known that the distribution of a stochastic quantity $V(t)$ is completely determined when all of its moments are known. By this, the averages $\langle V \rangle$, $\langle V^2 \rangle$, $\langle V^3 \rangle$, ... are understood. Often a certain distribution is known – or assumed – e.g. a Gaussian normal distribution for *random* events. Then it is sufficient to give the leading moments of the distribution, e.g. the average value $\langle V \rangle$ and the *variance* $\langle (V - \langle V \rangle)^2 \rangle$. The square root of the variance is called the *root-mean-square deviation* or in short r.m.s. value V_{rms},

$$V_{\mathrm{rms}}^2 = \frac{1}{T} \int_0^T [V(t) - \langle V(t) \rangle]^2 \, dt = \langle V^2(t) \rangle - \langle V(t) \rangle^2. \tag{A.1}$$

In experiments, nearly all measurable quantities are ultimately converted into electrical signals reflecting their properties. For processing dynamic electrical signals, *filters* play a very special role since their use allows desired and undesired parts of a signal to be separated from each other. The action of a filter or a combination of filters can be understood most simply by the effect on a sinusoidal or *harmonic* quantity with a varying frequency $f = \omega/2\pi$. Thus it is important for both theoretical and practical reasons to characterize the fluctuations of a measurable quantity not only in the time domain but also in the frequency domain, i.e. by a spectral analysis.

In physics and in the engineering sciences, the description of a time-dependent quantity by its frequency or *Fourier components* has proven to be invaluable for a long time. The complex voltage $V(t)$, for example, can be decomposed into partial waves and described in frequency space,

$$V(t) = \frac{1}{2\pi} \int_{-\infty}^{\infty} \mathcal{V}(\omega) \, e^{-i\omega t} \, d\omega = \int_{-\infty}^{\infty} \mathcal{V}(f) \, e^{-2\pi i f t} \, df. \tag{A.2}$$

Optics, Light and Laser. Dieter Meschede
Copyright © 2003 Wiley-VCH Verlag GmbH & Co. KGaA
ISBN: 3-527-40364-7

We can interpret $\mathcal{V}(f)\,df$ as the amplitude of a partial wave at the frequency f and with frequency bandwidth df. The *amplitude spectrum* has the unit $\mathrm{V\,Hz}^{-1}$ and as a complex quantity it also contains information about the phase angle of the Fourier components. The functions $V(t)$ and $\mathcal{V}(\omega)$ constitute a *Fourier transform pair* with the inverse transformation

$$\mathcal{V}(\omega) = \int_{-\infty}^{\infty} V(t)\,e^{i\omega t}\,dt. \tag{A.3}$$

The effect of a simple system of filters, e.g. low- or high-passes, on a harmonic excitation can often be given by a *transfer function* $T(\omega)$. The advantages of the frequency or Fourier decomposition according to Eq. (A.2) show up in the simple linear relation between the input and output of such a network,

$$V'(t) = \frac{1}{2\pi} \int_{-\infty}^{\infty} T(\omega)\mathcal{V}(\omega)\,e^{-i\omega t}\,d\omega.$$

This method delivers satisfactory results for numerous technical applications. This is especially valid in the case when the signal is periodic and the relation between time and frequency domain is exactly known. A noisy signal varies sometimes rapidly, sometimes slowly, and consequently it has contributions from both low and high frequencies. Thus the mathematical relation according to Eq. (A.2) cannot be given since an infinitely expanded measurement interval would be necessary. From a rigorous mathematical point of view, even a very large time interval cannot be considered a sufficiently good approximation since there is not even some information about the boundedness of the function and thus about the convergence properties of the integral transformation.

On the other hand, the Fourier component of an arbitrary signal can indeed be measured with an appropriate narrow-band filter by measuring its average transmitted power. In every *spectrum analyser* the signal strength V^2 transmitted through a filter with tunable centre frequency f and bandwidth Δf is measured. The square is generated by electronic hardware, e.g. by rectification and analogue quadrature.Let us take $P_V(t) = V^2(t)$ as the *generalized power* of an arbitrary signal $V(t)$. filter.

For the formal treatment, we introduce the Fourier integral transform of the function $V(t)$ on a finite measurement interval of length T,

$$\mathcal{V}_T(f) = \int_{-T/2}^{T/2} V(t)\,e^{i2\pi ft}\,dt. \tag{A.4}$$

The average total power in this interval is

$$\langle V^2 \rangle_T = \frac{1}{T} \int_{-T/2}^{T/2} V^2(t)\,dt.$$

We can introduce the Fourier integral transform according to Eq. (A.4) and exchange the order of integration (we leave out the index $\langle\ \ \rangle_T$ in the following since there cannot be any confusion),

$$\langle V^2 \rangle = \frac{1}{T} \int_{-T/2}^{T/2} \left[V(t) \int_{-\infty}^{\infty} \mathcal{V}_T(f) \, e^{-2\pi i f t} \, df \right] dt$$

$$= \frac{1}{T} \int_{-\infty}^{\infty} \left[\mathcal{V}_T(f) \int_{-T/2}^{T/2} V(t) \, e^{-2\pi i f t} \, dt \right] df.$$

The variable $\langle V^2 \rangle$ is very useful because with its help we can calculate the variance $\Delta V^2 = \langle V^2 \rangle - \langle V \rangle^2$ and thus the second moment of the distribution of the quantity $V(t)$, at least within the restricted interval $[-T/2, T/2]$. Since $V(t)$ is a real quantity, we have $\mathcal{V}_T(-f) = \mathcal{V}_T^*(f)$ according to (A.2) and we can write

$$\langle V^2 \rangle = \frac{1}{T} \int_{-\infty}^{\infty} [\mathcal{V}_T(f) \mathcal{V}_T(-f)] \, df = \frac{1}{T} \int_{-\infty}^{\infty} |\mathcal{V}_T(f)|^2 \, df.$$

Owing to the symmetry of $\mathcal{V}_T(f)$, it is sufficient to carry out the single-sided integration $0 \to \infty$. We define the power spectral density $S_V(f)$,

$$S_V(f) = \frac{2|\mathcal{V}_T(f)|^2}{T}, \tag{A.5}$$

obtaining a relation that may be interpreted as

$$\langle V^2 \rangle = \int_0^{\infty} S_V(f) \, df. \tag{A.6}$$

According to this, $S_V(f) \, df$ is exactly the contribution of the average power of a signal $V(t)$ transmitted by a linear filter with centre frequency f and bandwidth Δf. Towards higher frequencies the *power spectrum* $S_V(f)$ usually drops off with $1/f^2$ or faster so that the total noise power remains finite.

Often the formal and unphysical notation $\sqrt{S_V(f)}$ with units $V \, Hz^{-1/2}$ is used, which again gives a noise amplitude. This always refers to a noise power, however. For optical detectors the noise amplitudes of voltage and current in units of $(V^2 \, Hz^{-1})^{1/2}$ and $(A^2 \, Hz^{-1})^{1/2}$ respectively are most important and are thus given separately once again:

$$i_n(f) = \sqrt{S_I(f)}, \qquad e_n(f) = \sqrt{S_U(f)}. \tag{A.7}$$

Then, the r.m.s. values of noise current and voltage in a detector bandwidth B are $I_{rms} = i_n \sqrt{B}$ and $U_{rms} = e_n \sqrt{B}$, respectively. In a rather sloppy way they are often simply called 'current noise' and 'voltage noise', but one has to be aware of the fact that in calculations always only the squared values $i_n^2 B$ and $e_n^2 B$ are used, respectively. Also, for applications of this simple relation, one assumes that the noise properties are more or less constant within the frequency interval of width B.

A.1.1   Correlations

The fluctuations of measurable quantities can alternatively be described by means of correlation functions. With correlation functions, one investigates how the value of a quantity $V(t)$ evolves away from an initial value,

$$C_V(t, \tau) = \langle V(t)V(t+\tau) \rangle_T = \frac{1}{T} \int_{-T/2}^{T/2} V(t)V(t+\tau)\, dt.$$

In this case we have already assumed a realistic finite time interval T for the measurement. In general, we will investigate stationary fluctuations, which do not themselves depend on time, so that the correlation function does not explicitly depend on time either. Often useful physical information is given by the normalized correlation function,

$$g_V(\tau) = \frac{\langle V(0)V(\tau) \rangle}{\langle V \rangle^2} = 1 + \frac{\Delta V(\tau)^2}{\langle V \rangle^2}.$$

For $\tau \to 0$ the term $\Delta V(\tau)^2 = [V(\tau) - \langle V \rangle]^2$ exactly results in the variance. This directly allows one to assess the fluctuations.

We can build a valuable relation with the spectral power density by using the bounded Fourier transforms according to Eq. (A.4) and exchanging again the order of time and frequency integrations,

$$C_V(\tau) = \frac{1}{T} \int_{-\infty}^{\infty} \int_{-\infty}^{\infty} \int_{-T/2}^{T/2} \mathcal{V}_T(f')\mathcal{V}_T(f)\, e^{-i2\pi f't}\, e^{-i2\pi f(t+\tau)}\, df\, df'\, dt.$$

For very long times $T \to \infty$ we may replace the time integration by the Fourier transform of the delta function, $\delta(f) = \int_{-\infty}^{\infty} e^{i2\pi ft}\, dt$, yielding

$$C_V(\tau) = \frac{1}{T} \int_{-\infty}^{\infty} \int_{-\infty}^{\infty} \mathcal{V}_T(f')\mathcal{V}_T(f)\delta(f+f')\, e^{-i2\pi f\tau}\, df\, df'$$

$$= \int_0^{\infty} \frac{2|\mathcal{V}_T(f)|^2}{T}\, e^{-i2\pi f\tau}\, df.$$

With the help of Eq. (A.5) we can immediately justify the Wiener–Khintchin theorem, which establishes a relation between the correlation function and the power spectral density of a fluctuating quantity:

$$C_V(\tau) = \int_0^{\infty} S_V(f)\, e^{-i2\pi f\tau}\, df \tag{A.8}$$

and

$$S_V(f) = \int_0^{\infty} C_V(\tau)\, e^{i2\pi f\tau}\, d\tau. \tag{A.9}$$

A.1.2 Schottky formula

One of the most important and fundamental forms of noise is the so-called *shot noise*. It arises if a measurable quantity consists of a flow of particles being registered by the detector at random times, e.g. the photon flow of a laser beam or the photo-electrons in a photomultiplier or a photodiode.

Let us consider a flow of particles that are registered by a detector as needle-like sharp electrical impulses at random times. We are interested in the power spectrum of this current of random events. If N_T particles are registered during a measurement interval of length T, the current amplitude can be given as a sequence of discrete pulses registered at individual instants t_k:

$$I(t) = \sum_{k=1}^{N_T} g(t - t_k). \tag{A.10}$$

The function $g(t)$ accounts for the finite rise time τ of a real detector, which would give a finite length even to an infinitely sharp input pulse. At first we determine the Fourier transform

$$\mathcal{I}(f) = \sum_{k=1}^{N_T} \mathcal{G}_k(f),$$

with the Fourier transform of the kth individual event $\mathcal{G}_k(f) = e^{i2\pi f t_k} \mathcal{G}(f)$:

$$\mathcal{G}(f) = \int_{-\infty}^{\infty} g(t) \, e^{i2\pi f t} \, dt. \tag{A.11}$$

Any single event has to be normalized according to $\int_{-\infty}^{\infty} g(t) \, dt = 1$. If the events are shaped like pulses of typical length $\tau = f_G/2\pi$, then the spectrum has to be continuous at frequencies far below the cut-off frequency f_G, $\mathcal{G}(f \ll f_G) \simeq 1$.

By definition of the power spectrum (A.5) we have $S_I(f) = 2\langle |\mathcal{I}_T(f)|^2 \rangle / T$. Thus one calculates

$$|\mathcal{I}_T(f)|^2 = |\mathcal{G}(f)|^2 \sum_{k=1}^{N_T} \sum_{k'=1}^{N_T} e^{i2\pi f(t_k - t_{k'})}$$

$$= |\mathcal{G}(f)|^2 \left(N_T + \sum_{k=1}^{N_T} \sum_{k'=1, \neq k}^{N_T} e^{i2\pi f(t_k - t_{k'})} \right).$$

Averaging over an ensemble makes the second term in the lower row vanish, and N_T is replaced by the average value $\overline{N}$. Thus the power density of the noise is

$$S_I(f) = \frac{2\overline{N}|\mathcal{G}(f)|^2}{T}, \tag{A.12}$$

which depends only on the spectrum $\mathcal{G}(f)|^2$ of an individual pulse.

For 'needle-like' pulses with a realistic length τ, we anticipate an essentially flat spectrum, i.e. a *white spectrum* in the frequency range $f \leq \tau/2\pi$. For random uncorrelated pulses we expect not the amplitudes but the intensities to add. If it is also taken into account that $S_I(f)$ is obtained by single-sided integration (eq. (A.5)), we can interpret all factors in Eq. (A.12).

In the special case of an electric current the relation with the noise power spectral density is called the *Schottky formula*, which is valid for Fourier frequencies below the cut-off frequency of the detector f_G,

$$S_I(f) = 2e\overline{I},$$

(A.13)

where we have used $\overline{I} = e\overline{N}/T$.

If the amplitude of the individual event fluctuates as well, e.g. if we have $\int_{-\infty}^{\infty} g(t - t_k)\, dt = \eta_k$, then Eq. (A.12) is replaced by

$$S_I(f) = \frac{2\overline{N}\langle\eta^2\rangle|\mathcal{G}(f)|^2}{T}.$$

(A.14)

Now the average current is $\overline{I} = \overline{N}\overline{e}\,\overline{\eta}/T$, with an average charge $\overline{e}\,\overline{\eta}$. In the Schottky formula (A.13) an additional *excess noise* factor $F_e = \langle\eta^2\rangle/\langle\eta\rangle^2$ is introduced:

$$S_I(f) = 2\langle e\eta\rangle\langle I\rangle\frac{\langle\eta^2\rangle}{\langle\eta\rangle^2}.$$

(A.15)

This variant is important for photomultipliers and avalanche photodiodes subject to intrinsically fluctuating amplification.

Let us finally consider the special case of an amplitude distribution that has only the random values $\eta = 0$ and $\eta = 1$. In this case we have $F_e = 1$, so that events not being registered do not contribute to the noise.

A.2   Poynting theorem

The planar wave is the most important and most simple limiting case that is treated for the propagation of optical waves. There the field vector at a defined position is described by a harmonic function of time,

$$\mathbf{F} = \mathbf{F}_0\, e^{-i\omega t}.$$

Often averages of products of harmonically varying functions are required. For this, the Poynting theorem is very useful if physical quantities are described by the real part of a complex harmonic function. If $\mathbf{F}$ and $\mathbf{G}$ are two complex harmonic functions, then for arbitrary vector products $\otimes$ we have for the average taken over a period

$$\langle\Re\{\mathbf{F}\} \otimes \Re\{\mathbf{G}\}\rangle = \tfrac{1}{2}\langle\Re\{\mathbf{F} \otimes \mathbf{G}^*\}\rangle.$$

B Supplements in quantum mechanics

B.1   Temporal evolution of a two-state system

B.1.1   Two-level atom

A hypothetical two-level atom has only one ground state $|g\rangle$ and one excited state $|e\rangle$ to which the raising and lowering operators

$$\sigma^\dagger = |e\rangle\langle g| \qquad \text{and} \qquad \sigma = |g\rangle\langle e|$$

belong. They are known as linear combinations of the Pauli operators,

$$\sigma^\dagger = \tfrac{1}{2}(\sigma_x + i\sigma_y), \qquad \sigma = \tfrac{1}{2}(\sigma_x - i\sigma_y).$$

The Hamiltonian of the dipole interaction can be described by

$$H = \hbar\omega_0\sigma^\dagger\sigma + \hbar g\, e^{-i\omega t} + \hbar g^*\, e^{i\omega t}, \tag{B.1}$$

with $\omega_0 = (E_e - E_g)/\hbar$ and using the semi-classical approximation as well as the rotating-wave approximation (RWA). The dipole coupling rate g is derived from

$$V_{\mathrm{dip}} = (\mathbf{d}^{(+)} + \mathbf{d}^{(-)}) \cdot (\mathbf{E}^{(+)} + \mathbf{E}^{(-)}),$$

where the operator of the dipole matrix element is $q\mathbf{r} = \mathbf{d} = \mathbf{d}^{(+)} + \mathbf{d}^{(-)}$ and the electric field $\mathbf{E}(\mathbf{r}, t) = \mathbf{E}^{(+)}\, e^{-i\omega t} + \mathbf{E}^{(-)}\, e^{i\omega t}$, and in general geometric factors accounting for the vectorial nature have to be taken into account [97]. For instance, the rotating-wave approximation is of no relevance for a $\Delta m = \pm 1$ transition: because $\mathbf{d}^{(+)} = \langle d\rangle(\mathbf{e}_x + i\mathbf{e}_y)\, e^{-i\omega_0 t}$ and $\mathbf{E}^{(+)} = \mathcal{E}_0(\mathbf{e}_x + i\mathbf{e}_y)\, e^{-i\omega_0 t}$, we have exactly $\mathbf{d}^{(+)} \cdot \mathbf{E}^{(+)} = \mathbf{d}^{(-)} \cdot \mathbf{E}^{(-)} = 0$ in this case.

B.1.2   Temporal development of pure states

In the interaction picture of quantum mechanics, the temporal development of a state is described according to the equation

$$|\Psi_{\mathrm{I}}(t)\rangle = e^{-iH_I t/\hbar}|\Psi_{\mathrm{I}}(0)\rangle, \tag{B.2}$$

with the interaction Hamiltonian

$$\begin{aligned} H_{\mathrm{I}} &= \hbar g\sigma^\dagger + \hbar g^*\sigma \\ &= \hbar|g|(\cos\phi\,\sigma_x + \sin\phi\,\sigma_y). \end{aligned} \tag{B.3}$$

With the Rabi frequency $\Omega_R = 2|g|$, then

$$|\Psi_{\mathrm{I}}(t)\rangle = e^{-i(\Omega_R t/2)(\cos\phi\,\sigma_x - \sin\phi\,\sigma_y)}|\Psi_{\mathrm{I}}(0)\rangle. \tag{B.4}$$

The state development can be taken from the matrix equation

$$\exp(-i\alpha\boldsymbol{\sigma}\cdot\mathbf{n}) = \mathbf{1}\cos\alpha - i\boldsymbol{\sigma}\cdot\mathbf{n}\sin\alpha. \tag{B.5}$$

Optics, Light and Laser. Dieter Meschede
Copyright © 2003 Wiley-VCH Verlag GmbH & Co. KGaA
ISBN: 3-527-40364-7

B.2   Density-matrix formalism

For the expert reader, for convenience we here collect some results of quantum mechanics for the density operator leading to the optical Bloch equations. The density matrix formalism allows one to treat an ensemble of two-level atoms.

In a basis of quantum states $|i\rangle$ the density operator has the spectral representation

$$\hat{\rho} = \sum_{ij} \rho_{ij} |i\rangle\langle j|.$$

The equations of motion of the discrete elements can then be obtained from the Heisenberg equation with the Hamiltonian $\mathcal{H}$ under study,

$$i\hbar \frac{d}{dt}\hat{\rho} = [\mathcal{H}, \hat{\rho}].$$

For evaluation it is convenient to use the spectral representation of the Hamiltonian with elements $H_{ij} = \langle i|\mathcal{H}|j\rangle$,

$$\frac{d}{dt}\rho_{ij} = -\frac{i}{\hbar}\sum_k [H_{ik}\rho_{kj} - \rho_{ik}H_{kj}]. \tag{B.6}$$

According to this the density matrix of a two-level atom consists of the expectation values

$$\begin{pmatrix} \langle \sigma^\dagger \sigma\rangle & \langle \sigma^\dagger\rangle \\ \langle \sigma\rangle & \langle \sigma\sigma^\dagger\rangle \end{pmatrix}.$$

The Hamiltonian for the states $|g\rangle$ and $|e\rangle$ contains the undisturbed operator of the free atom and in semi-classical approximation the dipole term

$$\mathcal{V}_{\mathrm{dip}} = -(d_{eg}\sigma^\dagger + d_{ge}\sigma)(E^{(+)}e^{-i\omega t} + E^{(-)}e^{i\omega t}),$$

so that

$$\mathcal{H} = \tfrac{1}{2}\hbar\omega_0(\sigma^\dagger\sigma - \sigma\sigma^\dagger) + \tfrac{1}{2}(E_0^{(+)}e^{-i\omega t} + E_0^{(-)}e^{i\omega t})(d_{eg}\sigma^\dagger + d_{ge}\sigma).$$

We will see that the expectation values $\langle\sigma^\dagger\rangle$ and $\langle\sigma\rangle$ oscillate with $e^{i\omega_0 t}$ and $e^{-i\omega_0 t}$, respectively. In the vicinity of a resonance we use the 'rotating-wave approximation', for which the terms oscillating with $\omega+\omega_0$ are neglected. We abbreviate $g = -d_{eg}\mathcal{E}_0/2\hbar$ and find

$$\mathcal{H} = \hbar\omega_0\sigma^\dagger\sigma + \hbar g\, e^{-i\omega t}\sigma^\dagger + \hbar g^* e^{i\omega t}\sigma.$$

From this the equations of motions are obtained as

$$\dot{\rho}_{ee} = ig^* e^{-i\omega t}\rho_{eg} - ig\, e^{i\omega t}\rho_{ge} \qquad = -\dot{\rho}_{gg}$$

$$\dot{\rho}_{eg} = i\omega_0\rho_{eg} + ig\, e^{-i\omega t}(\rho_{ee} - \rho_{gg}) = \dot{\rho}_{ge}^*.$$

In the RWA it is moreover convenient to introduce 'rotating' elements of the density matrix $\rho_{eg} = \bar{\rho}_{eg}\, e^{-i\omega t}$ and $\rho_{ge} = \bar{\rho}_{ge}\, e^{i\omega t}$. Dropping the overbars for the sake of simplicity we obtain (detuning $\delta = \omega - \omega_0$)

$$\dot{\rho}_{ee} = -\dot{\rho}_{gg} = -ig\rho_{ge} + ig^*\rho_{eg},$$

$$\dot{\rho}_{eg} = -i(\omega - \omega_0)\rho_{eg} + ig(\rho_{ee} - \rho_{gg}) = -i\delta\rho_{eg} + ig(\rho_{ee} - \rho_{gg}).$$

From this system of equations the optical Bloch equations (6.28) can again be obtained by suitable replacements. After introducing phenomenological damping rates and $\rho_{eg} = u + iv$, for example, one obtains

$$
\begin{aligned}
\dot{u} &= \delta v - \tfrac{1}{2}\gamma u - 2\,\Im\{g\}w, \\
\dot{v} &= -\delta u - \tfrac{1}{2}\gamma v + 2\,\Re\{g\}w, \\
\dot{w} &= 2\,\Im\{g\}u - 2\,\Re\{g\}v - \gamma w.
\end{aligned}
\tag{B.7}
$$

B.3   Density of states

The calculation of the density of states (DOS) $\rho(E) = \rho(\hbar\omega)$ as a function of energy is a standard problem of the physics of many-particle systems. It depends on the dispersion relation,

$$E = E(\mathbf{k}),$$

and on the dimension of the problem. In the general case it can be anisotropic as well, though we here limit ourselves to the isotropic case. Two important examples are the dispersion relations of the electron gas and of the photon gas:

electrons: $E(\mathbf{k}) = \dfrac{\hbar^2\mathbf{k}^2}{2m}$,

photons: $E(\mathbf{k}) = \hbar\omega = \hbar c k.$

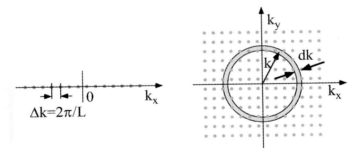

Fig. B.1: *Densities of states in 1D and 2D* **k***-spaces.*

The density of states $\rho(E)\,dE$ describes the number of states within an interval of width dE in energy space. It is calculated in n dimensions according to

$$
\begin{aligned}
\rho(E) &= 2\int_{V_\mathbf{k}} d^n k\, \rho_\mathbf{k}(\mathbf{k})\delta(E - E(\mathbf{k})) \\
&= 2\frac{1}{(2\pi)^n}\int_{V_\mathbf{k}} d^n k\, \delta(E - E(\mathbf{k})).
\end{aligned}
\tag{B.8}
$$

Tab. B.1: *Densities of states in one, two and three dimensions.*

1D	2D	3D
Electromagnetic radiation field: $\omega = ck$, $\rho(\omega)$		
$\dfrac{1}{\pi c}\, d\omega$	$\dfrac{\omega}{\pi c^2}\, d\omega$	$\dfrac{\omega^2}{\pi^2 c^3}\, d\omega$
Free electron gas: $E = \hbar^2 k^2 / 2m$, $\rho(E)$		
$\dfrac{m}{\pi\hbar}(2mE)^{-1/2}\, dE$	$\dfrac{m}{\pi\hbar^2}\, dE$	$\dfrac{m}{\pi^2\hbar^3}(2mE)^{1/2}\, dE$

In k-space we assume a constant density, $\rho_{\mathbf{k}}(\mathbf{k}) = (1/2\pi)^n$, within unit volume[1] and furthermore take into account the two-fold degeneracy due to the polarization of electromagnetic waves and the electron spin, respectively. Then we obtain the densities of states from Tab. B.1.

---

[1] For the calculation of physically measurable quantities, it has to be summed over the volume of the many-particle system. Thus here we set $L = 1$ for Eq. (B.8).

Bibliography

[1] Adams, C.; Riis, E.: Laser cooling and trapping of neutral atoms. Prog. Quantum Electron. **21** (1997) 1

[2] Allen, L.; Eberly, J.: Optical Resonance and Two-Level Atoms. Nachdruck 1987 Aufl. New York: Dover 1975

[3] Andreae, T.: Entwicklung eines kontinuierlichen Er:YAlO Lasers. Diplomarbeit, Sektion Physik der Ludwig-Maximilians-Universität München, München 1989

[4] Armstrong, J.; Boembergen, N.; Ducuing, J.; Perhan, P.: Interactions between light waves in a nonlinear dielectric. Phys. Rev. **127** (1962) 1918–1939

[5] Ashkin, A.; Boyd, G.; Dziedzic, J.: Resonant Optical Second Harmonic Generation and Mixing. IEEE J. Quant. Electronics **QE-2** (1966) 6 109–124

[6] Baer, T.: Large-amplitude Fluctuations Due to Longitudinal Mode Coupling in Diode-pumped Intracavity-doubled Nd:YAG Lasers. J.Opt.Soc.Am. B **3** (1986) 1175

[7] Bass, M.; van Stryland, E.; Williams, D.; Wolfe, W. (ed.): Handbook of Optics, Bd. I, II. New York: McGraw-Hill, Inc. 1995

[8] Baumert, T.; Grosser, M.; Thalweiser, R.; Gerber, G.: Femtosecond Time-Resolved Molecular Multiphoton Ionization: The Na_2 System. Phys. Rev. Lett. **67** (1991) 27 3753–3756

[9] Berman, P. (ed.): Cavity Quantum Electrodynamics. San Diego: Academic Press 1994

[10] Bloch, I.: Stimulierte Lichtkräfte mit Pikosekunden-Laserpulsen. Diplomarbeit, Mathematisch-naturwissenschaftliche Fakultät der Universität Bonn, Bonn 1996

[11] Born, M.; Wolf, E.: Principles of Optics. London: Pergamon Press 1975

[12] Börner, G.: Ist das kosmologische Standardmodell in Gefahr? Phys. i. u. Z. **28** (1997) 1 7–15

[13] Boyd, G.; Kleinman, D.: Parametric Interaction of Focused Gaussian Light Beams. J. Appl. Phys. **39** (1968) 8 3597–3639

[14] Boyd, R.: Nonlinear Optics. 2. Aufl. San Diego: Harlekijn 2002

[15] Boyle, W.; Smith, G.: Charge Coupled Semiconductor Devices. Bell Syst. Tech. J. **49** (1970) 587–593

[16] Brixner, T.; Gerber, G.: Quantum Control of Gas-Phase and Liquid-Phase Femtochemistry. Chem. Phy. Chem. **4** (2003) 418–438

[17] Byer, R.: Parametric Oscillators and Nonlinear Materials. In: P. Harper; B. Wherrett (ed.), *Nonlinear Optics*, Proceedings of the Sixteenth Scottish Universities Summer Scholl in Physics 1975. London: Academic Press, 1977 S. 47–160

[18] Carnal, O.; Mlynek, J.: Young's Double-Slit Experiment with Atoms: A Simple Atom Interferometer. Phys. Rev. Lett. **66** (1991) 2689

[19] Chang-Hasnain, J.: VCSELs. Advances and future prospects. Optics & Photonis News **5** (Mai 1998) 34–39

[20] Cohen-Tannoudji, C.; Diu, B.; Laloë, F.: Quantum Mechanics I,II. New York London Sydney Toronto: John Wiley & Sons 1977

[21] Corney, A.: Atomic and Laser Spectroscopy. Oxford: Clarendon Press 1988

[22] Crocker, J.: Engineering the COSTAR. Optics & Photonics New **4** (1993) 11 22–26

[23] Dehmelt, H.: Stored-Ion Spectroscopy. In: F. Arecchi; F. Strumia; H. Walther (ed.), *Advances in Laser Spectroscopy*. New York and London: Plenum Press, 1981 S. 153–188

[24] Demtröder, W.: Laser Spectroscopy: Basic Concepts and Instrumentation. 3. Aufl. Berlin-Heidelberg-New York: Springer 2003

[25] Desurvire, E.: Erbium Doped Fiber Amplifiers. New York: J. Wiley & Sons 1994

[26] Digonnet, M. J. F.: Rare Earth Doped Fiber Lasers and Amplifiers (Optical Engineering). New York: Marcel Dekker 2001

[27] Dmitriev, V.; Gurzadayan, G.; Nikogosyan, D.: Handbokk of Nonlinear Optical Crystals. Berlin-Heidelberg-New York: Springer 1991

[28] Drazin, P.; Johnson, R.: Solitons: An Introduction. New York: Cambridge University Press 1989

[29] Encyclopedia Britannica CD98. CD-ROM Multimedia Edition 1998. Table: The constant of the speed of light

[30] Fejer, M.; Magel, G.; Hundt, D.; Byer, R.: Quasi-phase-matched second harmonic generation: tuning and tolerances. IEEE J. Quantum Electron. **28** (1992) 2631–2654

[31] Feynman, R.; Leighton, R.; Sands, M.: The Feynman Lectures on Phyiscs, Bd. I-III. Reading, Ma: Addison-Wesley 1964

[32] Freitag, I.; Rottengatter, P.; Tünnermann, A.; Schmidt, H.: Frequenzabstimmbare, diodengepumpte Miniatur-Ringlaser. Laser und Optoelektronik **25** (1993) 5 70–75

[33] Fugate, R.: Laser Beacon Adaptive Optics. Optics & Photonics New **4** (1993) 6 14–19

[34] Gaeta, Z.; Nauenberg, M.; Noel, W.; C.R. Stroud, j.: Excitation of the Classical-Limit State of an Atom. Phys. Rev. Lett. **73** (1994) 5 636–639

[35] Gardiner, C.: Quantum Noise. Berlin-Heidelberg-New York: Springer 1991

[36] Goos, F.; Hänchen, H.: Ein neuer und fundamentaler Versuch zur Totalreflexion. Ann. Phys. (Leipzig) **1** (1947) 333–346

[37] Grimm, R.; Weidemüller, M.: Optical Dipole Traps for Neutral Atoms. Adv. At. Mol. Opt. Phys. **42** (2000) 95

[38] Grundmann, M.; Heinrichsdorff, F.; Ledentsov, N.; Bimberg, D.: Neuartige Halbleiterlaser auf der Basis von Quantenpunkten. Laser und Optoelektronik **30** (1998) 3 70–77

[39] Haken, H.: Laser Theory. Berlin-Heidelberg-New York: Springer 1983

[40] Happer, W.: Optical Pumping. Rev. Mod. Physics **44** (1972) 2 169–249

[41] Hariharan, P.: Optical Holography. Cambridge: Cambridge University Press 1987

[42] Hecht, E.: Optics. 4. Aufl. Boston: Addison-Wesley 2001

[43] Henderson, B.; Imbusch, G.: Optical Spectroscopy Of Inorganic Solids. Monographs on the Physics and Chemistry of Materials. Oxford: Oxford University Press 1989

[44] Henry, C.: Theory of the Linewidth of Semiconductor Lasers. IEEE Journal **QE-18** (1982) 259

[45] H.Ibach; Lüth, H.: Festkörperphysik. Eine Einführung in die Grundlagen. 4. Aufl. Berlin-Heidelberg-New York: Springer 1989

[46] Hinsch, K.: Lasergranulation. Phys. i. u. Z. **23** (1992) 2 59–66

[47] Houldcroft, P.: Lasers in Materials Process. Oxford: Pergamon 1991

[48] Hubble Space Telescope: Engineering for Recovery, Bd. 4 von *Optics & Photonics New* 1993. Heft 11, Special Issue

[49] Hänsch, T.; Walther, H.: Laser spectroscopy and quantum optics. Rev. Mod. Phys. **71** (1999) 242

[50] Javan, A.; W.R. Bennett, J.; Herriott, D.: Population Inversion and Continuous Optical Maser Oscillation in a Gas Discharge Containing a He-Ne Mixture. Phys. Rev. Lett. **6** (1961) 106

[51] Jessen, P.; Deutsch, I.: Optical Lattices. In: *Advances in Atomic, Molecular, and Optical Physics*, Bd. 37. Cambridge: Academic Press, 1996 S. 95–138

[52] Jung, C.; Jäger, R.; Grabherr, M.; Schnitzer, P.; Michalzik, R.; Weigl, B.; Müller, S.; Ebeling, K.: 4.8 mW single mode oxide confined top-surface emitting vertical-cavity laser diodes. Electron. Lett. **33** (1997) 1790–1791

[53] Kaenders, W.; Wynands, R.; Meschede, D.: Ein Diaprojektor für neutrale Atome. Phys. i. u. Z. **27** (1996) 1 28–33

[54] Kaminskii, A.: Laser Crystals, Bd. 14 von *Springer Ser.Opt. Sci.* Berlin-Heidelberg-New York: Springer 1990

[55] Kane, T.; R.L.Byer: Monolithic, unidirectional single-mode Nd:YAG ring laser. Opt. Lett. **10** (1985) 2 65–67

[56] Kapitza, H.: Mikroskopiern von Anfang an. Firmenschrift Zeiss, Oberkochen 1994

[57] Kittel, C.: Introduction to Solid State Physics. 7. Aufl. Hoboken: John Wiley & Sons 1995

[58] Kivshar, Y. S.; Stegeman, G. I.: Spatial Optical Solitons, Guiding Light for Future Technology. Optics & Photonics New **13** (2002) 59–63

[59] Klein, M.; Furtak, T.: Optics. 2. Aufl. Boston: John Wiley & Sons 1986

[60] Kogelnik, H.; Li, T.: Laser beams and resonators. Proc. IEEE **54** (1966) 10 1312–1329

[61] Laming, R.; Loh, W.: Fibre Bragg gratings; application to lasers and amplifiers. In: O. S. of America (ed.), *OSA Tops on Optical Amplification and Their Applications*, Bd. 5. Washington: Optical Society of America, 1996 S. XX

[62] Lauterborn, W.; Kurz, T.; Wiesenfeldt, M.: Kohärente Optik. Berlin-Heidelberg-New York: Springer 1993

[63] Letokhov, V.; Chebotayev, V.: Nonlinear Laser Spectroscopy, Bd. 4 von *Springer Ser. Opt. Sci.* Berlin-Heidelberg-New York: Springer 1977

[64] Lipson, H.; Lipson, S.; Tannhauser, D.: Optical Physics. 2. Aufl. Cambridge: Cambridge University Press 1981

[65] Lison, F.; Schuh, P.; Haubrich, D.; Meschede, D.: High brilliance Zeeman-slowed cesium atomic beam. Phys. Rev. A **A 61** (2000)

[66] Loudon, R.: The Quantum Theory of Light. Oxford: Clarendon Press 1983

[67] Louisell, W.: Quantum Statistical Properties of Radiation. New York: John Wiley & Sons 1973

[68] Magel, G.; Fejer, M.; Byer, R.: Quasi-phase-matched second harmonic generation of blue light in periodically poled $LiNbO_3$. Appl. Phys. Lett. **56** (1990) 108–110

[69] Maiman, T.: Stimulated optical radiation in ruby masers. Nature **187** (1960) 493

[70] Mandel, L.; Wolf, E.: Optical Coherence and Quantum Optics. Cambridge: Cambridge University Press 1995

[71] Meschede, D.: Radiating Atoms in Confined Space. From Spontaneous to Micromasers. Phys. Reports **211** (1992) 5 201–250

[72] Meschede, D.; Metcalf, H.: Atomic nanofabrication: atomic deposition and lithography by laser and magnetic forces. J. Phys. D: Appl. Phys. **36** (2003) R17–R38

[73] M.Niering; R.Holzwarth; J.Reichert; P.Pokasov; Th.Udem; M.Weitz; T.W.Hänsch; P.Lemonde; G.Santarelli; M.Abgrall; P.Laurent; C.Salomon; A.Clairon: Measurement of the Hydrogen 1S-2S Transition Frequency by Phase Coherent Comparison with a Microwave Cesium Fountain Clock. Phys. Rev. Lett. **84** (2000) 5496

[74] Mollenauer, L.; Stolen, R.; Islam, H.: Experimental demonstration of soliton propagation in long fibers: Loss compensated by Raman gain. Opt. Lett. **10** (1985) 229–231

[75] Moulton, P.: Spectroscopic and laser characteristics of $Ti:Al_2O_3$. J. Opt. Soc. Am. **B3** (1986) 125–133

[76] Möllenstedt, G.; Düker, H.: Beobachtungen und Messungen an Biprisma-Interferenzen mit Elektronenwellen. Z. Phys. **145** (1956) 377

[77] Nakamura, S.; Fasol, G.: The Blue Laser Diode. Berlin-Heidelberg-New York: Springer 1998

[78] NASA: http://aether.lbl.gov/www/projects/cobe/ 1997

[79] NASA: http://lisa.jpl.nasa.gov 1998

[80] Noeckel, J.: Mikrolaser als Photonen-Billards: wie Chaos ans Licht kommt. Phys. Bl. **54** (1998) 10 927

[81] Orloff, J.: Handbook of Charged Particle Optics. Boca Raton: CRC Press 1997

[82] Padgett, M.; Allen, L.: Optical tweezers and spanners. Physics World (September 1997) 35–38

[83] Panofsky, W.; Phillips, M.: Classical Electricity and Magnetism. Reading, USA: Addison Wesley 1978

[84] Paul, H.: Photonen. Eine Einführung in die Quantenoptik. Stuttgart: B. G. Teubner 1995

[85] Peréz, J.-P.: Optique fondements et applications. 3. Aufl. Paris: Masson 1996

[86] Physics, S.: Millenia-Laser. Technical description, Spectra Physics, Inc. 1995

[87] Reeves, W. H.; Skryabin, D. V.; Biancalana, F.; Knight, J. C.; Russell, P. S. J.; Omenetto, F. G.; Efimov, A.; Taylor, A. J.: Transformation and control of ultrashort pulses in dispersion-engineered photonic crystal fibres. Nature **424** (2003) 511–515

[88] Risset, C.; Vigoureux, J.: An elementary presentation of the Goos-Hänchen effect. Opt. Comm. **91** (1992) 155–157

[89] Saleh, B.; Teich, M.: Fundamentals of Photonics. New York Chichester Brisbane: John Wiley & Sons 1991

[90] SargentIII, M.; Scully, M.; Lamb, W.: Laser Physics. Reading, Ma: Addison Wesley 1974

[91] Schawlow, A.; Townes, C.: Infrared and optical masers. Phys. Rev. **112** (1958) 1940

[92] Schiller, S.; Mlynek, J. (ed.): Continuuos-wave optical parametric oscillators: materials, devices, and applications, Bd. 25 von *Appl. Phys. B* 1998. Heft 6, Special Issue

[93] Schmidt, O.; Knaak, K.-M.; Wynands, R.; Meschede, D.: Cesium stauration spectrsocopy revisited: How to reverse peaks and observe narrow resonances. Appl. Phys. B **59** (1994) 167–178

[94] Shen, Y. R.: The Principles of Nonlinear Optics. New York: John Wiley & Sons 1984

[95] Siegman, A.: Lasers. Mill Valley: University Science Books 1986

[96] Snitzer, E.: Optical Maser Action of Nd^+3 in a Barium Crown Glass. Phys. Rev. Lett. **7** (1961) 444

[97] Sobelman, I.: Atomic Spectra and Radiative Transitions. Berlin Heidelberg New York: Springer 1977

[98] Spence, D.; Kean, P.; Sibbett, W.: 60-fsec pulse generation from a self-modelocked Ti-sapphire laser. Opt. Lett **16** (1991) 42

[99] Strickland, D.; Mourou, G.: Compression of amplified chirped optical pulses. Opt. Commun. **56** (1985) 219–221

[100] Strong, C.: The Amateur Scientist: An unusual kind of gas laser that puts out pulses in the ultraviolet. Scientific American **230** (Juni 1974) 6 122–127

[101] Svelto, O.: Principles of Lasers. New York and London: Plenum Press 1998

[102] Szipöcs, R.; Ferencz, K.; Spielmann, C.; Krausz, F.: Sub-10-femtosecond. mirror-dispersion controlled Ti-sapphire laser. Opt. Lett. **20** (1994) 602

[103] Trillo, S.; Torruellas, W. E.: Spatial Solitons. Berlin-Heidelberg-New York: Springer 2001

[104] Ueberholz, B.: Ein kompakter Titan-Saphir-Laser. Diplomarbeit, Mathematisch-naturwissenschaftliche Fakultät der Universität Bonn, Bonn 1996

[105] von Neumann, J.: Notes on the photon-disequilibrium-amplification scheme (JvN), September 16, 1953. IEEE J. Quant. Electron. **QE-23** (1987) 659

[106] Wadsworth, W. J.; Ortigosa-Blanch, A.; Knight, J. C.; Birks, T. A.; Man, T.-P. M.; Russell, P. S. J.: Supercontinuum generation in photonic crystal fibers and optical fiber tapers: a novel light source. J. Opt. Soc. Am. **B19** (2002) 2148–2155

[107] Walls, D.; Milburn, G.: Quantum Optics. Berlin-Heidelberg-New York: Springer 1994

[108] Weisskopf, V.; Wigner, E.: Berechnung der natürlichen Linienbreite auf Grund der Diracschen Lichttheorie. Z. Phys. **63** (1930) 54

[109] Willardson, R.; Sugawara, M. (ed.): Self-Assembled Ingaas/GAAS Quantum Dots (Semiconductors & Semimetals). San Diego: Academic Press 1999

[110] Wineland, D.; Itano, W.; J.C.Bergquist: Absorption spectroscopy at the limit: detection of a single atom. Opt. Lett. **12** (1987) 6 389–391

[111] Woodgate, K.: Elementary Atomic Structure. Oxford: Clarendon Press 1980

[112] Wynands, R.; Diedrich, F.; Meschede, D.; Telle, H.: A compact tunable 60-dB Faraday optical isolator for the near infrared. Rev. Sci. Instrum. **63** (1992) 12 5586–5590

[113] Wöste, O.; Kühn, C.: Wie baue ich meinen Laser selbst? MNU-Zeitschrift **46/8** (1993) 471–473

[114] Yamamoto, Y. (ed.): Coherence, Amplification and Quantum Effects in Semiconductor Lasers. New York: John Wiley & Sons 1991

[115] Yariv, A.: Quantum Electronics. 4. Aufl. Fort Worth: Holt Rinehart and Winston 1991

[116] Yeh, C.: Applied Photonics. San Diego: Academic Press 1990

[117] Zellmer, H.; Tünnermann, A.; Welling, H.: Faser-Laser. Laser und Optoelektronik **29** (1997) 4 53–59

[118] Zernike, F.; Midwinter, J.: Applied Nonlinear Optics. J. Wiley & Sons 1973

[119] Zimmermann, C.; Vuletic, V.; Hemmerich, A.; Ricci, L.; Hänsch, T.: Design for a compact tunable Ti:sapphire laser. Opt. Lett. **20** (1995) 3 297

Index

Optics, Light and Laser. Dieter Meschede
Copyright © 2004 Wiley-VCH Verlag GmbH & Co. KGaA
ISBN: 3-527-40364-7